METAL IONS IN BIOLOGICAL SYSTEMS

Edited by

Helmut Sigel

Institute of Inorganic Chemistry
University of Basel
Basel, Switzerland

with the assistance of Astrid Sigel

VOLUME 11

Metal Complexes as Anticancer Agents

MARCEL DEKKER, INC. New York and Basel

MARCEL DEKKER, INC.
270 Madison Avenue, New York, New York 10016

ISSN: 0161-5149
ISBN: 0-8247-1004-5

Current printing (last digit):
10 9 8 7 6 5 4 3 2 1

PRINTED IN THE UNITED STATES OF AMERICA

PREFACE TO THE SERIES

Recently, the importance of metal ions to the vital functions of living organisms, hence their health and well-being, has become increasingly apparent. As a result, the long-neglected field of "bioinorganic chemistry" is now developing at a rapid pace. The research centers on the synthesis, stability, formation, structure, and reactivity of biological metal ion-containing compounds of low and high molecular weight. The metabolism and transport of metal ions and their complexes is being studied, and new models for complicated natural structures and processes are being devised and tested. The focal point of our attention is the connection between the chemistry of metal ions and their role for life.

No doubt, we are only at the brink of this process. Thus, it is with the intention of linking coordination chemistry and biochemistry in their widest sense that the series METAL IONS IN BIOLOGICAL SYSTEMS reflects the growing field of "bioinorganic chemistry." We hope, also, that this series will help to break down the barriers between the historically separate spheres of chemistry, biochemistry, biology, medicine, and physics, with the expectation that a good deal of the future outstanding discoveries will be made in the interdisciplinary areas of science.

Should this series prove a stimulus for new activities in this fascinating "field" it would well serve its purpose and would be a satisfactory result for the efforts spent by the authors.

Helmut Sigel

PREFACE TO VOLUME 11

In the preceding volume the interplay between carcinogenicity and metal ions was discussed. However, a further aspect to be considered is the use of metal complexes as anticancer agents. The alertness for this possibility was strongly stimulated by the discovery that certain platinum coordination compounds have a potent anticancer activity. It is hoped that this volume will promote further the efforts going on in this area because, as Barnett Rosenberg concludes, "a heavy responsibility devolves upon coordination chemists to advance this new field, for who else can do it?"

The introductory chapter on the antitumor properties of metal complexes sets the scene: it gives a historical review and it summarizes the properties of those metal ions (Co, Fe, Ni, Zn, Rh, Pd, Ir, Ga, etc.) and their complexes which are not covered in the remainder of this book. A large part of the volume is then devoted to the chemistry and the clinical aspects of platinum anticancer drugs. The following contributions examine carcinostatic copper complexes, the oncological potentialities of ruthenium, alkylating metal complexes as potential anticancer agents, and the metal binding to antitumor antibiotics. The closing chapter summarizes the interactions of metal-containing anticancer drugs with enzymes as well as the interactions of metalloenzymes and metal-ion-activated enzymes with other anticancer compounds.

Helmut Sigel

CONTENTS

PREFACE TO THE SERIES iii

PREFACE TO VOLUME 11 v

CONTRIBUTORS xi

CONTENTS OF OTHER VOLUMES xiii

Chapter 1

ANTITUMOR PROPERTIES OF METAL COMPLEXES 1

M. J. Cleare and P. C. Hydes

1. Introduction 2
2. Literature Prior to the Discovery of Antitumor Platinum Compounds 3
3. Discovery and Exploitation of Platinum Anticancer Drugs 9
4. Studies on Other Metals 28
5. Conclusions 52
6. Appendix: Screening Tests for Antitumor Activity 53
   Abbreviations 55
   References 56

Chapter 2

AQUEOUS PLATINUM(II) CHEMISTRY; BINDING TO BIOLOGICAL MOLECULES 63

Mary E. Howe-Grant and Stephen J. Lippard

1. Introduction 64
2. Thermodynamic and Kinetic Principles 68
3. Hydrolysis and Aqueous Substitution Chemistry 82
4. Binding Sites in Biological Systems 90
5. The Platinum Blues: A Case Study 108
6. Detecting Platinum in Biological Systems 114
   Abbreviations 118
   References 118

Chapter 3

CLINICAL ASPECTS OF PLATINUM ANTICANCER DRUGS 127

Barnett Rosenberg

1. Introduction 128
2. Anticancer Activity of Platinum Drugs in Animals 134
3. Toxic Side Effects in Animals 141
4. Drug Fate in Animals 143
5. Combination Chemotherapy in Animals 153
6. Development of Resistance to the Platinum Drugs in Animals 160
7. Mechanisms of Action 163
8. Mutagenesis and Carcinogenesis of Platinum Drugs 172
9. Anticancer Activity of Cisplatin in Humans 175
10. Active Combination Chemotherapies of Human Cancers with Cisplatin 179
11. General Conclusions 189
References 190

Chapter 4

CARCINOSTATIC COPPER COMPLEXES 197

David H. Petering

1. Introduction 198
2. Bis(thiosemicarbazonato) Copper Complexes 199
3. α-N-Heterocyclic Carboxaldehyde Thiosemicarbazonato Copper Complexes 212
4. Copper Bleomycin 218
Acknowledgments 225
Abbreviations 225
References 226

Chapter 5

ONCOLOGICAL IMPLICATIONS OF THE CHEMISTRY OF RUTHENIUM 231

Michael J. Clarke

1. Introduction 232
2. Chemical Properties of Ruthenium Complexes Pertinent to the Development of Anticancer Agents 233
3. Ruthenium(III) Complexes as Oncostatic Prodrugs 241
4. Coordination of Ru(II) and Ru(III) Ammine Complexes to Biologically Important Molecules and Functional Groups 244

5. Complexes with Nucleic Acids 256
6. Disposition of Ruthenium Complexes in Living Organisms 259
7. Conclusion 273
Acknowledgments 274
Abbreviations 275
References 276

Chapter 6

METAL COMPLEXES OF ALKYLATING AGENTS AS POTENTIAL ANTICANCER AGENTS 285

Melvin D. Joesten

1. Introduction 285
2. Alkylating Agents Used in Chemotherapy 287
3. Metal Complexes of Alkylating Agents 295
4. New Anticancer Drug Design 299
5. Recent Developments 301
Abbreviations 301
References 302

Chapter 7

METAL BINDING TO ANTITUMOR ANTIBIOTICS 305

James C. Dabrowiak

1. Introduction 306
2. Bleomycin 307
3. Tallysomycin 319
4. Phleomycin 322
5. Streptonigrin 324
6. Daunomycin and Adriamycin 326
7. Chromomycin, Olivomycin, and Mithramycin 328
8. Conclusions 330
References 331

Chapter 8

INTERACTIONS OF ANTICANCER DRUGS WITH ENZYMES 337

John L. Aull, Harlow H. Daron, Michael E. Friedman, and Paul Melius

1. Introduction 338
2. Interactions of Enzymes with Metal-Containing Anticancer Compounds 339

3. Interactions of Metalloenzymes and Metal-Ion-Activated Enzymes with Anticancer Compounds 353
Abbreviations 365
References 366

AUTHOR INDEX 377

SUBJECT INDEX 405

CONTRIBUTORS

Numbers in parentheses indicate the pages on which the authors' contributions begin.

J. L. AULL, Department of Chemistry, Auburn University, Auburn, Alabama (337)

M. J. CLARKE, Department of Chemistry, Boston College, Chestnut Hill, Massachusetts (231)

M. J. CLEARE, Johnson Matthey Research Centre, Reading, England (1)

J. C. DABROWIAK, Department of Chemistry, Syracuse University, Syracuse, New York (305)

H. H. DARON, Department of Animal and Dairy Sciences, Auburn University, Auburn, Alabama (337)

M. E. FRIEDMAN, Department of Chemistry, Auburn University, Auburn, Alabama (337)

M. E. HOWE-GRANT*, Department of Chemistry, Columbia University, New York, New York (63)

P. C. HYDES, Johnson Matthey Research Centre, Reading, England (1)

M. D. JOESTEN, Department of Chemistry, Vanderbilt University, Nashville, Tennessee (285)

S. J. LIPPARD, Department of Chemistry, Columbia University, New York, New York (63)

P. MELIUS, Department of Chemistry, Auburn University, Auburn, Alabama (337)

D. H. PETERING, Department of Chemistry, University of Wisconsin-Milwaukee, Milwaukee, Wisconsin (197)

B. ROSENBERG, Department of Biophysics, Michigan State University, East Lansing, Michigan (127)

*Current affiliation: Department of Chemistry, Polytechnic Institute of New York, Brooklyn, New York

CONTENTS OF OTHER VOLUMES

Volume 1. Simple Complexes*

1. STRUCTURE AND STABILITY OF METAL-NUCLEOSIDE PHOSPHATE COMPLEXES
   Anthony T. Tu and Michael J. Heller

2. KINETICS OF METAL ION INTERACTIONS WITH NUCLEOTIDES AND BASE-FREE PHOSPHATES
   Cheryl Miller Frey and John E. Stuehr

3. STEREOSELECTIVITY IN COMPLEXES OF AMINO ACIDS AND RELATED COMPOUNDS
   Klaus Bernauer

4. OPTICAL PROPERTIES OF TRANSITION METAL ION COMPLEXES OF AMINO ACIDS AND PEPTIDES
   R. Bruce Martin

5. KINETICS AND MECHANISMS OF METAL ION AND PROTON-TRANSFER REACTIONS OF OLIGOPEPTIDE COMPLEXES
   Dale W. Margerum and Gary R. Dukes

6. METAL ION-THIOETHER INTERACTIONS OF BIOLOGICAL INTEREST
   Donald B. McCormick, Rolf Griesser, and Helmut Sigel

AUTHOR INDEX-SUBJECT INDEX

Volume 2. Mixed-Ligand Complexes*

1. MIXED-LIGAND METAL ION COMPLEXES OF AMINO ACIDS AND PEPTIDES
   R. P. Martin, M. M. Petit-Ramel, and J. P. Scharff

2. STRUCTURAL ASPECTS OF MIXED-LIGAND COMPLEX FORMATION IN SOLUTION
   Helmut Sigel

3. KINETIC STUDY OF THE FORMATION OF MIXED-LIGAND COMPLEXES OF BIOLOGICAL INTEREST
   V. S. Sharma and D. L. Leussing

4. MULTIMETAL-MULTILIGAND EQUILIBRIA: A Model for Biological Systems
   D. D. Perrin and R. P. Agarwal

*Out of print

5. ARTIFICIAL ENZYMES
Arthur E. Martell

AUTHOR INDEX-SUBJECT INDEX

Volume 3. High Molecular Complexes

1. INTERACTIONS OF METAL IONS WITH NUCLEIC ACIDS
Michel Daune

2. METAL ION-PROTEIN INTERACTIONS IN SOLUTION
Ragnar Österberg

3. COMPLEX FORMATION BETWEEN METAL IONS AND COLLAGEN
Helmut Hörmann

4. THE INTERACTION BETWEEN METAL IONS AND RIBONUCLEASE
Esther Breslow

5. THE ROLE OF COPPER IN CYTOCHROME OXIDASE
David C. Wharton

6. THE ROLE OF COPPER IN HEMOCYANINS
René Lontie and Luc Vanquickenborne

7. MONOVALENT CATIONS IN ENZYME-CATALYZED REACTIONS
C. H. Suelter

AUTHOR INDEX-SUBJECT INDEX

Volume 4. Metal Ions as Probes

1. MAGNETIC CIRCULAR DICHROIC STUDIES OF THE ACTIVE SITE GEOMETRY IN METALLOENZYMES
Thomas A. Kaden

2. ENZYME, METAL ION, SUBSTRATE COMPLEXES
Joseph J. Villafranca

3. THE APPLICATION OF PARAMAGNETIC PROBES IN BIOCHEMICAL SYSTEMS
R. A. Dwek, R. J. P. Williams, and A. V. Xavier

4. THE THERMOCHEMISTRY OF BIOINORGANIC SYSTEMS
David R. Williams

AUTHOR INDEX-SUBJECT INDEX

Volume 5. Reactivity of Coordination Compounds

1. THE FORMATION OF SCHIFF BASES IN THE COORDINATION SPHERE OF METAL IONS
Daniel L. Leussing

2. THE ROLE OF DIVALENT METAL IONS IN PHOSPHORYL AND NUCLEOTIDYL TRANSFER
Barry S. Cooperman

3. METAL ION-CATALYZED DECARBOXYLATIONS OF BIOLOGICAL INTEREST
R. W. Hay

4. METAL ION-PROMOTED HYDROLYSIS OF AMINO ACID ESTERS AND PEPTIDES
R. W. Hay and P. J. Morris

5. CATALYTIC ACTIVITY OF POLY-L,α-AMINO ACID-METAL ION COMPLEXES: NEW APPROACHES TO ENZYME MODELS
Masahiro Hatano and Tsunenori Nozawa

6. REACTIONS OF MOLYBDENUM COORDINATION COMPOUNDS: MODELS FOR BIOLOGICAL SYSTEMS
Jack T. Spence

7. INTERACTION OF Cu(I) COMPLEXES WITH DIOXYGEN
Andreas D. Zuberbühler

AUTHOR INDEX-SUBJECT INDEX

Volume 6. Biological Action of Metal Ions

1. ZINC AND ITS ROLE IN ENZYMES
Jan Chlebowski and Joseph E. Coleman

2. VANADIUM IN SELECTED BIOLOGICAL SYSTEMS
Wilton R. Biggs and James H. Swinehart

3. THE CHEMISTRY OF BIOLOGICAL NITROGEN FIXATION
Peter W. Schneider

4. THE METAL ION ACCELERATION OF THE ACTIVATION OF TRYPSINOGEN TO TRYPSIN
Dennis W. Darnall and Edward R. Birnbaum

5. METAL CHELATES IN THE STORAGE AND TRANSPORT OF NEUROTRANSMITTERS
K. S. Rajan, R. W. Colburn, and J. M. Davis

6. THE ROLE OF DIVALENT METALS IN THE CONTRACTION OF MUSCLE FIBERS
F. Norman Briggs and R. John Solaro

AUTHOR INDEX-SUBJECT INDEX

Volume 7. Iron in Model and Natural Compounds

1. PREBIOTIC COORDINATION CHEMISTRY: The Possible Role of Transition Metal Complexes in the Chemical Evolution
M. T. Beck

2. BIOLOGICAL SIGNIFICANCE OF LOW MOLECULAR WEIGHT IRON(III) COMPLEXES
Peter M. May, David R. Williams, and Peter W. Linder

3. THE STORAGE AND TRANSPORT OF IRON
Thomas Emery

4. IRON-SULFUR PROTEINS AND THEIR SYNTHETIC ANALOGS
Bruce A. Averill and William H. Orme-Johnson

5. CATALASES AND IRON COMPLEXES WITH CATALASE-LIKE PROPERTIES
Peter Jones and Ian Wilson

6. MONOOXYGENASE HEMOPROTEINS: Cytochromes P-450
P. G. Debrunner, I. C. Gunsalus, S. G. Sligar, and G. C. Wagner

7. SYNTHETIC ANALOGS OF THE OXYGEN-BINDING HEMOPROTEINS
Christopher A. Reed

8. MECHANISMS FOR THE MODULATION OF HEMOGLOBIN OXYGENATION: A Statistical Mechanical Analysis of Equilibrium and Kinetic Data
H. Eugene Stanley, Rama Bansil, and Judith Herzfeld

9. HUMAN IRON METABOLISM
Edwin R. Hughes

AUTHOR INDEX-SUBJECT INDEX

Volume 8. Nucleotides and Derivatives: Their Ligating Ambivalency

1. X-RAY STRUCTURAL STUDIES OF METAL-NUCLEOSIDE AND METAL-NUCLEOTIDE COMPLEXES
Robert W. Gellert and Robert Bau

2. INTERACTIONS BETWEEN METAL IONS AND NUCLEIC BASES, NUCLEOSIDES, AND NUCLEOTIDES IN SOLUTION
R. Bruce Martin and Yitbarek H. Mariam

3. THE AMBIVALENT PROPERTIES OF SOME BASE-MODIFIED NUCLEOTIDES
Helmut Sigel

4. HEAVY METAL LABELING OF NUCLEOTIDES AND POLYNUCLEOTIDES FOR ELECTRON MICROSCOPY STUDIES
Thomas R. Jack

5. MACROMOLECULES OF BIOLOGICAL INTEREST IN COMPLEX FORMATION
S. L. Davydova

AUTHOR INDEX-SUBJECT INDEX

Volume 9. Amino Acids and Derivatives as Ambivalent Ligands

1. COMPLEXES OF α-AMINO ACIDS WITH CHELATABLE SIDE CHAIN DONOR ATOMS
R. Bruce Martin

2. METAL COMPLEXES OF ASPARTIC ACID AND GLUTAMIC ACID
Christopher A. Evans, Roger Guevremont, and Dallas L. Rabenstein

3. THE COORDINATION CHEMISTRY OF L-CYSTEINE AND D-PENICILLAMINE
Arthur Gergely and Imre Sóvágó

4. GLUTATHIONE AND ITS METAL COMPLEXES
Dallas L. Rabenstein, Roger Guevremont, and Christopher A. Evans

5. COORDINATION CHEMISTRY OF L-DOPA AND RELATED LIGANDS
Arthur Gergely and Tamás Kiss

6. STEREOSELECTIVITY IN THE METAL COMPLEXES OF AMINO ACIDS AND DIPEPTIDES
Leslie D. Pettit and Robert J. W. Hefford

7. PROTONATION AND COMPLEXATION OF MACROMOLECULAR POLYPEPTIDES: Corticotropin Fragments and Basic Trypsin Inhibitor (Kunitz Base)
Kálmán Burger

AUTHOR INDEX-SUBJECT INDEX

Volume 10. Carcinogenicity and Metal Ions

1. THE FUNCTION OF METAL IONS IN GENETIC REGULATION
Gunther L. Eichhorn

2. A COMPARISON OF CARCINOGENIC METALS
C. Peter Flessel, Arthur Furst, and Shirley B. Radding

3. THE ROLE OF METALS IN TUMOR DEVELOPMENT AND INHIBITION
Haleem J. Issaq

4. PARAMAGNETIC METAL IONS IN TISSUE DURING MALIGNANT DEVELOPMENT
Nicholas J. F. Dodd

5. CERULOPLASMIN AND IRON TRANSFERRIN IN HUMAN MALIGNANT DISEASE
Margaret A. Foster, Trevor Pocklington, and Audrey A. Dawson

6. HUMAN LEUKEMIA AND TRACE ELEMENTS
E. L. Andronikashvili and L. M. Mosulishvili

7. ZINC AND TUMOR GROWTH
Andre M. van Rij and Walter J. Pories

8. CYANOCOBALAMIN AND TUMOR GROWTH
Sofija Kanopkaitė and Gediminas Bražėnas

9. THE ROLE OF SELENIUM AS A CANCER-PROTECTING TRACE ELEMENT
Birger Jansson

10. TUMOR DIAGNOSIS USING RADIOACTIVE METAL IONS AND THEIR COMPLEXES
Akira Yokoyama and Hideo Saji

Volume 12. Properties of Copper (in press)

1. THE COORDINATION CHEMISTRY OF COPPER WITH REGARD TO BIOLOGICAL SYSTEMS
R. F. Jameson

2. COPPER(II) AS PROBE IN SUBSTITUTED METALLOPROTEINS
Ivano Bertini and A. Scozzafava

3. COPPER(III) COMPLEXES AND THEIR REACTIONS
Dale W. Margerum and Grover D. Owens

4. COPPER CATALYZED OXIDATION AND OXYGENATION
Harald Gampp and Andreas D. Zuberbühler

5. COPPER AND THE OXIDATION OF HEMOGLOBINS
Joseph M. Rifkind

6. TRANSPORT OF COPPER
Bibudhendra Sarkar

7. THE ROLE OF LOW-MOLECULAR-WEIGHT COPPER COMPLEXES IN THE CONTROL OF RHEUMATOID ARTHRITIS
Peter M. May and David R. Williams

AUTHOR INDEX-SUBJECT INDEX

Volume 13. Copper Proteins (in preparation)

1. THE EVOLUTION OF COPPER PROTEINS
Earl Frieden

2. PROPERTIES OF COPPER 'BLUE' PROTEINS
A. Graham Lappin

3. THE PROPERTIES OF BINUCLEAR COPPER CENTERS IN MODEL AND NATURAL COMPOUNDS
F. L. Urbach

4. CERULOPLASMIN: A MULTI-FUNCTIONAL METALLOPROTEIN OF VERTEBRATE PLASMA
Earl Frieden

5. COPPER MONOOXYGENASES: TYROSINASE AND DOPAMINE-β-HYDROXYLASE
Konrad Lerch

6. CYTOCHROME-C-OXIDASE: AN OVERVIEW OF RECENT WORK
Maurizio Brunori, Eraldo Antonini, and M. T. Wilson

7. THE ACTIVE SITES OF MOLLUSCAN AND ARTHROPODAN HEMOCYANINS
René Lontie and Rafaël Witters

8. THE COPPER/ZINC SUPEROXIDE DISMUTASE
James A. Fee

9. THE CHEMISTRY AND BIOLOGY OF COPPER-METALLOTHIONEINS
Konrad Lerch

10. METAL REPLACEMENT STUDIES OF BLUE COPPER PROTEINS
Bennett L. Hauenstein, Jr. and David R. McMillin

AUTHOR INDEX-SUBJECT INDEX

Volume 14. Inorganic Drugs in Deficiency and Disease (tentative)

1. DRUG-METAL ION INTERACTIONS IN THE GUT
P. F. D'Arcy and J. C. McElnay

2. CALCIUM DEFICIENCY AND ITS THERAPY
C. C. Ashley

3. THE USE OF ZINC IN ZINC DEFICIENCY AND AS A PHARMACOLOGICAL AGENT
George J. Brewer and Ananda S. Prasad

4. COPPER DEFICIENCY IN MAN: DIAGNOSIS, MANAGEMENT AND TREATMENT
William R. Collie

5. THE ANTI-INFLAMMATORY ACTIVITY OF COPPER COMPLEXES
John R. J. Sorenson

6. IRON CONTAINING DRUGS
David A. Brown and M. V. Chidambaram

7. GOLD COMPLEXES AS METALLO DRUGS
Kailash C. Dash and Hubert Schmidbaur

8. METAL COMPLEXES OF ANTIBIOTICS
R. E. Lenkinski

9. METAL IONS AND CHELATING AGENTS IN ANTIVIRAL CHEMOTHERAPY
D. D. Perrin and H. Stünzi

10. COMPLEXES OF HALLUCINOGENIC DRUGS
W. Hänsel

11. LITHIUM IN PSYCHIATRY
N. J. Birch

AUTHOR INDEX-SUBJECT INDEX

Other volumes are in preparation.

Comments and suggestions with regard to contents, topics, and the like for future volumes of the series would be greatly welcome.

Chapter 1

ANTITUMOR PROPERTIES OF METAL COMPLEXES

M. J. Cleare and P. C. Hydes
Johnson Matthey Research Centre
Reading, England

1. INTRODUCTION 2

2. LITERATURE PRIOR TO THE DISCOVERY OF ANTITUMOR PLATINUM COMPOUNDS 3

 2.1. Studies on Various Metals 4

 2.2. Chelation and Cancer 6

3. DISCOVERY AND EXPLOITATION OF PLATINUM ANTICANCER DRUGS 9

 3.1. Discovery 9

 3.2. Bacterial Studies 10

 3.3. Antitumor Studies 11

 3.4. Clinical Studies With Cis-$[PtCl_2(NH_3)_2]$ 24

 3.5. Mechanism of Antitumor Action of Metal Complexes. 26

4. STUDIES ON OTHER METALS 28

 4.1. General Considerations 28

 4.2. Rhodium 30

 4.3. Palladium 40

 4.4. Ruthenium 42

 4.5. Iridium 43

 4.6. Copper 44

 4.7. Cobalt 45

 4.8. Other Metals 47

4.9. Gallium and Group IIIA 49
5. CONCLUSIONS 52
APPENDIX: SCREENING TESTS FOR ANTITUMOR ACTIVITY 53
In Vitro 53
In Vivo 53
Screening Parameters 54
ABBREVIATIONS 55
REFERENCES 56

1. INTRODUCTION

Chemotherapy is a field of cancer treatment which has evolved over the last 30 or 40 years, and its use has escalated enormously in the last 10-15 years. The involvement of inorganic, metal-based compounds in cancer treatment was very limited until the discovery of potent anticancer activity in certain platinum coordination compounds by Rosenberg and Van Camp in 1969 [1]; this was perhaps not surprising, as medical scientists tend to regard all heavy metal compounds as non-selective poisons. The platinum discovery heralded the arrival of metal coordination compounds as a new class of potential antitumor drugs.

Although the idea of using certain heavy metals as selective cytotoxic agents had been mooted prior to 1969 [2], systematic investigations into this aspect of their biological activity have only been initiated over the last few years with a heavy emphasis on platinum amine complexes related to the initial active compounds reported by Rosenberg et al. [1]. The first platinum drug, cis-$[PtCl_2(NH_3)_2]$, has now been proved to be of significant value in treating several human tumors, particularly those of genitourinary origin. The drug has recently been approved for full medical availability by the Food and Drug Administration (FDA) in the United States and the Department of Health and Social Security (DHSS) in the United Kingdom.

This chapter attempts to summarize the anticancer studies on metal complexes, in general, in order to set the scene for the detailed chapters which follow. In addition, the history of the discovery and development of antitumor platinum compounds is covered at some length, as this has been the major catalyst in the emergence of this field of research.

2. LITERATURE PRIOR TO THE DISCOVERY OF ANTITUMOR PLATINUM COMPOUNDS

Prior to 1969, most cancer research interest was centered on the potential carcinogenicity of metals and their salts, rather than on any potential anticancer properties [3,4]. Another important research area involved the study of relative concentrations of essential metals in neoplastic and normal tissue with the aim of finding differences which might be exploitable in terms of cancer control [5-9]. The first metal-based remedy to undergo extensive clinical testing was Fowler's solution, which originally consisted of a suspension of lead arsenate in benzene [10]. Similarly, colloidal lead and lead phosphate have been tested with some reported success [11].

The use of such combinations indicates less concern for overall toxicity than exists today. However, sodium and potassium arsenite and various organoarsenicals are still used clinically to a limited extent, particularly for leukemias [12-14]. Elemental germanium has shown some activity against spontaneous mouse tumors but tin and vanadium were inactive [15]. Particularly after 1945, the efforts of most chemotherapists appear to have been directed towards the screening and testing of organic-based drugs. A discussion on animal tumor testing, including definitions of the screening parameters referred to in this text, can be found in the Appendix at the end of this chapter.

2.1. Studies on Various Metals

One of the earliest surveys [16] of the antitumor action of inorganic compounds was made by Collier and Krauss (1931). Sixty-four metal compounds, including oxides, halides, cyanides, and ammines (Cu, Pb, Cr, Mn, Fe, Co, Ni, Ru, Rh, Os) were tested against mice bearing an Ehrlich carcinoma. No quantitative data were given, but some salts of Pb, Cr, and Mn were described as slightly active. $Cs_2[RhCl_6]{\cdot}H_2O$ in oil was considered distinctly active while $K_4[Ru(CN)_6]$ and OsO_2 in oil registered marginal effects. Taylor and Carmichael (1953) assessed 37 metal chlorides and nitrates against the embryos and tumors of tumor-bearing eggs [17] and also on DBA mice bearing a sarcoma. The compounds were generally ineffective, showing a maximum of 41% inhibition of the sarcoma. Geschickter and Reid (1947) synthesized oil-soluble phthalate complexes [18] of Co(II), Ni(II), Fe(II), Cu(II), and Mn(II). No preparative or analytical details are given for these compounds, and their actual stoichiometry must be in doubt. However, Cu, Ni, and Co butylphthalate complexes were tested on human leukemia and some benefit was claimed although the response to therapy, when favorable, produced a regression which lasted only 3-6 weeks. The nickel compounds showed an effect against spontaneous lymphomas in DBA mice and estrogen-induced mammary carcinomas in rats, giving a considerable lifespan increase in each case. Subsequently nickel butylphthalate showed useful clinical effects on some solid human tumors. Recent testing against the ADJ/PC6A tumor at the Chester Beatty Institute in London showed no activity (Sec. 4.8.4), but these complexes and analogs may be worth screening against a broader range of tumors.

Balo and Banga (1957), working on the theory that the oxidative metabolism in tumor cells is suppressed, looked for autooxidizable complexes which might promote the oxidation mechanism and inhibit tumor growth [19]. They claimed to have prepared some 50 complexes of essential metals with ascorbic acid and the metabolites of the Krebs cycle as ligands. However, the only results given are for iron and cadmium ascorbate and manganese malate complexes against an Ehrlich mouse carcinoma and a Guerin rat carcinoma. The iron complex enhanced the growth of the mouse tumor, while the cadmium complex showed an inhibitory effect at identical doses, but was very toxic. The manganese malate was active against both tumors and was less toxic. Some tests were carried out to support the oxidative theory. The iron ascorbate inhibited the oxidation of p-phenylenediamine which had been added to the carcinoma tissue, whereas the cadmium complex enhanced the corresponding oxidation. The manganese malate formed a colored complex with p-phenylenediamine which decolorized quickly in normal organs but only slowly in tumor tissue.

Ru(II) and Cu(II) complexes with 3,4,7,8-tetramethyl-1,10-phenanthroline (phen) were evaluated against the Landschutz ascites tumor in mice [20]. $[Cu(phen)_2]Cl_2$ and $[Ru(phen)_3]Cl_2$ caused significant inhibition while $[Ru(phen)_2(acac)]Cl$ was less effective [21]. Similar tris-chelates have shown neuromuscular toxicity (Sec. 3.2.).

Topical application of $ZnCl_2$ [22,23] is sometimes used after surgical removal of skin tumors to kill remaining cancerous cells. Mercuric chloride is used as an irrigant solution for eradicating malignant cells from surgical wounds [24].

2.1.1. *Copper*

Simple copper salts have some cytostatic properties and the testing of copper chelates is still an active field. Thus for the ease of reference all of the copper studies are summarized in Sec. 4.6 (see also Chap. 4).

2.1.2. *Cobalt and Chromium*

Japanese researchers have tested a series of Co(III) and Cr(III) ammine complexes against the Yoshida sarcoma [25,26]. Some 40 compounds were examined, but none showed better than marginal activity.

The Co(III) hexammine and pentammine salts were very toxic (confirming earlier work on inert complexes with positive charges [27]; see Sec. 3.2) but $[CoCl_2(NH_3)_4]Cl$, $[Co(CO_3)(NH_3)_4]NO_3 \cdot \frac{1}{2}H_2O$, $[Cr(C_2O_4)(NH_3)_4]$, and $[Cr(C_2O_4)_2(NH_3)_2]$ showed some inhibition of of tumor growth. It is interesting to compare the configuration of these complexes (all cis except the chloride, which is not identified as cis or trans) with those of platinum, which are described in the next section. It is reported that a 1:1 complex between bis-histidinecobalt(II) and 8-azaguanine is somewhat effective against Ehrlich ascites in mice [28], while a cobalt complex with cysteine and phenylthiourea prolonged the lives of mice with an Ehrlich carcinoma [29]. $CoCl_2$ and $Na_3[Co(NO_2)_6]$ did not affect the growth of established epithelial tumors induced by methylcholanthrene, although both were inhibitory when given during methylcholanthrene administration [30]. Naturally occurring cobalt-containing molecules such as cyanocobalamin (vitamin B_{12}) [31] and cobalt protoporphyrin [32] have found occasional use against specific tumors.

2.2. Chelation and Cancer

No discussion of the earlier work in this field would be complete without mentioning the extensive review work of Furst [10,33]. He has postulated a relationship between chelation, carcinogens, and anticancer agents [10]. Furst points out that the majority, if not all, of nonmetallic carcinogens, or their proposed primary metabolic products, are potential chelating agents. Furst suggests that interactions with essential metals are involved, particularly with respect to removing them from specific cellular locations; inhibition of enzymes could occur in this manner. Similarly some metals are known to cause cancer while most chemotherapeutic agents are potential metal-binding agents. Metals may be best transported into cells in chelated forms, as most chelates have relatively high lipid solubility and the human body is capable of concentrating not only the essential metals but also abnormal ones. Thus a carcinogenic chelate may penetrate the cell wall and bring with it an abnormal metal which, on release, could perturbate the cell mechanisms and lead to its transformation into a neoplastic cell. These speculations are based on general observations on carcinogens and anticancer agents but some seem rather unlikely, particularly as many anticancer agents have been shown to react with nucleic acids and inhibit DNA synthesis. However, Furst had the foresight to suggest using Pt(II) and Pd(II) complexes (1962), particularly with sulphur mustards as ligands, although he seems to have been thinking in terms of using these complexes to bring about entry of the ligands into the cells.

Kirschner et al. partially adopted Furst's suggestions and synthesized complexes with either naturally occurring biological molecules or known antitumor agents as the ligands [34]. Three of the five active complexes (Table 1) consisted of Pd(II) and Pt(IV) bound to an S-donor ligand (6-mercaptopurine), and such strongly bound ligands would only be replaced with great difficulty (probably by another S-donor ligand), unless some type of enzymatic action were involved. Attachment of an active ligand to a metal may greatly aid cell transport, but unless the activity is substantially better than that of the free ligand, nothing has been achieved. This appears to be the case for the complexes in Table 1.

The use of chelating agents to regulate the concentration of metal ions in biological systems has been practiced for some three decades [35]. There are several excellent reviews on the application of chelation to medicine which include discussions on drug design of chelating agents [36]. The most extensive work in this area is the monograph by Albert [37], entitled *Selective Toxicity*.

TABLE 1

Screening Results for Some Complexes with Biologically Active Ligands

Complex	Tumor	Dose range[a]	Optimum T/C	Dose[b]	Toxic Level[c]
$Na_2[Pt(mp)_2Cl_4]\cdot 2H_2O$	S180	44-150	7	150	>150
	Ca755	4-280	<1	16	>32
$Na_2[Pd(mp)_2Cl_2]\cdot H_2O$	S180	100[d]	13	100	100
	Ca755	9-36	<1	18	9-36[f]
$Na_2[(BiO)(mp)_2]\cdot 3H_2O$	S180	14-112	16	112	>112
	Ca180	0.5-46	<1	23	45
$[Pd(butp)_3Cl]Cl$	Ca755	3-1000	4	1000	>1000
$[Bi(tgn)_2H_2O]\cdot 3.5H_2O$	Ca755	5-10[e]	<1	5	5-10

[a] Daily dose.
[b] Dose at which optimum T/C was obtained.
[c] Dose at which survivors are >83%.
[d] Only one dose tested.
[e] Only two doses tested.
[f] Variable toxicity.
Source: Reprinted with permission from Ref. 2.

3. DISCOVERY AND EXPLOITATION OF PLATINUM ANTICANCER DRUGS

3.1. Discovery

As has been the case in many other medical developments, the discovery of antitumor activity in platinum ammine complexes was somewhat fortuitous [38,39]. Professor Barnett Rosenberg of Michigan State University, a physicist by training, was fascinated by the appearance of spindle cell formation during cell mitosis as they appeared to him to resemble lines of magnetic force as seen with iron filings around a magnet. Thus a study was initiated to see if an electromagnetic field would influence cell division. The setting up of these experiments involved two pieces of fortune. Firstly, ac current was passed through *Escherichia coli* bacteria in a growth chamber via a set of platinum electrodes, and secondly, the bacteria were supported in a nutrient medium containing ammonium chloride as the nitrogenous source. *E. coli*, and procaryotic cells in general, do not show mitotic figures in division and were only being used to test the equipment; this was another piece of fortune. Under the influence of the current, the bacteria underwent filamentous growth. Bacterial rods, normally some 2-5 μm in length, formed strands up to some 300 times that length. Thus cell division was inhibited while cell growth was unaffected. Further tests showed that gram-negative rods were most sensitive with gram-positive rods much less so; spherical bacilli (cocci) were unaffected.

A long series of control experiments showed that the current was not causing the filamentous growth but was causing some 10 ppm of platinum to dissolve from the electrodes. The species formed was identified as $[PtCl_6]^{2-}$, which is present in part as the ammonium salt. Fresh solutions of $(NH_4)_2[PtCl_6]$ are bacteriostatic and inhibit cell growth at these concentrations (approximately 10 ppm); however, aged solutions (2-3 days) were very effective in producing filaments at low platinum concentrations. Further studies confirmed the photochemical reaction [40]:

$$[PtCl_6]^{2-} + nNH_4^+ \rightarrow cis\text{-}[PtCl_{6-n}(NH_3)_n]^{(2-n)-} + nH^+ + nCl^-$$

which in the time period studied did not proceed much beyond the n = 2 stage. The neutral species cis-$[PtCl_4(NH_3)_2]$ is a potent inhibitor of cell division while having only a small inhibitory effect on the growth rate. Testing of synthesized cis and trans isomers showed that only the former was biologically active. The corresponding cis and trans Pt(II) species, $[PtCl_2(NH_3)_2]$, were also tested and again only the cis compound caused filamentation.

3.2. Bacterial Studies

Although metal complexes have been used in the treatment of bacterial infections from early times (e.g., As, Sb, Hg, and Ag) they have been largely superseded in modern times by organic drugs, such as sulphonamides and the antibiotics. Fully substituted inert, chelated complexes of ruthenium and osmium with phenanthroline and bipyridyl ligands, for instance, $[RuL_3]^{3+}$, [27,41] were shown to be effective bacteriocidal agents against gram-positive microorganisms. These charged complexes appeared to affect the neuromuscular junction in animals, giving rise to severe curare-like toxicity limiting their use to topical administration.

Clinical trials did establish a usefulness in the treatment of some skin infections, such as dermatosis and dermatomycosis, but they were not pursued. However, Rosenberg's work was the first to identify filamentation effects at concentrations below bacteriocidal levels. Tests on a variety of platinum amine species showed that neutral, as opposed to charged, species tended to inhibit cell division and cause filamentation. Cis configurations were active and trans, inactive. Forming a filament was not a terminal event; on removal of the platinum complex from solution (or transfer of the filaments to a normal medium suitable for growth), the filaments divide into normal bacteria [38] in contrast to the effects observed with nitrogen mustards and penicillin.

Renshaw and Thomson [42] studied the distribution of platinum from ultraviolet-irradiated $(NH_4)_2[PtCl_6]$ in *E. coli* and two gram-positive bacteria. In the latter case most of the platinum was bound to metabolic intermediates, whereas in *E. coli* it was distributed among the cytoplasmic proteins and nucleic acids. A clue to the mode of action was given by a comparison with the distribution of platinum from a fresh $(NH_4)_2[PtCl_6]$ solution, which is bacteriocidal, where nearly all the platinum was in the cytoplasmic protein and very little was associated with the nucleic acid.

Other Group-VIII metal compounds were tested against bacteria and some Ru and Rh compounds were found to cause filamentous growth in *E. coli* [87] (Table 2), but not to the same extent as that for the platinum compounds. Gillard and coworkers have since extended the Rh studies [43,44] (Sec. 4.2).

Another aspect of bacterial activity was reported by Reslova et al. [45,46], who found that lysogenic strains of *E. coli* (i.e., *E. coli* which have previously been infected by bacteriophage) can be induced by the platinum compounds to develop partial or complete viruses leading to lysis of the cell. Ultraviolet radiation, x-rays, and other known antitumor agents, also have this effect [47]. The ability to induce lysis in such bacterial strains correlates well with anticancer activity.

3.3. Antitumor Studies

The property of inhibiting cell division, but not cell growth, suggested that these compounds might have antitumor properties; this was emphasized by the fact that known antitumor agents caused elongation and lysis in lysogenic bacteria. Initially four compounds--cis-$[PtCl_4(NH_3)_2]$, cis-$[PtCl_2(NH_3)_2]$, $[PtCl_4(en)]$, and $[PtCl_2(en)]$--were tested against Sarcoma 180 in the ICR strain of mice and were found to be effective in inhibiting the tumor growth [1]. As had been predicted by the bacteriological results, neither of the trans isomers showed any appreciable activity. The two cis compounds

TABLE 2

Effects of Some Group VIII Complexes on Bacterial Growth (*E. Coli* B)

Complex	Bacteriostatic concentration ($\mu g\ ml^{-1}$)	Filament-inducing concentration ($\mu g\ ml^{-1}$)	Filaments (%)	Elongation
Rhodium				
$(NH_4)_3[RhCl_6]$		20-30	75	5-25×
$K_3[Rh(NO_2)_6]$		20-60	10	3-5×
$[RhCl_3](aq)$[a]		30-100	75	5-25×
trans-$[RhCl_2(NH_3)_4]NO_3$		25	10	0-5×
mer-$[RhCl_3(NH_3)_3]$		25	[b]	5-10×
trans-$[RhCl_2(py)_4]Cl \cdot H_2O$[c]	20	2.5	[b]	5-10×
cis- and trans-$[RhCl_2(en)_2]NO_3$		Inactive		
Ruthenium				
$(NH_4)_3[RuCl_6]$		15-30	10	5-20×
$K_2[RuCl_5(NO)]$		40-100	50	5-10×
$[RuCl_3(NH_3)_3]$	>25	5	[b]	>25×
$[RuCl_3](aq)$		Inactive		
Iridium				
mer-$[IrCl_3(NH_3)_3]$		Inactive		
$K_3[Ir(NO_2)_6]$		Inactive		
$(NH_4)_2[IrCl_6]$		Bacteriocidal		

Palladium				
$[Pd(NH_3)_2ox]$	>0.25			
$[Pd(en)ox]$	>0.25			
$(NH_4)_2[PdCl_4]$		Bacteriocidal		
Others				
$[UO_2(CH_3CO_2)_2]$		25-75	5	3-5×
$(NH_4)_2[OsCl_6]$		Bacteriocidal		
$[NiCl_2]$		Bacteriocidal		
$[CoCl_2]$		Bacteriocidal		
$[Co(NH_3)_6]Cl_3$		Inactive		
$[Ni(NH_3)_6]Cl_2$		Inactive		

[a] Widely differing results have been obtained for various samples of hydrated rhodium trichloride.

[b] Not estimated.

[c] For more data, see Ref. 2.

Source: Reprinted with permission from Ref. 2.

were submitted to the U. S. National Cancer Institute and screened against L1210 leukemia in mice [1]. The compounds showed potent antitumor activity and effected several cures with single injections at the therapeutic dose of 8 $mg \cdot kg^{-1}$ of body weight. Rosenberg and Van Camp went on to show that cis-$[PtCl_2(NH_3)_2]$ was capable of regressing large solid Sarcoma-180 tumors (8 days old) in Swiss white mice [48]. Cis-$[PtCl_2(NH_3)_2]$ appeared to be the more potent of the original compounds and has since been tested against many transplanted animal tumors and has proved to have a wide spectrum of activity. It has been under extensive toxicological and clinical trials leading to recent governmental approval for certain human tumors (see Sec. 3.4).

Since the original discovery of anticancer effects of Pt complexes, studies have been undertaken to determine what relationships exist between chemical structure and antitumor activity with a view to finding more active, less toxic drugs [49-55]. This work has been discussed in several reviews and a summary account is given below [2,56].

Platinum complexes of the type $[PtX_2A_2]$ (X_2 = two monodentate anionic ligands or one bidentate anionic ligand; A_2 = two monodentate amine ligands or one bidentate amine ligand) have attracted the most attention while limited data on other systems have been reported. It has been established that where activity exists it is only found in neutral species and in cis rather than trans isomers [49,55]. The X and A ligands have been systematically varied and the chemistry of active compounds has been investigated, particularly in kinetic terms [56].

Several compounds have been identified which show better or comparable activity against several animal tumors, to cis-$[PtCl_2(NH_3)_2]$. Some of these compounds have the advantage, from the viewpoint of mode of administration, of relatively high water solubilities although these often require higher doses to achieve maximum effects.

3.3.1. Structure-Activity Studies

Initial studies concentrated on variation of the anionic ligands (X) in species of the type cis-$[PtX_2(NH_3)_2]$, using the solid Sarcoma-180 tumor in Swiss white mice [49]. The results indicated that in the ammine system, labile groups such as NO_3^- and H_2O give rise to highly toxic species, while strongly bound ligands such as SCN^- and NO_2^- formed inactive complexes. Activity was found with monodentate ligands of intermediate lability (e.g., Cl^-, Br^-). Since this early work, several groups of researchers have reported on the complex nature of cis-$[Pt(NH_3)_2(H_2O)_2]^{2+}$ solutions which tend to polymerize on standing via the formation of hydroxo bridges [57,58]; the monomer and oligomers have widely different toxicities, and activities with the dimer being particularly toxic.

A major outcome of this phase of the work was the activity of complexes with the chelating dicarboxylate ligands, oxalate, malonate, and substituted malonates. Further testing against the ADJ/PC6A solid plasma cell tumor in BALB/$\bar{C}$ mice at the Chester Beatty Institute confirmed these results (Table 3). Subsequent synthesis and testing on other amine systems have shown that complexes with these ligands usually have comparable or superior activity to those containing chloride groups.

Tobe and Connors et al. studied the effect of varying the amine ligands (A_2) in species of the type cis-$[PtCl_2A_2]$ and found them to have a primary effect on the antitumor properties [55]. The ADJ/PC6A tumor appears to be very sensitive to platinum compounds and relatively minor structural changes can lead to major changes in the therapeutic index (TI). Most of the changes in TI are associated with toxicity rather than potency. Heterocyclic and alicyclic amines (Table 4), and straight and branched chain alkylamines all give compounds with appreciable activity and selectivity. Unfortunately the highest TIs were associated with compounds of extremely low aqueous solubility (injected intraperitoneally as a suspension in arachis oil), which means that they are probably acting as slow release systems, thus making a true

TABLE 3

Screening Data for Platinum(II) Malonate Derivatives Against the ADJ/PC6A Tumor

Malonato complexes	LD_{50} ($mg \cdot kg^{-1}$)	ID_{90} ($mg \cdot kg^{-1}$)	TI	Aqueous solubility (mM)
$[Pt(NH_3)_2mal]$	225	18.5	12 2	1.0
$[Pt(NH_3)_2Memal]$	112	4.5	24.9	7.0
$[Pt(NH_3)_2Etmal]$	132	12	11	160.0
$[Pt(NH_3)_2OHmal]$	150	4.9	30.6	-
$[Pt(NH_3)_2Benzmal]$	150	1.85	81.1	-
$[Pt(NH_3)_2(1,1\text{-}CBDCA)]$	180	14.5	12.4	50.0
[Pt(en)mal]	220	18.5	12	-
[Pt(en)Memal]	200	50	4	-
[Pt(en)Etmal]	450	49	9.2	-
$[Pt(MeNH_2)_2mal]$	670	56	12	-

Source: Reprinted with permission from Ref. 56.

comparison of toxicity impossible. Chelating amines such as 1,2-diaminocyclohexane and o-phenylenediamine were also effective. Results on the plasma cell tumor indicated that toxicity may not be so closely related to activity as to prevent the existence of active complexes with relatively low toxicities. The S180 tumor was much less sensitive to variations in amine ligands although primary alkylamines showed more activity than secondary [49].

Results on the L1210 tumor in BDF_1 mice [56,59] have confirmed the activity of straight- and branched-chain alkylamines and alicyclic amines, particularly at the C_3-C_4 level (Table 5). The most impressive results were obtained on a daily 1-9 dose schedule. Preliminary estimates of nephro- and myelotoxicity on mice by measurement of blood urea nitrogen (BUN) levels in urine and white blood cells (WBC) in blood indicate that some compounds may be less toxic in both respects than cis-$[PtCl_2(NH_3)_2]$.

TABLE 4

Screening Data for Alicyclic Amine Complexes Against the ADJ/PC6A Tumor

A	Solvent	Dose range $(mg \cdot kg^{-1})$	Dose response	LD_{50} $(mg \cdot kg^{-1})$	ID_{90} $(mg \cdot kg^{-1})$	TI
ADJ/PC6A plasma cell tumor						
NH_3	A	0.1-40	+	13.0	1.6	8.1
CH_3NH_2	A		-	18.5	18.5	1.0
$ClC_2H_4NH_2$	A		+	45.0	17.5	2.6
NH	A	2.5-160	+	56.5	2.6	21.7
NH	A	3-200	+	141	10.8	13.1
C_2H_4OH NH	A		-	90	>90	<1.0
O NH	A		-	18	>18	<1.0
NH_2	A	1-80	+	56.5	2.3	24.6
NH_2	A	6-750	+	90	2.9	31.0
NH_2	A	1-3200	+	565.6	2.4	235.7
NH_2	A	1-3200	+	>3200	12	>267
NH_2	A	5-625	+	>625	18	>35

TABLE 5

Screening Data for Platinum Complexes Against L1210 Tumor (10^6 Cells) in BDF_1 Mice

Complex	Optimum dose ($mg\ kg^{-1}$)	Schedule (days)	Median survival (% T/C)	Therapeutic ratio (MTD-MED)	Toxicity	
					(BUN)	(WBC)
cis$[PtCl_2(CH_3NH_2)_2]$	16	1	129	1	+[a]	±[b]
	2	1-9	121	1		
cis$[PtCl_2(n\text{-}C_3H_7NH_2)_2]$	8	1	157	2	+	+
	8	1-9	157	2		
cis$[PtCl_2(i\text{-}C_3H_7NH_2)]_2$	32	1	171	1	+	±
	8	1-9	179	2		
cis$[PtCl_2(C_3H_5NH_2)_2]$	16	1	157	4	+	+
	8	1-9	164	2		
cis$[PtCl_2(OH)_2(i\text{-}C_3H_7NH_2)_2]$[c]	32	1	171	2	+	±
	16	1-9	207	2		
cis$[PtCl_2(i\text{-}C_4H_9NH_2)_2]$	64	1	171	4	+	±
	16	1-9	193	8		
cis$[PtCl_2(t\text{-}C_4H_9NH_2)_2]$	64	1	Inactive			
	32	1-9	Inactive			

cis[$PtCl_2(C_4H_7NH_2)_2$]	32	1	157	4	+	±
	16	1-9	221	4		
[Pt(mal)(1,2-DAC)]	32	1	154	-		
	16	1-9	254	4		
[Pt(mal)($C_5H_9NH_2)_2$]	128	1	-	-		
	128	1-9	-	-		
[Pt(OHmal)($NH_3)_2$]	64	1	150	1	±	+
	32	1-9	200	2	+	+
[Pt(OHmal)(i-$C_5H_{11}NH_2)_2$]	128	1	-	-		
	128	1-9	-	-		
[Pt(Etmal)($NH_3)_2$]	128	1	171	2	+	±
	64	1-9	186	4	+	+
[Pt(1,1-CBDCA)($NH_3)_2$]	128	1	150	2	+	±
	64	1-9	157	2	+	+

[a] + Denotes toxicity lower than for cisPt(II).

[b] ± Denotes toxicity comparable with cisPt(II).

[c] trans (OH) groups.

Source: Reprinted with permission from Ref. 56.

Some malonato and substituted malonato complexes have also shown promise in this system (Table 5).

Tobe and Connors et al. went on to exploit the (IV) oxidation state of platinum in order to synthesize more water-soluble compounds. Trans dihydroxo Pt(IV) species were synthesized and the hydrophilic properties of the OH^- ligands gave improved aqueous solubility while maintaining the activity in some cases [55]. In general the order of activity was

$$[Pt(II)Cl_2A_2] > [Pt(IV)Cl_4A_2] \sim [Pt(IV)Cl_2A_2(OH)_2]$$

Although variations in lipid solubility were noticed, these did not correlate with the antitumor effect.

Present efforts appear to be concentrated on a comprehensive study of the more effective amine systems, reported by Tobe and Connors et al. for chloro complexes, with a range of different leaving groups in an attempt to optimize activity and solubility [56]. The ligands under investigation include bi- and monodentate carboxylates and O-bonded anionic groups such as sulphate, nitrate, and phosphate. Gale [53] and Speer [54] and their coworkers have reported similar studies which have concentrated on the 1,2-diaminocyclohexane system. Little data is available but initial results indicate that at least some compounds will match or improve upon the activity shown in the chloro species (Table 6).

Several of the promising compounds have been subjected to animal testing against a wide variety of transplanted tumors, mainly via the U. S. National Cancer Institute. The compounds $[Pt(Etmal)(NH_3)_2]$, $[Pt(CBDCA)(NH_3)_2]$ (CBDCA = 1,1-cyclobutanedicarboxylate), and cis-$[PtCl_2(OH)_2(isopropylamine)_2]$ show good activity against most systems; a selection of results for the latter is shown in Table 7 [56]. All of these compounds have relatively high aqueous solubilities.

TABLE 6

Screening Data for Platinum Complexes Against the L1210 Tumor (10^6) Cells in BDF_1 Mice

Complex	Optimum dose ($mg \cdot kg^{-1}$)	Schedule (days)	Median survival (% T/C)	Therapeutic ratio (MTD-MED)
[Pt(mal)(1,2-DAC)][a]	32	1	154	-
	16	1-9	254	4
$[Pt(ClAc)_2(i\text{-}C_3H_7NH_2)_2]$[a]	32	1	179	4
	16	1-9	207	8
$[Pt(ClAc)_2(C_5H_9NH_2)_2]$[a]	128	1	-	-
	32	1-9	143	1
$[Pt(NO_3)_2(i\text{-}C_3H_7NH_2)_2]$[a]	64	1	171	4
	16	1-9	193	4
$[Pt(SO_4)(H_2O)(1,2\text{-DAC})]$[b]	4	1	119	-
	5	1	116[c]	-
	3.33	1,5,9	285[c]	-
	0.6	1-9	239[c]	-

[a] Testing data of Bristol Laboratories.

[b] Data of Gale et al., U.S. Patent Appl. 769,888.

[c] 10^5 cells injected.

Source: Reprinted with permission from Ref. 56.

TABLE 7

Multiple Tumor Screening Data for $[PtCl_2(OH)_2(i\text{-}C_3H_7NH_2)_2]$

Tumor	Host	Parameter	Schedule (days)	Optimum dose ($mg \cdot kg^{-1}$)	% T/C
B16 melanoma	BDF_1	MDST	1-9	12.5	164
P388	CDF_1 (a)	MDST	1-9	18.0	202
	CDF_1 (a)	MDST	1,5,9	50	154
L1210	BDF_1 (a)	MDST	1	32	171
	CDF_1 (b)	MST	1	50	137
	BDF_1 (b)	MST	1-5	20	149
	BDF_1 (a)	MDST	1-9	16	207
	CDF_1 (b)	MST	1-9	25	191
Lewis lung carcinoma	BDF_1 (b)	MDST	1-9	6.25	127
Colon 26	CDF_1	MDST	1,5,9	25	161
Colon 38	BDF_1	MTW	1,8,15	25	52
Mammary	CDF_1	MTW	1,8,15, etc.	50	8
L1210/cytoxan (c)	BDF_1 (b)	MST	1-5	30	166
L1210/BCNU (c)	BDF_1 (b)	MST	1-5	20	139
L1210/L-sarcolysine (c)	BDF_1 (b)	MST	1-5	20	118

(a) 10^6 cells.

(b) 10^5 cells.

(c) Resistant to drug shown.

Source: Reprinted with permission from Ref. 56.

3.3.2. Reactivity of Antitumor-Active Species

The chemistry of platinum(II) ammine species of the type $[PtX_2A_2]$ is dominated by the high affinity of NH_3 for the Pt(II) center. Affinities for common ligands vary [60]:

$$CN^- > NH_3 \sim OH^- > I^- > SCN^- > Br^- > Cl^- > F^- \sim H_2O$$

The strength of the Pt-N bond tends to override the trans effect factors, which often control the substitution kinetics of Pt(II) complexes (along with smaller cis effects). Organic amines, especially the simpler alkyl- and alicyclic amines are expected to behave similarly to NH_3. Thus chemical studies on both isomers of $[PtX_2(NH_3)_2]$ have clearly shown that X are the reactive or leaving groups, while the NH_3 ligands are relatively inert [61,62]. The order of leaving ability has been established for the reaction [62]

$$[Pt(dien)X]^+ + py \rightarrow [Pt(dien)py]^{2+} + X^-$$

where the order of decreasing rate constants is

$$X = NO_3^- > H_2O > Cl^- > Br^- > I^- > SCN^- > NO_2^-$$

The spread in rates for the reactions is about 10^6, showing that the leaving group X and the consequent breaking of the Pt-X bond have a substantial effect on the reaction site. The testing results against the S180 system reflected this order, with strongly bound (poor-leaving) ligands giving rise to inactive species, while reactive aquo ligands were very toxic. However, aquo species from bulkier organic amines were far less toxic [49] and recent studies have shown that that X ligands (such as sulphate, which is expected to be very reactive) can give rise to active species. Much of the aquo species toxicity appears to be due to polymerization by formation of hydroxo bridges, which seems to be a slow process for some of the organic amine systems.

The amine ligands (A) have a primary effect on the antitumor properties, while they are expected to have only a secondary effect on reactivity via different steric, electronic, and basic properties. Aqueous solutions of organic amine complexes containing various leaving ligands have been examined by conductivity and ultraviolet/visible spectroscopy techniques in order to compare their reactivities against a common incoming group [56]. The results may be summarized as follows. Variation in the amine (A) group has little effect on the solvolysis of the complexes by water and DMSO. The wide variations in activity do not correlate with reactivity and appear to be due to biophysical, rather then chemical, factors.

Although the empirical rules concerning neutrality and a cis configuration still hold, we can now identify three classes of active compounds on a kinetic basis:

1. Reactive species, such as sulphato and nitrato complexes, which are rapidly hydrolyzed and also converted to chloro species in the presence of physiological levels of saline.
2. Species with intermediate reactivity towards water and chloride. Reactions of chloro complexes themselves are, of course, suppressed in the presence of chloride ion which serves to protect them in the serum. Compounds containing halogenoacetate ligands will undergo chloride replacement at an intermediate rate.
3. Bidentate carboxylate complexes are the only kinetically inert species to show activity so far. These are so nonreactive in comparison to other antitumor-active species that we have previously mentioned that an in vivo activation mechanism might be operating, possibly involving enzymes [2].

3.4. Clinical Studies With Cis-$[PtCl_2(NH_3)_2]$

Since the commencement of Phase 1 clinical trials in 1972 under the auspices of the U. S. National Cancer Institute, cis-$[PtCl_2(NH_3)_2]$ (cisplatin) has shown significant activity against a range of tumor types, particularly those of the genitourinary region. Extensive data have been compiled for the treatment of testicular and ovarian cancer patients. The compound is now recommended as first-line chemotherapy for these tumors, particularly in combination with other approved chemotherapeutic agents. Other tumors have shown notable responses, but the data are not yet extensive enough for official approval; these include bladder, head and neck, cervix, prostate, and lung tumors.

Human studies with radioactive platinum confirmed animal work and showed high uptake in the kidney, liver, and intestine [63-66]. Evidence from brain scans and brain tissue samples suggested poor penetration of the drug into the central nervous system [66,67]. After initial injection, Pt was rapidly cleared from the blood and more than 90% of that remaining in the post-distribution phase (after approximately 30 min) was protein-bound [68,69].

Cis-$[PtCl_2(NH_3)_2]$ is excreted primarily in the urine but only 30-50% of the injected dose is excreted in the first 5 days [68]; the remaining platinum, much of which is in the liver, is only removed slowly. Nephrotoxicity is the major dose-limiting toxicity for the drug, but this has been largely overcome by the use of prehydration with diuresis and infusion techniques. Whereas renal toxicity was originally significant with doses around 50 mg$\cdot$M^{-2},* doses around 120 mg$\cdot$M^{-2} can be safely given, using these techniques. The ability of the drug to act synergically in combination with other established chemotherapeutic drugs has led to the use of regimens involving low non-nephrotoxic doses. Bone marrow toxicity occurs, largely associated with a reduction in circulating white blood cells,although this is lower than that for most other antitumor drugs. Other side effects include nausea, vomiting, and high-frequency hearing loss (ototoxicity). Peripheral neuropathy has been observed on repeated treatment.

In December 1978 cis-$[PtCl_2(NH_3)_2]$ received U. S. Food and Drug Administration approval as an anticancer drug for testicular and ovarian cancers. The formulated drug containing sodium chloride and mannitol is marketed under the name of Platinol. British approval of the drug came in March 1979; it is known in the U. K. as Neoplatin. License applications are pending in many other countries.

---

*Dosage given as milligram per square meter surface area.

3.5. Mechanism of Antitumor Action of Metal Complexes

Filamentous growth in bacteria is indicative of the ability of an agent to react with DNA, giving selective inhibition of DNA synthesis but no accompanying effect on other biosynthetic pathways. Induction of bacteriophage from lysogenic bacteria and the mutogenicity of some active complexes are also important evidence for direct DNA attack [70,71,88].

Biochemical studies on cells in culture have shown that cis-$[PtCl_2(NH_3)_2]$ selectively and persistently inhibits the rate of DNA synthesis, as compared to RNA and protein synthesis [72-75] (Fig. 1). It is postulated that the primary chemical lesion is in the DNA, inhibiting it as a template for replication but not affecting transcription or translation.

Studies on cis and trans isomers of $[PtCl_2(NH_3)_2]$ on cells in vitro showed that trans binds to cell macromolecules as effectively as cis [74]. There are more platinum moieties bound per molecule of DNA than to either RNA or protein. Interstrand cross-linking has been demonstrated to occur for Pt compounds, but to a much lesser extent than for alkylating agents. The balance of evidence highly favors the proposal that this form of binding to DNA is not an important cytotoxic event [75]. Linking between bases on the same strand has been proposed [76] and some evidence for this has been obtained, particularly from work on the inactivation of bacteriophage [77].

Guanine, adenine, and cytosine react with both $[PtCl_2(NH_3)_2]$ isomers, the rate being fastest for guanine [78-80]; the N-7 sites of adenine and guanine and N-3 sites of cytidine are favored. A very slow reaction occurs with thymine. Studies on DNAs with varying GC-AT ratios suggest guanosine as a major reaction site [81,82]. Intrastrand linking between neighboring guanine bases has been proposed [83]. Chelation between the N-7 and O-6 on the same guanine has been claimed for the cis and not the trans isomer [84] from ESCA studies and is a current point of controversy.

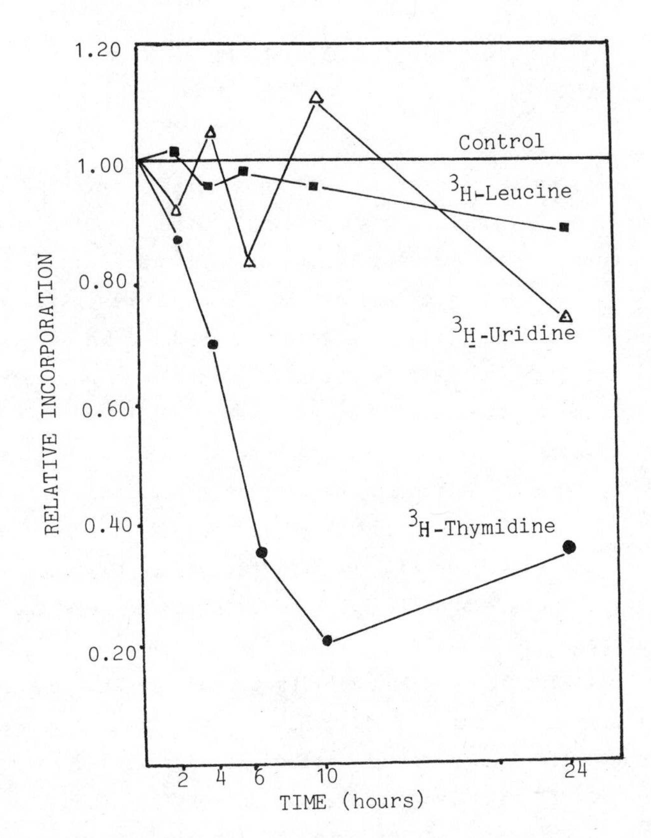

FIG. 1. Selective inhibition of DNA synthesis in human AV_3 cells grown in tissue culture with exposure to 5 μM cis-$[PtCl_2(NH_3)_2]$. Reprinted with permission from Ref. 47.

It seems that only a proportion of the lesions caused by cis-$[PtCl_2(NH_3)_2]$ can be recognized by a DNA excision repair process although certain mammalian cells have effective postreplication repair systems. Inability to synthesize past some Pt lesions eventually leads to cell death [69]. An interesting question which remains unanswered is whether the many different active platinum compounds now reported produce the same type of lesion. Are all platinations equal, or do some compounds give a higher proportion of a more lethal lesion, leading to less Pt on the DNA at the therapeutic dose? This could give less overall toxicity.

4. STUDIES ON OTHER METALS

4.1. General Considerations

The platinum studies have stimulated considerable interest in other metal complexes although, not surprisingly, the amount of work completed to date is far less than that for platinum. Cis-$[PtCl_2(NH_3)_2]$ differs from many of the compounds discussed in Sec. 2, in that it contains simple inorganic ligands which have no carcinostatic ability in their own right. Platinum is not normally found in cells (unlike Cu, Co, and Zn) and, once bound to a macromolecule, is likely to form a strong bond which cells will find difficult to repair. Although the kinetic properties of the active platinum compounds vary considerably, it seems logical to suppose that very inert compounds will have little chance of success while very reactive species are likely to react prior to reaching the site of action.

General kinetics and equilibrium trends are well established for metal compounds, particularly those of the transition metals. Bond strengths for complexes involving the same ligand increase on going from first- to third-row transition metals. Thus heavier metals should form less easily repaired lesions although excision repair may invalidate this point. Regardless of the metal's oxidation state, the ligands will affect the stability of the complex. The heavier transition metals are generally referred to as being class "b" (soft) in character or are intermediate, tending toward class b [85]. Thus these metals form their most stable bonds with heavier donor atoms. Generally the following order of decreasing stability applies:

$$S \sim C > I > Br > Cl > N > O > F$$

This is particularly important when considering the possible sites of reaction with biological macromolecules. Relatively inert metal systems, such as those of Cr(III) and Co(III) are of class "a" and tend toward hard ligands; hence the reverse order will generally apply.

Inertness to ligand substitution for transition metals increases in the order first row < second row < third row for a given set of ligands. This property of transition metals is presumably reflected in the observation that of the essential transition metals found in biological systems, only Mo is not in the first row. It is likely that second- and third-row metal complexes are too inert to perform metabolic functions. Within a particular row, independent of mechanism, the contribution of crystal field energy predicts that for octahedral complexes, the following order of inertness will generally occur [60]: d^6(low) > d^3 > d^4(low) > d^5(low); d^0, d^1, d^2, d^5(high-spin), d^6(high-spin), and d^7(high-spin) are relatively labile systems compared with the above.

Many first-row metals may be too labile, except for Co(III) (d^6 low-spin) and Cr(III) (d^3), although Cu(II) shows activity (Sec. 4.6.). In the second and third rows, most of the potentially useful systems lie among the platinum group metals, such as Rh(III), Ru(II), Ir(III), Pt(IV) (d^6); Ru(III) (d^5); and Os(IV) (d^4). Mo(III), W(III), and Re(IV) (d^3) are relatively unstable and are likely to hydrolyze under physiological conditions. As Thomson et al. have pointed out [46], the biological system acts at a low redox potential and many metals will tend to exist in their lower oxidation states, e.g., Os(II) and Ru(II) rather than Os(III) and Ru(III). This may well be the explanation for the activity of certain Pt(IV) cases, as all active Pt(IV) compounds have an active Pt(II) analog. It is also possible that introduction of the metal in a higher oxidation state could be advantageous in terms of cell transport ability. Apart from Pt(II), the d^8 square planar possibilities are Pd(II), Au(III), Rh(I), and Ir(I) [assuming Ni(II) is too labile]. Under physiological conditions Au(III) is likely to be reduced to Au(I) (linear or tetrahedral) and Rh(I) and Ir(I) will tend to undergo oxidative addition to the Rh(III) and Ir(III) state. For comparable compounds the reaction rates of Ni(II), Pd(II), and Pt(II) are in the approximate ratio $5 \times 10^6:10^5:1$, respectively [60]; Au(III) complexes are somewhat slower than Pd(II).

The labilizing effect of strongly bound trans ligands is especially important in these square planar systems.

These general comments are intended to highlight the importance of varying stability, and particularly the reactivity, of a drug system in achieving the difficult task of reaching and reacting at the site of antitumor action. Many other properties are, of course, important. Aqueous and lipid solubility will affect both transport through membranes into cells and removal from the system via the kidneys. Of the other metals which have been the subject of anticancer study, by far the most work has been carried out on rhodium, largely due to observations on bacterial activity. Other platinum-group metals have been the subject of preliminary investigations. Copper salts and complexes were identified as potential agents prior to the platinum discovery and studies have been continuing since the mid-sixties. The major study outside of the transition group has involved the Group-IIIA metals, with gallium nitrate showing sufficient promise for a limited clinical trial sponsored by the National Cancer Institute (Sec. 4.9.).

4.2. Rhodium

4.2.1. Bacterial Studies

The study of biological activity in Rh complexes was pioneered by Gillard and coworkers [43,44], who reported on the bacterial activity of a large number of complexes of the type trans-$[RhX_2L_4]Y$, where L_4 = pyridine, substituted pyridine, or other heterocyclic amine ligands or two bidentate N donors such as ethylenediamine, 2,2'-bipyridyl, or 1,10-phenanthroline (Table 8). Many of these have antibacterial activity which is more pronounced for gram-positive than gram-negative bacteria with massive filamentation occurring at sublethal doses, indicating an interference with cell division. The Rh(I) state may be involved, as only the most easily reduced compounds showed activity as indicated by measurement of reduction potentials. Filamentation also appeared to be related

TABLE 8

Bacterial Activity of Rh(III) Amine Complexes

Complex	Filamentation (amount %/elongation)		$E_{1/2}$	log K_D [a]
cis-$[RhCl_2(en)_2]NO_3$	-	-	-520	-
trans-$[RhCl_2(en)_2]NO_3$	-	-	-460	1.78
cis-$[RhCl_2(udmen)_2]ClO_4$	-	-	-420	2.19
trans-$[RhCl_2(udmen)_2]ClO_4$	-	-	-230	1.68
trans-$[RhCl_2(trimen)_2]ClO_4$	3	5×	-240	2.21
trans-$[RhCl_2(tetmen)_2]Cl$	100	60×	-110	2.31
trans-$[RhCl_2(py)_4]Cl$	85	10-30×	+30	1.71
trans-$[RhCl_2(4\text{-}Mepy)_4]Cl$	98	20-100×	-150	0.68
trans-$[RhCl_2(4\text{-}Etpy)_4]Cl$	98	50-100×	-30	-0.19
trans-$[RhCl_2(iquin)_4]Cl$	100	100-300×	+20	-2.12
trans-$[RhCl_2(4\text{-}Prpy)_4]Cl$	50	50-100×	-120	-2.24
trans-$[RhCl_2(4\text{-}tBupy)_4]Cl$	25	3-30×	-190	-2.70

[a] $K_D = \frac{\text{concentration of complex in aqueous phosphate (pH7)}}{\text{concentration of complex in octanol}}$.

Source: Reprinted with permission from Ref. 44.

to lipophilic character. Neither the position of substitution nor the counterion seem to be important. In these complexes the Rh(I) (d^8) state is generated comparatively easily and is thought (in catalytic quantities) to be responsible for the enhanced chemical reactivity of amine complexes in the presence of reducing agents.

Some of the above complexes have been tested for antitumor activity but failed to show activity against a screen which is very sensitive to platinum compounds (ADJ/PC6A, Chester Beatty Institute, London) [86]. Despite the fact that Pt(II) and Rh(I) are both d^8 square planar systems, there are several reasons to believe that different in vivo mechanisms could be involved. There is microscopic evidence for a membrane effect in the filamentation and it has been suggested that Rh(I) may be generated within the cell

TABLE 9

Screening of Rhodium Amine Complexes

Tumor	Complex vehicle[a]	Dose range	Dose response	Toxic level	T/C	Dose
Sarcoma 180	mer-$[Rh(NH_3)_3Cl_3]$ (S.S.)	12-100	+	70-100	17	100[b]
	(S.)	5-30[c]	+	20	31	20
	mer-$[Rh(dien)Cl_3]$(S.S.)	25-100	±	>100	54	50-100
	trans-$[Rh(NH_3)_3Cl_2(H_2O)]NO_3$(W)	50-150	±	125-150	48	150
	mer-$[Rh(NH_3)_3(Cl)ox]$(W.S.)	50-125	-	>125	73	125
	trans-$[Rh(NH_3)_4Cl_2]NO_3$(W)	25-100	-	>100	82	25-100
	trans$[Rh(py)_4Cl_2]Cl$(S)	20-50	±	40-50	45	50[b]
	mer-$[Rhpy_3Cl_3]$(S.S.)	50-200	±	100	51	100[b]
	cis-$[Rh(en)_2Cl_2]NO_3$(S)	50-200	±	>200	63	50-200
	trans-$[Rh(en)_2Cl_2]NO_3$(S)	5-50	±	>50	46	50

Sarcoma 180 (ascites)[d]	mer-$[Rh(NH_3)_3Cl_3]$(S.S.)	50-100	+	>100	23[e]	100
				LD_{50}	ID_{90}	TI
ADJ/PC6A[f]	mer-$[Rh(NH_3)_3Cl_3]$(A)	4-500	+	235	86	2.6
	mer-$[Rh(NH_3)_3(NO_3)_3]$(A)	12-1500	+	135	59	2.3
Walker 256 carcinosarcoma[g]	mer-$[Rh(NH_3)_3Cl_3]$	5-160	+	~170	~40	~4

[a] S = saline; S.S. = saline slurry; W = water; W.S. = aqueous slurry, A = Arachis oil.
[b] 66% survivors only.
[c] Daily dose for 8 days.
[d] Swiss white mice.
[e] % increase in life span.
[f] BALB/C̄ mice.
[g] Rats.

Source: Reprinted with permission from Ref. 51.

envelope, where the reductive enzymes of the electron transport chain are situated. These compounds are poor inducers of phage from lysogenic bacteria [44] in contrast to active Pt(II) species. The trans structure contrasts with the required cis configuration in platinum and binding studies to cell components would be of interest.

Rosenberg and his coworkers also demonstrated filamentation effects in *E. coli* for some rhodium compounds as indicated in Sec. 3.2 [87] (Table 2).

4.2.2. Rhodium Amine Complexes

A series of Rh(III) amine complexes were tested on the basis of the criteria from structure-activity studies on the platinum compounds. Thus some emphasis was placed on complexes with cis leaving groups and/or neutrality (Table 9) [51]. Activity was found in species of the type mer-$[RhX_3A_3]$ (A = NH_3 or three monodentate amines or one tridentate amine ligand; X = anionic ligands). One particular complex, mer-$[RhCl_3(NH_3)_3]$, has shown definite activity against four different animal tumor systems (Table 9) and is the best tested example of an active nonplatinum complex in the second and third rows of the transition series. It is, however, inferior to most of the active platinum compounds against all but the Walker 256 carcinosarcoma tumor. A limited amount of structure-activity work has been attempted (Table 10) with limited success, possibly due to low aqueous solubility [88]. However, this remains an area which is well worth further attention both in synthetic structure-activity terms and in studies on cell binding.

4.2.3. Rhodium(I) Complexes

Using the rationale of a cis configuration in a square planar d^8 system, Giraldi and coworkers have studied some Rh(I) complexes containing 1,5-cyclooctadiene (1,5-COD) {[Rh(1,5-COD)A]Cl, where A = bipy or o-phen and [RhCl(1,5-COD)A] where A = NH_3 or piperidine} [89]. These showed little or no activity against the S180 or L1210

TABLE 10

Screening Data for $[RhX_3(NH_3)_3]$ Derivatives Against the ADJ/PC6A Tumor[a]

	LD_{50}	ID_{90}	TI
$[RhCl_3(C_6H_5CH_2NH_2)_3]$	>1000		
$[RhCl_3(C_2H_5CN)_3]$	>250		
$[Rh(C_6H_5CO_2)_3(NH_3)_3]$	67	25	2.7
$[Rh(C_6H_5CH{=}CHCO_2)_3(NH_3)_3]$	121		
$[RhCl_3(morpholine)_3]$	280		

[a]Doses in $mg \cdot kg^{-1}$.

Source: Reprinted with permission from Ref. 88.

mouse systems but some activity against the Ehrlich ascites; again the activity is far less than for Pt compounds. The authors suggest that oxidative addition to Rh(III) may be important, as Rh(III) is a less reactive system than Rh(I); the Cl ligands in Rh(I) systems are some 10^3 times more reactive than those in Pt(II). Tests with [RhCl(1,5-COD)(pip)] on the incorporation of labeled biological macromolecule precursors into Ehrlich ascites cells indicated selective inhibition of protein synthesis {^{3}H-leucine incorporation}, which is an interesting contrast with cis-$[PtCl_2(NH_3)_2]$. However, tests with [Rh(1,5-COD)(bipy)]Cl on *E. coli* bacteria showed a reduction in both DNA and RNA synthesis with protein unaffected. The compound [Rh(acac)(1,5-COD)] also showed activity against the Ehrlich ascites test system (superior to cis-$[PtCl_2(NH_3)_2]$ in a comparison test) and seemed to show selective inhibition of uridine (RNA) uptake; it was also inactive against S180 and L1210 tumor systems [90]. Histological damage was claimed to be less pronounced than for cis-$[PtCl_2(NH_3)_2]$. These Rh(I) studies indicate some potential and should perhaps be extended to a more representative selection of tumors. The confused situation in the cell binding studies should be resolved.

4.2.4. Rhodium Carboxylates

Early reports on the antitumor activity of platinum complexes indicated that binding of these complexes to macromolecules in vivo inhibited DNA, RNA, and protein synthesis. Tetra-μ-acetato-dirhodium (II) (Fig. 2) readily forms adducts by coordination with nucleophiles at the axial positions of the molecule and it was thought that this complex could also inhibit biological processes in vivo by this method [91]. Initial trials used the Ehrlich ascites tumor in Swiss white mice and the L1210 tumor in BDF_1 mice. Although the acetate complex was ineffective against the latter tumor, significant activity was demonstrated with the Ehrlich ascites tumor, especially when the rhodium complex was used in combination with arabinosyl-cytosine. Comparative studies with the acetate and propionate complexes on the Ehrlich ascites tumor demonstrated higher activity for the propionate by a factor of 10. A much lower dose was used for the propionate in this experiment (2 $mg \cdot kg^{-1}$ propionate, 16 $mg \cdot kg^{-1}$ acetate) due to the higher toxicity of the complex. In vitro experiments revealed that both complexes bind to DNA and RNA polymerases, but the difference in degree of inhibition was only a factor of 2 in contrast to the antitumor activity detailed above.

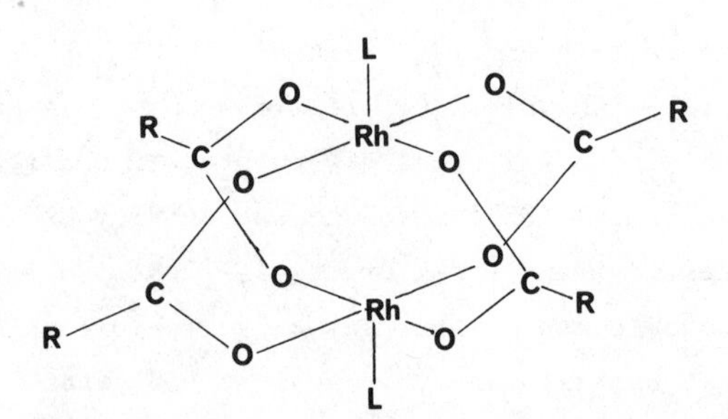

FIG. 2. Structure of tetra-μ-carboxylatodirhodium(II) complexes [rhodium(II) carboxylates], where L represents the axial ligands and R, the carbon chain of the carboxylate groups. When R = $-CH_3$, the complex is rhodium(II) acetate; R = $-CH_2CH_3$, rhodium(II) propionate; R = $-CH_2OCH_3$, rhodium(II) methoxyacetate. Reprinted with permission from Ref. 92.

Subsequent trials compared the antitumor activity of the acetate, methoxyacetate, propionate, and butyrate complexes, where it was found that the butyrate was the most active with an increase in lifespan (ILS) of 197% for mice bearing the Ehrlich ascites tumor (Table 11) [92]. The interaction of these carboxylate complexes with biological macromolecules was also investigated using equilibrium dialysis techniques and from the characteristic color changes observed (blue-green → pink), it was concluded that binding occurs to nitrogen donor ligands at the axial positions as predicted earlier [93]. The stability of complexes formed by the acetates and propionate with 5'-AMP, 5'-ADP, and 5'-ATP in aqueous solution was also determined in an attempt to correlate the observed activity of the rhodium species with the thermodynamic stability of the complexes formed between rhodium carboxylates and macromolecules [94]. Both monomolecular and bimolecular adducts were formed with the nitrogen donor ligands and the relative stabilities matched the observed trend of antitumor activity, i.e.,

propionate > acetate > methoxyacetate

although the difference in stability for these adducts was insufficient to account for the variation in biological effects.

The observation of a synergistic effect between the acetate and arabinosyl-cytosine (ara-C) led to an investigation of the effect of the carboxylate complex on cellular deaminases [95]. The latter are known to deactivate ara-C, and rhodium acetate was subsequently found to inhibit this class of enzymes, particularly cytidine deaminase, indicating that this may be the mechanism by which the acetate improves the clinical activity of ara-C. The acetate and propionate complexes were shown to inhibit DNA synthesis in vitro [91]. In vivo studies on the propionate and butyrate were carried out using Ehrlich ascites cells taken from mice at different times after drug administration (Fig. 3) [93]. Cytosine arabinoside was used as a positive control and this inhibited thymidine incorporation into DNA by more than 90% of the control level at 1 hr after administration. Neither of the rhodium

TABLE 11

Screening Data for Rhodium Acetate Derivatives Against the Ehrlich Ascites Tumor

Compound	Saline			Isopropyl Myristate			% Rh bound to tumor cells	Partition Coefficients[a]	
	LD_{50} ($mol \cdot kg^{-1}$ × 6 days)	% ILS[b]	TI	LD_{50} ($mol \cdot kg^{-1}$ × 6 days)	% ILS	TI		Chloroform	1-Octanol
$RhCl_3$	4.2×10^{-3}			Insoluble			12.0	<0.0005	1.32×10^{-4}
$Rh_2[OOCCH_2OCH_3]_4$	5.9×10^{-4}	50	1.1	Insoluble			2.5	<0.005	5.21×10^{-2}
$Rh_2[OOCCH_3]_4$	2.6×10^{-4}	88	1.8	Insoluble			25.0	0.071	6.54×10^{-2}
$Rh_2[OOCCH_2CH_3]_4$	1.9×10^{-5}	149	4.2	1.2×10^{-5}	175	2.0	68.0	0.540	8.17
$Rh_2[OOC(CH_2)_2CH_3]_4$	6.0×10^{-6}	197	4.7	1.4×10^{-6}	179	2.4	Insoluble	15.6	898
$Rh_2[OOC(CH_2)_3CH_3]_4$	Insoluble			1.7×10^{-6}	236	3.1	Insoluble	>300	>3000
$Rh_2[OOC(CH_2)_4CH_3]_4$	Insoluble			2.2×10^{-6}	136	2.0	Insoluble	>300	>3000

[a] All partition coefficients were determined at 37°C.

[b] % ILS and TI are based on the LD_{10} dose.

Source: Reprinted with permission from Ref. 92.

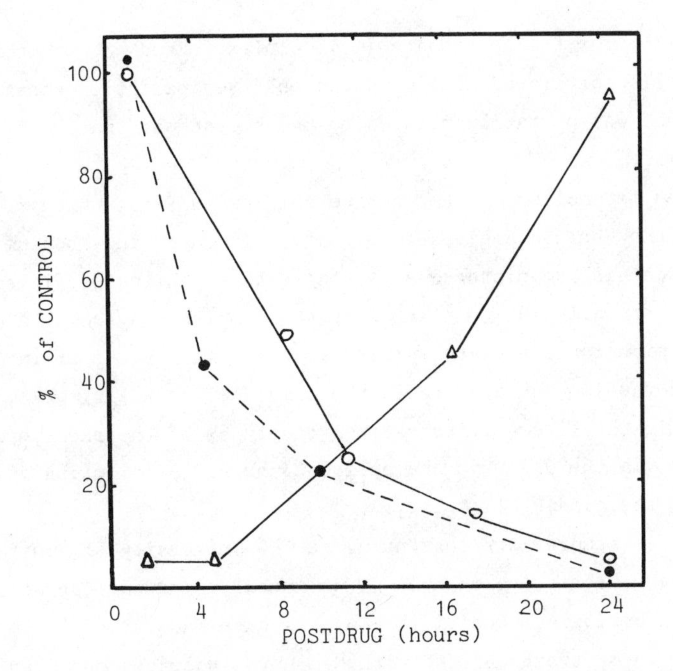

FIG. 3. In vivo inhibition of DNA synthesis by rhodium(II) carboxylates. O—O—O = rhodium(II) butyrate (0.6 mg kg^{-1}); ●--●--● = rhodium(II) propionate (2 mg kg^{-1}); △--△--△ = cytosine arabinoside (15 mg kg^{-1}). A 30-min labeling time of thymidine-5-methyl-^{3}H (specific activity, 10 μCi·$mmol^{-1}$) was not included in postdrug intervals. Each point represents the average of three mice per group. Reprinted with permission from Ref. 93.

compounds showed appreciable inhibition of DNA synthesis after 1 hr, but inhibition increased as a function of time to greater than 90% at 24 hr. The prolonged effect on DNA synthesis is reminiscent of that observed with platinum compounds.

In an attempt to rationalize the differential toxicity and antitumor activity for rhodium carboxylates, a series of complexes were synthesized with C_1-C_5 alkyl chain lengths. The saline-chloroform and saline-1-octanol partition coefficients were determined together with the activity against the Ehrlich ascites tumor and the extent of rhodium binding to tumor cells (Table 11) [92]. The C_4, C_5 analogs were insoluble in saline and were administered in isopropyl myristate but the maximum activity for the series was an

ILS of 236% for the pentanoate (C_4) complex. Rhodium uptake by the tumor cells correlated well with the observed partition coefficients and was primarily related to the hydrophobic nature of the complexes.

The metabolism of the acetate complex in Swiss white mice bearing the Ehrlich ascites tumor was followed, using rhodium(II) (-^{14}C) acetate administered as a single therapeutic dose injected intraperitoneally [96]. After 2 hr the complex was shown to break down to give ionic acetate species with a small degree of incorporation of rhodium and acetate in the tumor cells and primary concentration in the liver. Interestingly, only 5% of the rhodium was excreted via the urine in the period 24 hr after injection compared with 50% for cis-$[PtCl_2(NH_3)_2]$.

These studies are continuing at the University of Houston and provide an interesting example of how metal compounds can be involved in biological activity. In this case the use of a compound known to form strong adducts with N-donors indicates one line of approach to designing effective metal-based drugs, even though there is evidence to indicate that enzymes rather than DNA bases may be the important binding sites.

4.3. Palladium

As Pd(II) is the second-row d^8 system equivalent to Pt(II), a series of Pd(II) analogs of active Pt(II) complexes were tested against the S180 and ADJ/PC6A tumors at an early stage of the platinum structure-activity studies [51]. No significant activity was observed. Some of these compounds were tested against *E. coli* bacteria and although they were bacteriostatic at low concentrations, no filamentation was observed. Durig et al. studied $[PdCl_2(en)]$, $[PdCl_2(Me_4en)]$, and $[PdCl_2(glyH)_2]$ and also failed to observe filamentation of *E. coli* in "C" medium at sublethal doses; marginally significant filamentation was only observed for a near-lethal dose of $[PdCl_2(en)]$ [97].

Williams et al. studied the effects of a number of chloroamine palladium(II) and (IV) species against a variety of biological systems [98]. Cis-$[PdCl_4(NH_3)_2]$ was reported to reduce S180 and Landschutz ascites tumors by 82%, an interesting result which should be verified. Some doubt may be cast on the tumor system, as trans-$[PdCl_2(NH_3)_2]$ also reduced them by 45% and this compound had no activity in another S180 system [99].

Furst (see Sec. 2.2) suggested the use of Pd(II) complexes containing sulphur mustard ligands as a means of transporting the cytotoxic ligands into cells, and several groups have followed this approach. Livingstone et al. synthesized and screened Pd(II) complexes containing 6-mercaptopurine ($[Pd(MP)_2]\cdot 2H_2O$; MP = 6-mercaptopurine) and thioguanine ($[Pd_3Cl_2(AMP)_4(AMPH)]\cdot 4H_2O$, AMPH = thioguanine), both ligands being established as having antimetabolite properties [100]. Screening was carried out against an L1210 tumor and although % T/C (ILS) values of 160 were recorded, control experiments showed these to be inferior to tests using the free ligands alone (% T/C 180-MPH, 240-AMPH). However, both Pd(II) complexes were nontoxic at doses up to 400 mg kg^{-1}. Kirschner and coworkers have continued the studies discussed in Sec. 2.2 and during this work synthesized cis-$[PdCl_2(N_2H_4)_2]$ and cis-$[PdCl_2(piperidine)_2]$ [101]. The latter was shown to induce filamentation in *E. coli* and also act as a potent inhibitor of DNA synthesis using murine leukemia L1210 tumor cells in culture. $[Pd(dtp)_2]$ (dtp = 0,0'-dimethyldithiophosphate) has been reported as active against a Walker 256 carcinosarcoma, reducing test tumors to 6% of the controls; no other details are given [102].

L-asparaginase and glutaminase both exhibit antitumor activity and the corresponding amino acids, together with L-serine, have been considered for use as transport agents to deliver metals to tumor cells [103]. Complexes of amino acids with palladium were not isolated for L-glutamine and L-serine and studies on aqueous mixtures of $[PdCl_4]^{2-}$ and the free amino acids revealed only marginal filamentation activity (L-serine) with *E. coli*. A 1:1 molar ratio of L-glutamine and $K_2[PdCl_4]$ was reported to be active against the P388 leukemia tumor in mice with a low ILS (%) of 39% at a dose of 6.25 mg kg^{-1}; unfortunately control data for L-glutamine alone were not presented.

Charlson studied palladium complexes with thiosemicarbazone derivatives of sugars, presumably by analogy with the successful copper compounds (see Sec. 4.6. and Chap. 4, in general). The Pd chelate of D-glucosone bis(thiosemicarbazone) was inactive against the L1210 leukemia up to 200 mg kg^{-1}, and KB cells in culture. However, the dimethylglyoxime Pd(II) chelate was active against the latter culture medium (ED_{50} 0.9 μg·ml^{-1}) [104].

Although a considerable amount of work has been carried out on palladium compounds, they do not as a whole seem to be very promising. Any activity which has been observed in simple compounds without carcinostatic ligands is marginal at best and vastly inferior to the Pt(II) complexes. This is perhaps explained by the fact that Pd(II) systems are much more labile than Pt(II) compounds (Sec. 4.1.) and if studies are attempted, they should concentrate on more inert Pd(II) species. Where compounds with carcinostatic ligands are concerned, this may be an advantage in releasing these within the tumor cell, providing that the complex remains intact prior to penetrating the cell membrane.

4.4. Ruthenium

Several ruthenium compounds have shown bacterial activity with $K_2[RuCl_5(NO)]$, $(NH_4)_3[RuCl_6]$, $K_2[RuCl_5(H_2O)]$ and fac-$[RuCl_3(NH_3)_3]$ [44,97], inducing filamentation in *E. coli*; the last of these was particularly effective and comparable to the platinum compounds. However, antitumor tests on this compound against both the S180 [2] and ADJ/PC6A tumors showed no activity [105]. $K_2[RuCl_5(H_2O)]$ and fac-$[RuCl_3(NH_3)_3]$ both gave negative results for filamentation against gram-positive bacteria.

Studies on a variety of Ru(II) and Ru(III) amine species and their reactions with DNA and its precursors have been carried out by Kellman and his coworkers [106]. A range of nitrosyl and purine Ru complexes have been synthesized and their effects on protein and nucleic acid synthesis studied in vitro. Some compounds showed DNA inhibition; however, studies are continuing including systematic antitumor screening of ruthenium amine complexes. A detailed account of this work is presented in Chap. 5.

The only other ruthenium compound reported in the literature to date is cis-$[RuCl_2(DMSO)_4]$, which was studied by Giraldi et al. on the grounds of its cis configuration and neutrality [90]. This compound induces prophage and filamentous growth in *E. coli* [107] and has been screened for anticancer activity against the Ehrlich ascites carcinoma and L1210 leukemia in a comparative study with cis-$[PtCl_2(NH_3)_2]$ and [Rh(acac)(1,5-COD)] (see Sec. 4.2.3.). The Ru compound was given in very high doses (100-800 mg kg^{-1}) and recorded a comparable result to cis-$[PtCl_2(NH_3)_2]$ (T/C's of 125 and 129) against the L1210. However, this data may be very misleading as cis-$[PtCl_2(NH_3)_2]$ is far more effective against other varieties of the L1210 tumor. It is important to test active compounds against a representative selection of tumors and this compound should probably be retested against two or three other animal tumors. Interestingly this comparative study indicated less general histological damage with the Rh complex, as compared to Ru and Pt.

There seems little doubt that ruthenium complexes are of considerable interest where biological activity is concerned. As yet no really significant antitumor results have been reported, but some systematic structure-activity studies should be undertaken.

4.5. Iridium

Gale et al. [108] have reported activity against the Ehrlich ascites tumor (BALB/$\bar{C}$ mice) for a violet solution prepared by irradiating with ultraviolet light a solution of $(NH_4)_2[IrCl_6]$ in ammonia for 7 hr. The activity was limited to a maximum 140% increase in lifespan using a daily dose of 100 mg kg^{-1} for 8 days. Further testing against L1210 (BDF_1 mice) indicated a low activity with a maximum increase in lifespan of 20%. No evidence was obtained that this product is a selective inhibitor of DNA synthesis. It is suggested that the irradiated solution contains the Ir(IV) complex cis-$[IrCl_4(NH_3)_2]$, purely on the basis of analogy with the $(NH_4)_2[PtCl_6]$ system. However, a polynuclear mixed oxidation state [Ir(III), Ir(IV)] species would seem more likely. This solution does cause filamentous growth in *E. coli* [108].

Some Ir(III) amine complexes have been examined for biological activity against *E. coli* and S180, but no activity was observed [51]. Ir(III) complexes are known to be extremely inert and this is likely to be the reason for their inactivity and particularly their lack of toxicity.

4.6. Copper

Simple copper salts appear to have some cytostatic properties. $CuSO_4$ has produced significant inhibition of the tumor growth rate of S180 and various experimental carcinomas (including Ehrlich) [109]. However, no significant prolongation of survival time was recorded. Various oximes were tested against the Ehrlich ascites and S180 tumors with little effective increase in lifespans (ILS $<$ 30%) [110,111]. Dimethylglyoxime (DMG) showed no inhibition of the test tumors, but gave good results in the presence of Cu(II), although not with Ni(II), Co(II), Zn(II), Mn(II), Mg(II), or Hg(II). The $[Cu(DMG)_2]$ chelate gave an increased lifespan of 200-300%. These compounds were initially tested on the basis that the metal might mask the hydrophilic group of the oxime and increase its permeability through the cell membrane. Once in the cell, the antitumor activity of the free oxime (which appears to be quite small in vivo) might be displayed.

Antitumor activity has also been recorded for copper chelated with thiosemicarbazones. The most investigated system involves 3-

ethoxy-2-oxobutyraldehyde bis(thiosemicarbazone) known as kethoxal bis(thiosemicarbazone) or KTS [112,113]. The activity of KTS in animal systems was found to be enhanced by the presence of Cu^{2+} ions and more careful investigations showed that the antitumor activity and the toxicity of KTS are directly dependent on the dietary intake of Cu^{2+} ion. Cu^{2+} is critical for KTS activity whether the drug is given orally or intraperitoneally, but Cu^{2+} has no or little effect in the absence of KTS. Similarly the Cu(II) KTS chelate has antitumor activity, regardless of the presence of dietary or environmental Cu. These studies are continuing and will be considered in detail in Chap. 4.

Copper has also been used as a radiation sensitizer and the cytotoxic effect of radiation has been shown to be proportional to the cell copper content [114]. The mechanism of action has been postulated as radiation catalyzed reduction of Cu(II) to Cu(I) and subsequent oxidative destruction of nucleic acids by the Cu(I) species. The synergistic effect could be due to catalysis of the reduction reaction in hypoxic cells in the cores of solid tumors, which are resistant to radiotherapy.

4.7. Cobalt

Schiff-base complexes of Co(II) have shown limited activity against the Walker 256 carcinosarcoma in rats (Table 12) [115,116]. A similar range of complexes was screened against the Lewis lung, L1210, and S180A tumors, but none showed significant activity. Some Schiff bases are known to have anticancer activity in their own right and control experiments should be included. It is interesting to note that the above compounds were administered by both intraperitoneal and gavage (stomach-tube) routes although the latter required very high doses of 200-400 mg kg^{-1} daily for 5-9 days. The development of an orally active metal drug would be of great significance to the pharmaceutical industry.

TABLE 12

Screening Results for Co(II) Schiff-base Complexes Against the Walker 256 Tumor in Rats

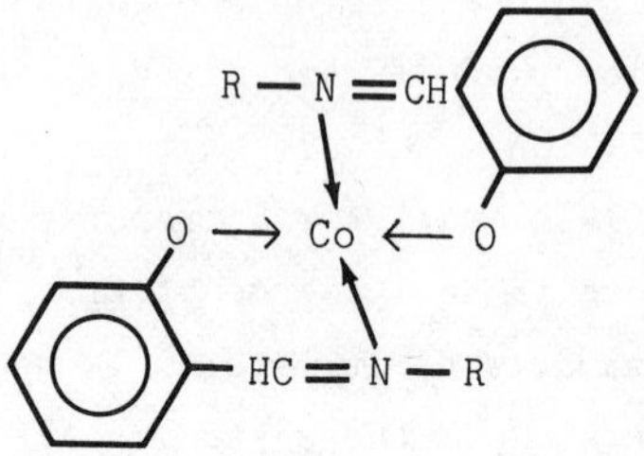

R	Dose range[a,b,c]	Dose response	Best T/C[d]	Dose
$(CH_2)_2CH_2OH$	9.4-37.5 (3)	+	21	37.5
$C(CH_3)_2CH_2OH$	12.5-100 (4)	+	24	100
$C(CH_3)(CH_2OH)_2$	50-100 (2)	+	56	100
$C(C_2H_5)(CH_2OH)_2$	50-200 (3)	+	38	200
$CH_2 \cdot CH_2 \cdot OC_2H_5$	50-100 (2)	+	43	100
$CH_2 \cdot (CH_2)_2 \cdot NH(CH_2)_2OH$	25-100 (3)	+	43	100
$CH_2(CH_2)_2N(CH_2 \cdot CH_2)_2O$	12.5-50 (3)	+	29	50

[a]Number of dose levels in parentheses.

[b]One dose daily for 4 days; administered ip as a suspension in an unspecified medium.

[c]Doses in $mg \cdot kg^{-1}$.

[d]No indication of toxic levels is given and better results may be obtainable.

Source: Reprinted with permission from Ref. 116.

4.8. Other Metals

4.8.1. Iron

A series of ferrocenyl polyamine derivatives were prepared with a view to their selectively binding to cell-surface nucleic acids [117] which had been proposed as present only on tumorgenic cells [118]. There is considerable doubt as to the validity of this proposition. The aim was for the polyamine to bind at the nucleic acid site, after which the ferrocenyl hapten portion of the molecule could conjugate with cell-surface protein and elicit an antigenic response. The target compounds failed to show activity against the P388 tumor in mice but the amide intermediates in these preparations showed marginal activity (% T/C 120-132).

4.8.2. Zinc

Zinc is an essential component of several enzymes [119], including some involved in the synthesis of nucleic acids. Experimentally induced zinc deficiency has been used to inhibit the growth of several animal tumors including L1210 and P388 leukemias [120,121]. However, excess zinc can also have a cytotoxic effect (Sec. 2.1.) and zinc sulphate injections have been used to inhibit the growth of S180 in mice [122]. Aqueous solutions of zinc acetate were evaluated against both the ascitic L1210 leukemia (ip) and a solid lymphocytic leukemia BW5147 in mice. Growth of the L1210 tumor was effectively inhibited, although the mean survival time was only marginally increased for the solid-tumor animals [123]. Several theories have been postulated for the effects of Zn on tumor growth, including enzyme inhibition [124] and support of an immune response due to increased lymphocyte transformation [125]. Zinc demand should be higher in tumor cells due to their faster growth rate; therefore zinc treatment could be selectively cytotoxic to tumor cells while replenishing Zn-depleted body tissues [123].

4.8.3. Nickel

$[Ni(dtp)_2]$ (dtp = dimethyldithiophosphate) is reported to be active against a Walker 256 carcinosarcoma tumor system but no data are given and the compound appears to be inactive against the L1210 leukemia [102].

Ni and Fe chelates of D-glucosone bis(thiosemicarbazone) were screened against L1210 leukemia in mice and KB cell cultures and found to be inactive at the high doses of 200-400 mg kg^{-1}. Similar results were reported for nickel dimethylglyoxime complexes [104].

4.8.4. Miscellaneous

Piperazinium (and substituted analogs) complex salts of Os, Cu, and Pt are reported to have antitumor activity (e.g., $LH_2[OsCl_6]$; L = piperazine, etc.) [126]. Control data for uncomplexed piperazines were not presented and these are important as the amine is not coordinated.

Au(III) remains the major heavy metal d^8 square planar system to have escaped investigation. The only result known to the authors is for $[AuCl_3(C_2H_4NH)]$ against the ADJ/PC6A tumor in BALB/C̄ mice, where the compound was inactive with an LD_{50} of >6.4 mg kg^{-1} [127]. Under physiological conditions, Au(III) may be reduced to Au(I) (linear or tetrahedral). Comparable Au(III) complexes generally react far more rapidly than Pt(II), but slightly slower than Pd(II).

Complexes of Ni(II), Cu(II), Co(II), and Zn(II) with sec-butylphthalate and n-butylphthalate were synthesized and tested against the ADJ/PC6A tumor at the Chester Beatty Institute. All were inactive with LD_{50}s ranging from 20 mg kg^{-1} (Co(II)-n-Bu) to 280 mg kg^{-1} (Zn-sec-Bu) [128]. The order of toxicity was Co > Cu > Ni > Zn (see Sec. 2.1.).

4.9. Gallium and Group IIIA

The rationale for screening elements of this group originated with the use of gallium-67 (^{67}Ga) as a tumor-imaging nuclide for solid tumors, such as melanomas and lymphomas in the body. It was found to be localized in bone tissue and also in solid, slow-growing non-osseous human tumors. Rodent studies demonstrated preferential uptake in carcinomas and sarcomas compared with normal tissue, and preliminary trials with $Ga(NO_3)_3$ revealed inhibition of the growth of the i.m. Walker 256 tumor in rats [129].

Extensive screening of the Group IIIA salts against animal tumors by the National Cancer Institute was initiated with the determination of their toxicity in CDF_1 mice and Sprague-Dawley rats over the dose range 0-400 $mg \cdot kg^{-1}$. Toxicity was shown to increase with atomic weight, with LD_{50}s ranging from 5-320 $mg \cdot kg^{-1}$, i.e., $In(NO_3)_3 \cdot 5H_2O > TlCl_3 > Ga(NO_3)_3 \cdot 9H_2O > Al(NO_3)_3 \cdot 9H_2O$. In vitro cytotoxicity studies for these salts with L1210 and Walker 256 cells revealed a similar order of toxicity to that demonstrated earlier in vivo, although the latter cells were noticeably more sensitive. This may reflect the lower degree of uptake for these compounds by leukemia cells, although ^{67}Ga distribution studies in mice and humans present conflicting evidence [129].

Antitumor activity studies against six tumor systems used a day 1-10 schedule administering the above compounds i.p. All four were active against the ascitic Walker 256 carcinosarcoma in the order of increasing effectiveness:

$$Tl^{3+} > Al^{3+} > Ga^{3+} > In^{3+}$$

with up to 42% long-term survivors. However, no activity was observed in the remaining five tumors, i.e., Ehrlich ascites, YPC-1 plasma cell and the rapidly dividing leukemias L1210, K1964, and P388 (Table 13). Subsequent screening of compounds of the metals and other anions, e.g., Cl^-, SO_4^{2-}, and $CH_3CO_2^-$, on the Walker 256 tumor showed similar effects on MST and % ILS to the nitrate salts (Table 14). Additional screening of the original salts against

TABLE 13

Effect of Gallium nitrate on the Growth of Various Experimental Animal Tumors[a]

Tumor	Optimal anti-tumor dose ($mg \cdot kg^{-1}$)	Increase in MST (% of controls)	Number of long-term survivors/number of animals treated[b]	Growth inhibition[c] (% of controls)
I.p. transplantation				
Walker 256 carcinosarcoma	60	138	4/12 (33)	
Leukemia L1210	50-60	0	0/12 (0)	
Leukemia K1964	50-60	0	0/12 (0)	
Leukemia P388	50-60	0	0/12 (0)	
Plasma cell YPC-1	50-60	0	0/12 (0)	
Ehrlich ascites	50-60	3	0/12 (0)	
S.c. transplantation				
Fibrosarcoma M-89	50			95
Leukemia K1964	50			54
Adenocarcinoma 755	60			93
Mammary carcinoma YMC	50			24
Reticulum cell sarcoma A-RCS	50			94
Lymphosarcoma P1798	30			98
Walker 256 carcinosarcoma	60			92
Osteosarcoma 124F	40			96

[a]Animals were injected ip daily for 10 days, beginning 1 day after tumor transplant.

[b]Numbers in parentheses = % of long-term survivors.

[c]Three-dimensional measurements of sc transplanted tumors were taken 20 days after transplant, with the exception of the Walker 256 carcinosarcoma, which was measured 10 days after transplantation.

Source: Reprinted by permission from Ref. 129.

TABLE 14

Effect of Anion Substitution on Antitumor Activity of Group-IIIA Metal Salts Against Ascites Walker 256 Carcinosarcoma in Rats[a]

Salt	Optimal antitumor dose ($mg \cdot kg^{-1}$)	Increase in MST (% of controls)	Number of long-term survivors/ Number of animals treated[b]
Aluminium nitrate	150	300	5/12 (42)
$AlCl_3 \cdot 6H_2O$	129	550	4/12 (33)
$Al_2(SO_4)_3 \cdot 18H_2O$	89	200	3/12 (25)
Gallium nitrate	60	138	4/12 (33)
$GaCl_2$	25	180	1/12 (9)
$Ga_2(SO_4)_3 \cdot 18H_2O$	45	138	2/12 (17)
Indium nitrate	1.0	124	0/18 (0)
$InCl_3$	1.7	50	0/12 (0)
$In_2(SO_4)_3 \cdot 5H_2O$	0.8	27	0/12 (0)
$In(OOCCH_3)_3$	1.6	32	0/12 (0)
Thallium chloride	3.0	769	3/12 (25)
$Tl(OOCCH_3)_3$	3.9	427	2/12 (17)

[a] Rats were given daily ip injections of the appropriate salt beginning 1 day after tumor transplantation. At least 3 doses of each salt < the LD_{10} for that salt were tested.

[b] Animals surviving 4 months with no histologic evidence of tumor are termed "long-term survivors." Number in parentheses = % of long-term survivors.

Source: Reprinted with permission from Ref. 129.

eight s.c. transplanted leukemias and solid tumors demonstrated little activity for $TlCl_3$, greater than 90% inhibition for six tumors with $Ga(NO_3)_3$ and 20-60% inhibition for the Al and In salts. ^{67}Ga has been shown to be concentrated in three of these tumors [130] and to a greater extent than with ^{114m}In, which reflected the observed trend in tumor inhibition.

Preclinical toxicology studies have been completed for gallium nitrate indicating primary toxicity at renal and hepatic centers, and a Phase I clinical trial including four melanoma patients has been initiated by the U. S. National Cancer Institute.

5. CONCLUSIONS

Studies on metal complexes as antineoplastic agents have been dominated by platinum compounds, which is hardly surprising, as these compounds have shown good results in the clinic against human tumors. While it should be emphasized that a cytotoxic drug such as cisplatin is not a panacea for cancer control, its effectiveness against several tumors highlights the need for more studies on metal-based compounds. The success of cisplatin should help to redress the balance between the screening of inorganic and organic compounds. Sufficient metal compounds with differing structures and chemical properties have shown activity to make this a worthwhile proposition.

Metals are capable of reacting with a wide variety of biological macromolecules, precursors, and intermediates, and thus have numerous possible modes of cytotoxic action. In the case of cisplatin, the evidence favors direct lesions in DNA although considerable binding to protein and RNA has been shown to occur. Preliminary evidence for rhodium acetate indicates an enzyme-inhibition mode. It is very unlikely that the active metal compounds act in the same way. Several studies have shown that the application of inorganic principles of binding and kinetics can be a worthwhile and successful approach.

However, the true problem in cancer chemotherapy remains one of selectivity. The present drugs, both organic and inorganic, are largely cytotoxic drugs with rather poor selectivity towards tumor cells. Much of the selectivity they have relies on the fact that they tend to be able to kill rapidly dividing cells--this includes normal as well as malignant ones, hence toxicity and side effects. Much work in the organic cancer-drug field at present concerns the design and synthesis of drugs containing a component which could make them more selective, and this should be extended to the metal systems. General approaches include the incorporation of molecules known to be taken up or used by tumor cells and the incorporation of large molecules such as DNA. Tumor cells are known to have a higher pinocytotic rate than normal cells and may selectively take up a large molecule with a cytotoxic entity (e.g., a metal) built in. The cytotoxic portion could be released within the tumor cell by lytic enzymes. Utilization of tumor properties to design more selective metal-based drugs should be a major objective for fundamental research in this area.

APPENDIX: SCREENING TESTS FOR ANTITUMOR ACTIVITY

In Vitro

Most antitumor agents have proved to be selectively toxic for cells undergoing DNA synthesis and, not surprisingly, there is some correlation between antitumor properties and effects on rapidly growing organisms. Rosenberg's discovery for platinum hinged on an observation of bacteriological activity. However, in general, these systems are most useful for detailed mechanistic studies after detection of antitumor activity. Human or animal cells in culture are used as screens, but these give only cell toxicity data and no information on the drug's overall effectiveness or on whole-body toxicity.

In Vivo

Ideally, animals bearing spontaneous tumors should be used, but these tumors tend to arise late in lifespan and are inconsistent from animal to animal. Most spontaneous or chemically induced tumors can be transplanted to other animals by tumor fragments, which grow immediately; this is the basis for almost all animal screens. A large number of tumor lines have been transplanted, but there are a few which are most commonly used (see Ref. 2). The National Cancer Institute has set up protocols for at least 20 in vivo test tumor systems. Mouse tumors are most popular for primary screens as relatively small quantities of drug are required, especially by comparison with rats. Particular tumors often appear to be more sensitive to a particular type of drug (see Ref. 2). No particular screen seems to be sufficiently sensitive to all known drugs to be used as a universal primary screen (although leukemias L1210 and P388 are most widely used and would have detected the majority of the drugs in clinical use). However, a variety of screens should be employed even on the same family of compounds.

Screening Parameters

Screening should determine whether any antitumor effect is present up to the toxic dose, and should enable comparisons to be made with other compounds. Important parameters are defined as follows:

1. *Lethal doses* are represented by LD_n, where n is the percentage of animals killed at that dose. LD_{50} is most commonly used.
2. Similarly *effective* or *inhibitory* doses are represented by ED_n or ID_n respectively. ID_{90}'s are most commonly quoted.
3. For solid tumors, inhibition is measured by a comparison of the weights of treated and untreated (control) tumors. This is expressed as a percentage and termed T/C. Values of less than 50 are generally considered significant.
4. For ascites and leukemias the mean survival time is compared to that of the controls. Any percentage *increase in life span* (ILS) is a measure of antitumor activity.
5. In mice, the drugs are usually administered intraperitoneally dissolved or suspended in a suitable solvent,

either as a single dose or on a daily basis. Drugs are administered either on the day after transplantation or sometimes when the tumor has reached an advanced state. At the end of a fixed time, the test animals and controls are killed and the tumors are dissected out and weighed. Screens are routinely checked against compounds of known TI, which act as a positive control against histological changes in the tumor line.

ABBREVIATIONS

Chemical

acac	acetylacetonate
aq	H_2O
5' AMP	5'-adenosine monophosphate
5' ADP	5'-adenosine diphosphate
5' ATP	5'-adenosine triphosphate
ara-C	cytosine arabinoside arabinosyl cytosine
Benz mal	2-benzylmalonate
bipy	bipyridyl
butp	butylphthalate
4t-Bupy	4-tert-butylpyridine
1,1-CBDCA	1,1-cyclobutanedicarboxylate
cisplatin	cis-$[PtCl_2(NH_3)_2]$ (official name)
ClAc	chloroacetate
1,5-COD	1,5-cyclooctadiene
1,2-DAC	1,2-diaminocyclohexane
dien	diethylenetriamine
DMG	dimethylglyoxime
DMSO	dimethylsulphoxide
dtp	O,O'-dimethyldithiophosphate
en	ethylenediamine
Et mal	2-ethylmalonate
4 Et py	4-ethylpyridine
ESCA	electron spectroscopy for chemical analysis
glyH	glycine
iquin	isoquinoline
KTS	kethoxal bis(thiosemicarbazone)
mal	malonate
Me	methyl
Me_4en	tetramethylethylenediamine
Me mal	2-methylmalonate
4-Me py	4-methylpyridine
mp	6-mercaptopurine
OH mal	2-hydroxymalonate
ox	oxalate
phen	3,4,7,8-tetramethyl-1,10-phenanthroline

pip	piperidine	tgn	thioguanine
4 Pr py	4-propylpyridine	trimen	N,N'-dimethylethylene-diamine
py	pyridine	udmen	N,N-dimethylethylene-diamine
sdmen	N,N'-dimethylethylene-diamine		
tetmen	N,N,N',N'-tetramethyl-ethylenediamine		

Biological

A	arachis oil	MTD	minimum toxic dose
BUN	blood urea nitrogen	MTW	mean tumor weight
DNA	deoxyribonucleic acid	RNA	ribonucleic acid
ILS	increase in life span	S	saline
i.m.	intramuscular	SS	saline slurry
i.p.	intraperitoneally	s.c.	subcutaneously
ID_{90}	dose that causes 90% inhibition of a solid tumor	TI	therapeutic index (LD_{50}/ID_{90})
LD_{50}	dose that kills 50% of a group of animals	% T/C	% treated/controls
		W	water
MDST	median survival time	WBC	white blood cell count
MED	minimum effective dose	WS	aqueous slurry
MST	mean survival time		

REFERENCES

1. B. Rosenberg, L. Van Camp, J. E. Trosko, and V. H. Mansour, *Nature* (London), *222*, 385 (1969).

2. M. J. Cleare, *Coord. Chem. Revs.*, *12*, 349 (1974).

3. A. Furst and R. T. Haro, in *Progress in Experimental Tumour Research*, Vol. 12, Karger, Basel, 1969, p. 120 ff.

4. A. Furst, in *Environmental Geochemistry in Health and Disease*, Memoir 123, H. L. Cannon and H. C. Hopps, eds., Geological Society of America, Boulder, Colo., 1971, p. 109 ff.

5. C. A. Tobias, R. Wolfe, R. Dunn, and I. Rosenfeld, *Acta Unio Int. Contra Cancrum*, 7, 874 (1951).

6. K. B. Olson, G. E. Heggen, and C. F. Edwards, *Cancer* (Philadelphia), *11*, 554 (1958).

7. S. Murakami, *Cancer Chemother. Abstr.*, *2*, 1344 (1961).

8. M. Arnold and D. Sasse, *Cancer Res.*, *21*, 761 (1961).

9. F. Svec, *J. Physiol.* (Paris), *49*, 387 (1957).

10. A. Furst, in *The Chemistry of Chelation in Cancer*, C. C. Thomas, Springfield, Ill., 1963.

11. W. B. Bell, *Brit. Med. J.*, *431* (1929).

12. G. Tarchiani and S. Vitale, *Clin. Ther.*, *31*, 101 (1964).

13. W. Luhrs and A. Reincke, *Hippokrates*, *35*, 463 (1964).

14. T. J. Bardos, N. Datta-Gupta, and P. Hebborn, *J. Med. Chem.*, *9*, 221 (1966).

15. M. Kanisawa and H. A. Schroeder, *Cancer Res.*, *27*, 1192 (1967).

16. W. A. Collier and F. Krauss, *Z. Krebsforsch.*, *34*, 526 (1931).

17. A. Taylor and N. Carmichael, Univ. Texas Publ., No. 5314, *Biochem. Inst. Studies*, *5*; *Cancer Studies*, *2*, 36 (1953).

18. C. F. Geschickter and E. E. Reid, in *Approaches to Tumor Chemotherapy*, E. R. Moulton, ed., AAAS, Washington, D.C., 1947, p. 431 ff.

19. J. Balo and I. Banga, *Acta Unio Int. Contra Cancrum*, *13*, 463 (1957).

20. F. P. Dwyer, E. Mayhew, E. M. F. Roe, and A. Shulman, *Brit. J. Cancer*, *19*, 195 (1965).

21. E. Mayhew, E. M. F. Roe, and A. Shulman, *Brit. Emp. Cancer Campaign, 40th Ann. Rep.*, Part II, 1962, p. 106 ff.

22. J. T. Phelan and H. Milgrom, *Surg. Gynecol. Obstet.*, *125*, 549 (1967); *Cancer Chemother. Abstr.*, *8*, no. 835 (1967).

23. T. A. Trainovitch, G. Beirne, and C. Beirne, *Cancer* (Philadelphia), *19*, 867 (1966).

24. H. B. Devlin, *Proc. Roy. Soc. Med.*, *61*, 341 (1968).

25. K. Kajiwara, *Gann.*, *41*, 168 (1951).

26. K. Kajiwara, *Gann.*, *42*, 272 (1951)

27. F. P. Dwyer, E. C. Gyarfas, R. D. Wright, and A. Shulman, *Nature* (London), *179*, 425 (1957).

28. K. Yui, *Jap. J. Bacteriol.*, *14*, 411 (1959); *Cancer Chemother. Abstr.*, *1*, 3556 (1960).

29. K. Yui, *Jap. J. Bacteriol.*, *14*, 339 (1959); *Cancer Chemother. Abstr.*, *1*, 3557 (1960).

30. G. P. O'Hara, D. E. Mann, Jr., and R. F. Gautieri, *J. Pharm. Sci.*, *60*, 473 (1971).

31. A. Sawitsky and F. Deposito, *J. Pediat.*, *67*, 99 (1965).

32. Y. Ishikawa, M. Fukushima, A. Horimai, T. Sato, and S. Machida, *Igaku To Seibutsugaku (Med. Biol. Tokyo)*, *74*, 112 (1967); *Cancer Chemother. Abstr.*, *8*, No. 1623 (1967).

33. A. Furst, in *Metal Binding in Medicine*, M. J. Seven and L. A. Johnson, eds., Lippincott, Philadelphia, 1960, Chap. 14.

34. S. Kirschner, Y. K. Wei, D. Francis, and J. G. Bergman, *J. Med. Chem.*, *9*, 369 (1966).

35. G. A. Zentmyer, *Phytopathol.*, *33*, 1121 (1943).

36. *Fed. Proc., Fed. Amer. Soc. Exp. Biol.*, *20*, Suppl. No. 10 (1961).

37. A. Albert, Selective Toxicity, Methuen, London, 1960.

38. B. Rosenberg, *Naturwissenschaften*, *9*, 399 (1973).

39. B. Rosenberg, L. Van Camp, and T. Krigas, *Nature* (London), *205*, 698 (1965).

40. B. Rosenberg, L. Van Camp, E. B. Grimley, and A. J. Thomson, *J. Biol. Chem.*, *242*, 1347 (1967).

41. F. P. Dwyer, E. C. Gyarfas, W. P. Rogers, and J. H. Koch, *Nature* (London), *170*, 490 (1952).

42. E. Renshaw and A. J. Thomson, *J. Bacteriol.*, *94*, 1915 (1967).

43. R. J. Bromfield, R. H. Dainty, R. D. Gillard, and B. T. Heaton, *Nature* (London), *223*, 735 (1969).

44. R. D. Gillard, in Recent Results in Cancer Research, Vol. 48, T. A. Connors and J. J. Roberts, eds., Springer-Verlag, New York, 1974, p. 29 ff.

45. S. Reslova, *Chem.-Biol. Interact.*, *4*, 66 (1971).

46. A. J. Thomson, R. J. P. Williams, and S. Reslova, *Struc. Bonding* (Berlin), *11*, 1 (1972).

47. B. Rosenberg, *Plat. Met. Rev.*, *15*, 3 (1971).

48. B. Rosenberg and L. Van Camp, *Cancer Res.*, *30*, 1799 (1970).

49. M. J. Cleare and J. D. Hoeschele, *Bioinorg. Chem.*, *2*, 187 (1973).

50. M. J. Cleare and J. D. Hoeschele, *Plat. Met. Rev.*, *17*, 2 (1973).

51. M. J. Cleare, in Recent Results in Cancer Research, Vol. 48, T. A. Connors and J. J. Roberts, eds., Springer-Verlag, New York, 1974, p. 2 ff.

52. K. P. Beaumont, C. A. McAuliffe, and M. J. Cleare, *Chem.-Biol. Interact.*, *14*, 179 (1976).

53. G. R. Gale and S. Meischen, U.S. Patent, Appl. No. 769, 888.

54. H. J. Ridgeway, R. J. Speer, L. M. Hall, D. P. Steward, A. D. Newman, and J. M. Hill, *J. Clin. Hematol. Oncol.*, 7, 1, 220, 231 (1977).

55. T. A. Connors, M. Jones, W. C. J. Ross, P. D. Braddock, A. R. Khokhar, and M. L. Tobe, *Chem.-Biol. Interact.*, *5*, 415 (1972); ibid., *11*, 145 (1975).

56. M. J. Cleare, P. C. Hydes, B. W. Malerbi, and D. M. Watkins, *Biochim.*, *60*, 835 (1978).

57. B. Lippert, *J. Clin. Hematol. Oncol.*, *7*, 1 (1977), p. 26 ff.

58. J. A. Stanko, L. S. Hollis, J. A. Schneifels, and J. D. Hoeschele, *J. Clin. Hematol. Oncol.*, *7*, 1 (1977), p. 138 ff.

59. A. W. Prestayko, W. T. Bradner, J. B. Huftaler, W. C. Rose, J. E. Schurig, M. J. Cleare, P. C. Hydes, and S. T. Crooke, *Cancer Treat. Rep.*, *63*, 1503 (1979).

60. F. Basolo and R. G. Pearson, in Mechanisms of Inorganic Reactions (2d ed.), Wiley, New York, 1967, p. 359 ff.

61. D. Banerjea, F. Basolo, and R. G. Pearson, *J. Amer. Chem. Soc.*, *79*, 4055 (1957).

62. F. Basolo, H. B. Gray, and R. G. Pearson, *J. Amer. Chem. Soc.*, *82*, 4200 (1960).

63. R. C. Lange, R. P. Spencer, and H. C. Harder, *J. Nucl. Med.*, *13*, 328-330 (1972).

64. R. C. Lange, R. P. Spencer, and H. C. Harder, *J. Nucl. Med.*, *14*, 191-195 (1973).

65. C. L. Litterst, T. E. Gram, R. L. Dedrick, A. F. Leroy, and A. M. Guarino, *Cancer Res.*, *36*, 2340 (1976).

66. P. H. S. Smith and D. M. Taylor, *J. Nucl. Med.*, *15*, 349 (1974).

67. J. M. Hill, E. Loeb, A. MacLellan, N. O. Hill, A. Khan, and J. J. King, *Cancer Chemother. Rep.*, *59*, 647 (1975).

68. R. C. DeConti, B. R. Toftness, R. C. Lange, and W. A. Creasey, *Cancer Res.*, *33*, 1310-1315 (1973).

69. J. J. Roberts and A. J. Thomson, in Progress in Nucleic Acids Research and Molecular Biology, Vol. 22, Academic, New York, 1979, p. 71 ff.

70. D. J. Beck and R. R. Brubaker, *Mut. Res.*, *27*, 181 (1975).

71. P. Lecointe, J.-P. Macquet, J.-L. Butour, and C. Paoletti, *Mut. Res.*, *48*, 139 (1977).

72. H. C. Harder and B. Rosenberg, *Int. J. Cancer*, *6*, 207 (1970).

73. J. A. Howle and G. R. Gale, *Biochem. Pharmacol.*, *19*, 2757 (1970).

74. J. M. Pascoe and J. J. Roberts, *Biochim. Pharmacol.*, *23*, 1345 (1974a).

75. J. M. Pascoe and J. J. Roberts, *Biochim. Pharmacol.*, *23*, 1359 (1974b).

76. H. C. Harder, *Chem.-Biol. Interact.*, *10*, 27 (1975).

77. K. V. Shooter, R. Howse, R. K. Merrifield, and A. B. Robbins, *Chem.-Biol. Interact.*, *5*, 289 (1972).

78. P. Horacek and J. Drobnik, *Biochem. Biophys. Acta*, *254*, 341 (1971).

79. S. Mansy, B. Rosenberg, and A. J. Thomson, *J. Amer. Chem. Soc.*, *95*, 1633 (1973).

80. A. B. Robbins, *Chem.-Biol. Interact.*, *6*, 35 (1973).

81. P. J. Stone, A. D. Kelman, and F. M. Sinex, *Nature*, *251*, 736 (1974).

82. L. Munchausen and R. O. Rahn, *Biochem. Biophys. Acta*, *414*, 242 (1975).

83. A. D. Kelman, H. J. Peresie, and P. J. Stone, *J. Clin. Hematol. Oncol.*, *7*, 440 (1977).

84. M. M. Millard, J. P. Macquet, and T. Theophanides, *Biochem. Biophys. Acta*, *402*, 166 (1975).

85. S. Ahrland, J. Chatt, and N. R. Davies, *Quart. Rev. Chem. Soc.*, *12*, 265 (1958).

86. T. A. Connors, personal communication, 1975.

87. B. Rosenberg, E. Renshaw, L. Van Camp, J. Hartwick, and J. Drobnik, *J. Bacteriol.*, *93*, 716 (1967).

88. M. J. Cleare, C. A. McAuliffe, and R. Pollock, unpublished data, 1975.

89. T. Giraldi, G. Zassinovich, and G. Mestroni, *Chem.-Biol. Interact.*, *9*, 389 (1974).

90. T. Giraldi, G. Sava, G. Bertoli, G. Mestroni, and G. Zassinovich, *Cancer Res.*, *37*, 2662 (1977).

91. A. Erck, L. Rainen, J. Whileyman, I.-M. Chang, A. P. Kimball, and J. L. Bear, *Proc. Soc. Exp. Biol. Med.*, *145*, 1278 (1974).

92. R. A. Howard, E. Sherwood, A. Erck, A. P. Kimball, and J. L. Bear, *J. Med. Chem.*, *20*, (7), 943 (1977).

93. J. L. Bear, H. B. Gray, Jr., L. Rainen, I.-M. Chang, R. Howard, G. Serio, and A. P. Kimball, *Cancer Chemother. Rep.*, *59*, 611 (1975).

94. L. Rainen, R. A. Howard, A. P. Kimball, and J. L. Bear, *Inorg. Chem.*, *14*, 2752 (1975).

95. S. H. Lee, D. L. Chao, J. L. Bear, and A. P. Kimball, *Cancer Chemother. Rep.*, *59*, 661 (1975).

96. A. Erck, E. Sherwood, J. L. Bear, and A. P. Kimball, *Cancer Res.*, *36*, 2204 (1976).

97. J. R. Durig, J. Danneman, W. D. Behnke, and E. E. Mercer, *Chem.-Biol. Interact.*, *13*, 287 (1976).

98. D. R. Williams and R. D. Graham, in Recent Results in Cancer Research, Vol. 48, T. A. Connors and J. J. Roberts, eds., Springer-Verlag, New York, 1974, p. 27.

99. M. J. Cleare, unpublished results, 1973.

100. M. Das and S. E. Livingstone, *Brit. J. Cancer*, *38*, 325 (1978).

101. S. Kirschner, A. Maurer, and C. Dragulescu, *J. Clin. Hematol. Oncol.*, *7*, (1), 190 (1977).

102. S. E. Livingstone and A. E. Mihkelson, *Inorg. Chem.*, *9*, 2545 (1970).

103. A. J. Charlson, R. J. Banner, R. P. Gale, N. T. McArdle, K. E. Trainor, and E. C. Walton, *J. Clin. Hematol. Oncol.*, *7*, (1), 294 (1977).

104. A. J. Charlson, in Recent Results in Cancer Research, Vol. 48, T. A. Connors and J. J. Roberts, eds., Springer-Verlag, New York, p. 34.

105. M. J. Cleare and T. A. Connors, unpublished data, 1973.

106. A. D. Kelman, M. J. Clarke, S. D. Edmonds, and H. J. Peresie, *J. Clin. Hematol. Oncol.*, *7*, (1), 274 (1977).

107. C. Monti-Bragadin, M. Tamaro, and E. Banfi, *Chem.-Biol. Interact.*, *11*, 469 (1975).

108. G. R. Gale, E. M. Walker, Jr., A. B. Smith, and A. E. Stone, *Proc. Soc. Exp. Biol. Med.*, *136*, 1197 (1971).

109. I. V. Savitskii, *Gig. Tr.*, *42* (1970).

110. K. Takamiya, *Nature* (London), *185*, 190 (1960).

111. K. Takamiya, *Gann.*, *50*, 265 (1959).

112. D. H. Petering, *Biochem. Pharmacol.*, *23*, 567 (1974) and references therein.

113. A. C. Sartorelli and B. A. Booth, *Cancer Res.*, *27*, 1614 (1967).

114. J. M. Thomson, Y. Maruyama, S. Schwartz, and E. Hahn, *Radiol.*, *101*, 187 (1971).

115. E. M. Hodnett, C. H. Moore, and F. A. French, *J. Med. Chem.*, *14*, (11), 1121 (1971).

116. E. M. Hodnett and W. J. Dunn, III, *J. Med. Chem.*, *15*, (3), 339 (1972).

117. V. J. Fiorina, R. J. Dubois, and S. Brynes, *J. Med. Chem.*, *21*, (4), 393 (1978).

118. B. Rosenberg, *Cancer Chemother. Rep.*, Part 1, *59*, 589 (1975).

119. B. L. Vallee and W. E. C. Wacker, in The Proteins, Vol. V, H. Neurath, ed., Academic Press, New York, 1970.

120. W. DeWys and W. J. Pories, *J. Nat. Cancer Inst.*, *48*, 375 (1972).

121. D. H. Barr and J. W. Harris, *Proc. Soc. Exp. Biol. Med.*, *144*, 284 (1973).

122. A. D. Woster, M. L. Failla, and M. W. Taylor, *J. Nat. Cancer Inst.*, *54*, 1001 (1975).

123. J. L. Phillips and P. J. Sheridan, *J. Nat. Cancer Inst.*, *57*, (2), 361 (1976).

124. M. Mustafa, C. Cross, and R. Munn, *J. Lab. Clin. Med.*, *77*, 563 (1971).

125. H. Kirchner and H. Ruhl, *J. Nat. Cancer Inst.*, *54*, 1001 (1975).

126. A. Doadrio, D. Craciunescu, and G. Ghirvu, *An. Quim.*, *73*, (7-8), 1042 (1977).

127. M. L. Tobe and T. A. Connors, unpublished data, 1974.

128. M. J. Cleare, T. A. Connors, and C. A. McAuliffe, unpublished data reported at the 15th ICCC, Dublin, 1974.

129. R. H. Adamson, G. P. Canellos, and S. M. Sieber, *Cancer Chemother. Rep.*, Part 1, *59*, (3), 599 (1975).

130. M. H. Hart, C. F. Smith, S. T. Yancey, and R. H. Adamson, *J. Nat. Cancer Inst.*, *47*, (5), 1121 (1971).

Chapter 2

AQUEOUS PLATINUM(II) CHEMISTRY; BINDING TO BIOLOGICAL MOLECULES

Mary E. Howe-Grant* and Stephen J. Lippard
Department of Chemistry
Columbia University
New York, New York

1. INTRODUCTION 64
   1.1. Introduction to Platinum(II) Chemistry 65
2. THERMODYNAMIC AND KINETIC PRINCIPLES 68
   2.1. Thermodynamic Stabilities 68
   2.2. The Chelate Effect 71
   2.3. Redox Chemistry of Platinum(II) 71
   2.4. Trans and Cis Influences 73
   2.5. Kinetics of Substitution Reactions 74
   2.6. Kinetic Trans and Cis Effects 79
   2.7. Kinetics Related to Chelation 81
   2.8. Additional Kinetic Factors 82
3. HYDROLYSIS AND AQUEOUS SUBSTITUTION CHEMISTRY 82
   3.1. Effects of Buffers and Salts on Complex Stabilities 82
   3.2. Hydrolysis of Platinum(II) Complexes: Substitution Reactions 86
   3.3. Acidity of Coordinated Water 87
   3.4. The Hydroxide Ion as Ligand 88
   3.5. Proposed Hydrolysis of Coordinated Ligands 89

*Current affiliation: Department of Chemistry, Polytechnic Institute of New York, Brooklyn, New York

4. BINDING SITES IN BIOLOGICAL SYSTEMS 90
4.1. Nucleic Acids: Potential Binding Sites 91
4.2. Binding Sites on Nucleic Acid Constituents 93
4.3. Binding to Polynucleotides 96
4.4. Proteins: Potential Binding Sites 98
4.5. Platinum(II)-Amino Acid Binding 100
4.6. Binding to Proteins and Polypeptides 103
4.7. Other Biological Molecules 108
5. THE PLATINUM BLUES: A CASE STUDY 108
5.1. Statement of the Problem 108
5.2. Molecular Structure of Platinum Blues 110
5.3. Electronic Structure of Platinum Blues 111
5.4. Solution and Redox Chemistry 113
5.5. Binding to DNA 114
6. DETECTING PLATINUM IN BIOLOGICAL SYSTEMS 114
6.1. Methods of Detection 115
6.2. Nature and Amount of Bound Platinum 117
ABBREVIATIONS 118
REFERENCES 118

1. INTRODUCTION

The chemistry of platinum has been studied widely for over two centuries. Most of the work published through 1971 is nicely summarized by F. R. Hartley [1]. Platinum exhibits oxidation states ranging 0 to 6. The divalent state is the most common in aqueous systems and forms compounds with members of nearly every group in the periodic table. In this chapter the aqueous solution chemistry of divalent platinum, especially in the presence of biological molecules, is discussed. A comprehensive review of even the most recent literature has not been attempted, owing to space limitations.

The chapter is organized in the following manner. The first portion covers broad principles of aqueous platinum(II) chemistry. An introduction to the divalent oxidation state and to basic thermodynamic and kinetic principles of platinum(II) reactions is followed by a more comprehensive examination of aqueous substitution reactions including both hydrolyses and the effects of salts on aqueous equilibria. Biological applications are discussed in the latter portion of the chapter. One section each is devoted to discussing the known and suspected platinum binding sites on biopolymers, the platinum blues, and the analytical methods used to detect platinum in its interactions with biological molecules.

1.1. Introduction to Platinum(II) Chemistry

Platinum(II) has the electronic configuration $[Xe]4f^{14}5d^{8}$. Although both five- and six-coordinate compounds are known, the majority of its compounds exhibit coordination number 4 with square planar geometry [1-5]. The stability of square coordination for a d^8 configuration is apparent upon examination of the relative energies of the d orbitals in different coordination environments, as calculated according to crystal field theory (see Fig. 1) [6]. The orbital splittings shown in Fig. 1 are only approximate for a given coordination environment, however, since the actual orbital energy differences depend upon the nature of both the metal and its ligands. For example, although comparison of five-coordinate environments for a d^8 system (Fig. 1) indicates a slight advantage for square pyramidal over trigonal bipyramidal geometry, the known pentacoordinate platinum(II) compounds are trigonal bipyramidal [1,3-5]. These five-coordinate compounds all contain at least one ligand capable of accepting π-electron density from the metal, thereby lowering the energy of the metal's π-donating d orbitals. This π interaction renders the trigonal bipyramidal configuration more stable than the square pyramidal one [1]. Both four- and five-coordinate platinum(II) complexes are diamagnetic.

Square planar platinum(II) complexes are relatively kinetically inert, which is a feature of considerable importance in their chemistry. Although the radius of square planar platinum(II) is exactly the same as that of the d^8 palladium(II) ion, compounds of

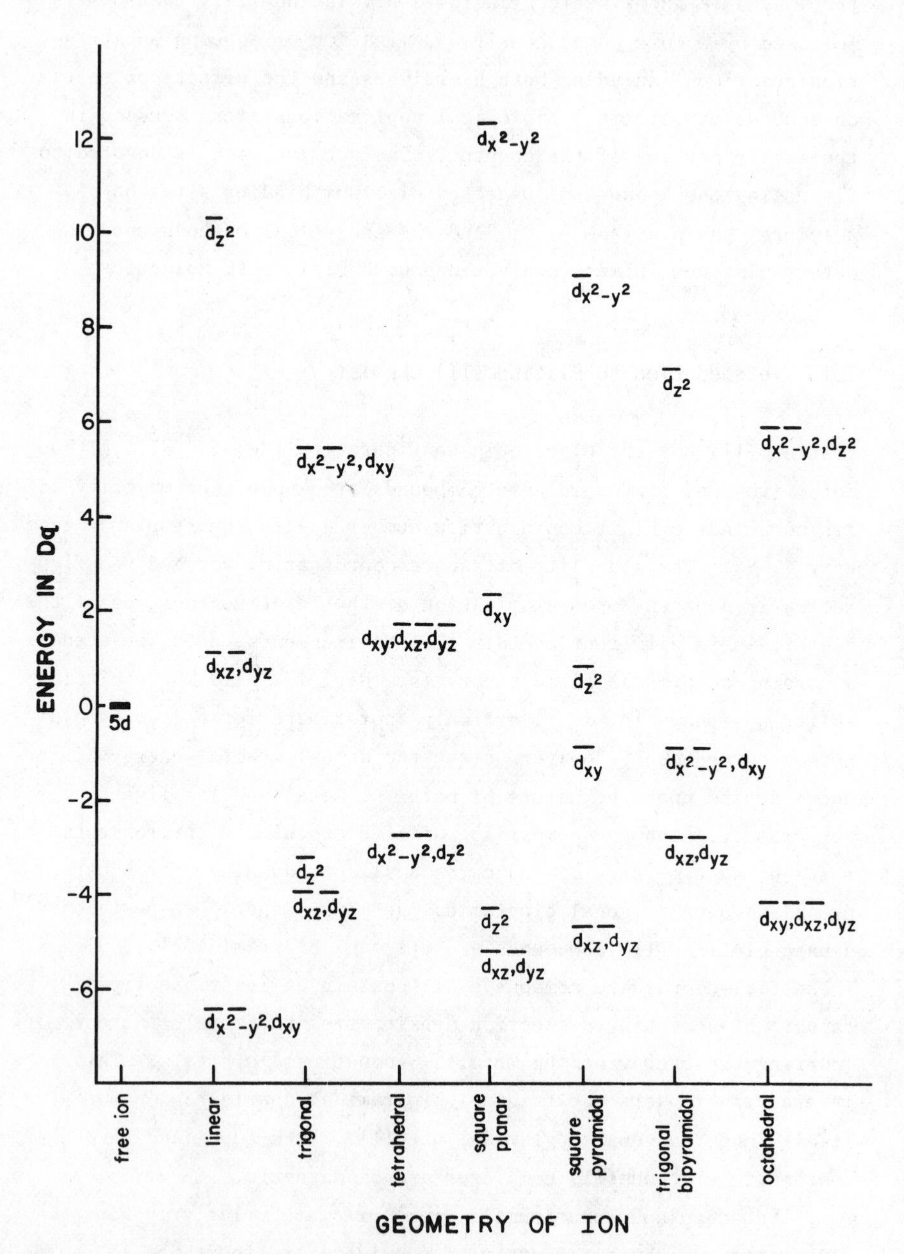

FIG. 1. Relative energies of metal d orbitals in various coordination environments. Splittings are according to Ref. 6.

the latter react up to 10^6 times faster than the corresponding platinum(II) complexes [1,2]. The major electronic difference between the two elements is the existence of 4f electrons for platinum. Because of the effective nuclear shielding of these 4f electrons, the divalent ions are the same size. The 5d orbitals of platinum are more extended spatially than the 4d orbitals of palladium, however, inhibiting axial coordination and retarding the rate of substitution reactions. Platinum(II) complexes are therefore more inert.

Platinum(II) forms stable compounds with both σ- and π-electron donating ligands. Differences in the nature of the electron donation and in the ability of a ligand to accept metal π-electron density influence the thermodynamic and kinetic behavior of platinum(II) complexes [1,2,7,8]. The ligands are generally anionic or neutral and the list of donor atoms encompasses nearly every nonmetallic element. Because compounds formed with the heavier nonmetallic elements are generally more stable, platinum(II) is classified as a b-type metal [1,9,10]. Class a metals form their most stable compounds with the first member of a group: nitrogen, oxygen, and fluorine. Class b metals form their most stable compounds with the heavier members of the chemical family. Although the usual order for stability of a class b metal with donor atoms is $S \sim C > I > Br > Cl > N > O > F$ [9,10], the environment of the donor atom must be considered. In the case of platinum(II), thioether (R_2S) and thiolate (RS^-) complexes are much more stable than those with R_2O and RO^-, whereas H_2O and ROH complexes are more stable than those with H_2S and RSH. Complexes containing OH^- and SH^- are of comparable stability [1]. Classification as a class b metal does not, therefore, preclude the existence of stable complexes with the first member of a chemical group, nor does the classification imply anything about the kinetic properties of a particular metal complex.

Ligand donor atoms with the appropriate orbital geometry and available electron density can bridge two or more platinum atoms, forming dimeric or polymeric species. For one series of anionic ligands, the tendency to form bridges was found to decrease as $PR_2^- > SR^- > R_2PO_2^- > I^- > Br^- > Cl^- > RSO_2^- > SnCl_3^-$ [2,11]. In aqueous solution of moderate or high pH, the hydroxide ion often serves as a bridging ligand. Hydroxo-bridged complexes are discussed in some detail in Sec. 3.4.

Apart from ligand-bridged systems, square planar platinum(II) compounds such as $[(en)PtCl_2]$ form polymeric species in the solid state through axial metal-metal stacking interactions [12]. The polymeric, stacked structure of the platinum blues appears to be maintained even in solution [13]. Platinum compounds that undergo stacking interactions were recently reviewed [14]. The majority of these complexes display unusual conductivity properties [12-15].

2. THERMODYNAMIC AND KINETIC PRINCIPLES

2.1. Thermodynamic Stabilities

The thermodynamic stability of a coordination complex may be expressed either in terms of a series of stepwise formation constants, K_1, K_2, ..., K_n, where n is the number of maximum coordination, or in terms of the overall formation constant β_n, defined in Eq. (1).

$$\beta_n = \prod_{i=1}^{n} K_n \tag{1}$$

The stepwise formation constants for a square planar platinum(II) system and the corresponding chemical processes are shown in Fig. 2. Table 1 gives the overall formation constants of platinum(II) complexes formed from some common ligands in aqueous solution. As expected for a class b metal, $\beta_4(I^-) > \beta_4(Br^-) > \beta_4(Cl^-)$. β_4 values for $[PtCl_4]^{2-}$ and $[PtBr_4]^{2-}$ were recently determined at 25°C in a 0.5 to 1.00 M $HClO_4$ medium to be 9.77×10^{13} and 1.3×10^{16}, respectively [16]. Although the values at this higher ionic strength are less than those reported in Table 1, the order of the stabilities remains the same.

$$PtS_4^{2+} + L^{-x} \rightleftharpoons PtLS_3^{(2-x)} \qquad K_1 = \frac{[PtLS_3^{(2-x)}]}{[PtS_4^{2+}][L^{-x}]}$$

$$PtLS_3^{(2-x)} + L^{-x} \rightleftharpoons PtL_2S_2^{(2-2x)} \qquad K_2 = \frac{[PtL_2S_2^{(2-2x)}]}{[PtLS_3^{(2-x)}][L^{-x}]}$$

$$PtL_2S_2^{(2-2x)} + L^{-x} \rightleftharpoons PtL_3S^{(2-3x)} \qquad K_3 = \frac{[PtL_3S^{(2-3x)}]}{[PtL_2S_2^{(2-2x)}][L^{-x}]}$$

$$PtL_3S^{(2-3x)} + L^{-x} \rightleftharpoons PtL_4^{(2-4x)} \qquad K_4 = \frac{[PtL_4^{(2-4x)}]}{[PtL_3S^{(2-3x)}][L^{-x}]}$$

FIG. 2. Platinum(II) stepwise formation constants. S represents the solvent, and L^{-x} is the ligand.

TABLE 1

Overall Formation Constants for Platinum(II) Complexes in Water at 25°C[a]

Ligand	Complex	log β_4
CN^-	$[Pt(CN)_4]^{2-}$	41
NH_3	$[Pt(NH_3)_4]^{2+}$	35.3
OH^-	$[Pt(OH)_4]^{2-}$	35
I^-	$[PtI_4]^{2-}$	29.6
Br^-	$[PtBr_4]^{2-}$	20.5
Cl^-	$[PtCl_4]^{2-}$	16.6

[a]Equilibrium constants for the reaction $Pt^{2+} + 4L^{-x} \rightleftharpoons PtL_4^{(2-4x)}$ are taken from Ref. 10.

TABLE 2

Equilibrium Constants, K, for Some Platinum(II) Substitution Reactions in Water at 25°C[a]

$$[Pt(C_2H_4)Cl_3]^- + X^- \rightleftharpoons trans\text{-}[Pt(C_2H_4)Cl_2X]^- + Cl^-$$

X^-	NH_3	OH^-	SCN^-	I^-	Br^-	F^-	H_2O
log K	7.8	6	>2	2	0.5	-1.5	-2.5

$$trans\text{-}[Pt(NH_3)_2(OH)_2] + 2X^- \rightleftharpoons trans\text{-}[Pt(NH_3)_2X_2] + 2OH^-$$

X^-	I^-	SCN^-	Br^-	Cl^-
log K	-6.3	-6.6	-8.6	-10

$$[Pt\{(CH_3)_2S\}Cl_3]^- + CH_3OH \rightleftharpoons trans\text{-}[Pt\{(CH_3)_2S\}(CH_3OH)Cl_2] + Cl^-$$

log K -2.85[b]

[a]Data are taken from Ref. 10, except where noted.
[b]Data from Ref. 17 at 30°C in methanol.

The relative affinities of various ligands for Pt(II) may also be ascertained by examining equilibrium constants for reversible ligand substitution reactions. Table 2 contains some of these data. Information from Tables 1 and 2 reveals that the relative affinity of ligands for Pt(II) decreases as $CN^- > NH_3 \sim OH^- > I^- > SCN^- > Br^- > Cl^- >> F^- \sim H_2O \sim$ MeOH. Thermodynamic stability is always measured relative to other components of the system. A complex will be thermodynamically stable depending upon the nature and concentration of other potential ligands. The stability of complexes in aqueous systems, where water is a potential ligand present in high concentration, is discussed in the next section. Moreover, square planar complexes often can exist as cis and trans isomers. Although the formation of the initial complex is kinetically controlled, the trans isomers are generally more stable than the corresponding cis compounds. For example, cis-$[Pt(NH_3)_2Cl_2]$ isomerizes to the trans species with an estimated $\Delta H°$ of isomerization of -3.0 kcal mol^{-1} [10].

2.2. The Chelate Effect

A complex that contains a chelated ligand is generally more stable than a similar complex having no chelate rings, especially if the ring size is 5 or 6. The enhanced stability of chelated systems is called the chelate effect. The chelate effect has both entropic and enthalpic origins [2,10]. Because of the chelate effect, dichloroethylenediamineplatinum(II) is expected to be more stable than cis-dichlorodiammineplatinum(II), both of which are shown below.

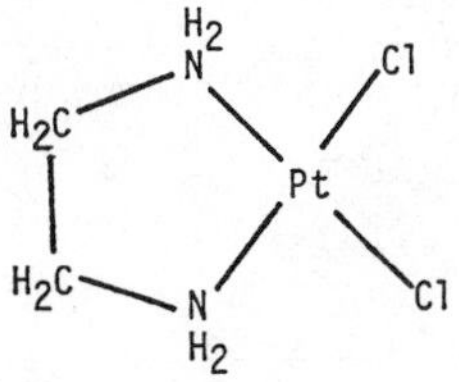

[(en)PtCl] 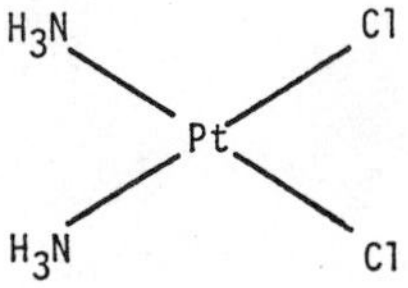 cis-$[Pt(NH_3)_2Cl_2]$

2.3. Redox Chemistry of Platinum(II)

The stability of platinum(II) with respect to disproportionation to Pt(0) and Pt(IV) or to oxidative addition is of interest. Table 3 contains some platinum reduction potentials. Whereas $[PtCl_4]^{2-}$ is stable to disproportionation [Eq. (2)] at 1 M concentrations and 25°C, it is unstable at 60°C. The equilibrium constant for Eq. (2)

$$2PtCl_4^{2-} \rightleftharpoons Pt(0) + PtCl_6^{2-} + 2Cl^- \quad (2)$$

at 60°C in the presence of 3 M HCl is 50 M^{-1} versus 0.21 M^{-1} under standard conditions.

Oxidative addition reactions such as that shown in Eq. (3) may occur with even the most stable of platinum(II) complexes,

$$Pt(CN)_4^{2-} + I_2 \rightleftharpoons trans\text{-}Pt(CN)_4I_2^{2-} \quad (3)$$

TABLE 3

Selected Platinum(II) Reduction Potentials[a]

Half-reaction	E° (V)
$Pt^{2+} + 2e^- \rightleftharpoons Pt$	1.2
$Pt(OH)_2 + 2e^- \rightleftharpoons Pt + 2OH^-$	0.14
$PtCl_4^{2-} + 2e^- \rightleftharpoons Pt + 4Cl^-$	0.75
$PtBr_4^{2-} + 2e^- \rightleftharpoons Pt + 4Br^-$	0.67
$PtI_4^{2-} + 2e^- \rightleftharpoons Pt + 4I^-$	0.40
$PtCl_6^{2-} + 2e^- \rightleftharpoons PtCl_4^{2-} + 2Cl^-$	0.77
$PtBr_6^{2-} + 2e^- \rightleftharpoons PtBr_4^{2-} + 2Br^-$	0.64
$PtI_6^{2-} + 2e^- \rightleftharpoons PtI_4^{2-} + 2I^-$	0.39

[a]Data are taken from Ref. 1.

TABLE 4

Examples of Oxidative Addition Reactions of Pt(II) Complexes in Aqueous Media

Reaction	Ref.
$[Pt(CN)_4]^{2-} + I_2 \rightleftharpoons$ trans-$[Pt(CN)_4I_2]^{2-}$	18
$[Pt(NH_3)_4]^{2+} + H_2O_2 \rightleftharpoons [Pt(NH_3)_4(OH)_2]^{2+}$	19
$2[Pt(NH_3)_4]^{2+} + 2S_2O_8^{2-} + H_2O \rightleftharpoons [Pt(NH_3)_4(OH)(SO_4)]^+ + [Pt(NH_3)_4(SO_4)_2] + SO_4^{2-} + H^+$	19
trans-$[Pt(NH_3)_2Cl_2] + Cl_2 \rightleftharpoons$ trans-$[Pt(NH_3)_2Cl_4]$	20

$\beta_4(CN^-) = 10^{41}$. The equilibrium constant reported for this reaction at 25°C in 0.50 M $HClO_4$ is $1.29 \times 10^4\ M^{-1}$ [18]. Table 4 lists some additional reactions for which thermodynamic data are not available.

2.4. Trans and Cis Influences

The trans and cis influences are two thermodynamic phenomena observed for square planar platinum(II) complexes. The trans influence [21] is the tendency of a ligand to weaken the bond trans to itself in the ground state of a metal complex. The cis influence is the analogous effect on a cis ligand. The experimental evidence and theoretical rationale for the trans influence have been thoroughly reviewed [7]. Most of the information supports the theory that a rehybridization of metal σ orbitals occurs whereby the ligand exhibiting a strong trans influence competes more effectively for the platinum 6s orbital than do the other ligands. The bond in the trans position is therefore weakened. The net transfer of electron density from ligand to metal is also of importance, however, and under certain circumstances, e.g., π bonding, may outweigh the degree of metal rehybridization.

The trans influence is revealed by comparing bond lengths determined by x-ray crystallography for similar compounds [7,22,23]; by nmr chemical shifts and coupling constants, since these parameters reflect the hybridization of a metal ion [7,24,25]; by infrared spectroscopy [7]; and in the case of complexes containing Pt-I bonds, by Mössbauer parameters, which reflect both differences in s character and overall σ and π interactions in the metal-iodide bonds [26]. While the trans influence is partly dependent upon the total composition of a complex, it generally decreases as $PF_3 > PEt_3 > C_2H_4 > CO > H_2S > NH_3 > H_2O$ for neutral ligands and $H^- > SiH_3^- > CN^- > CH_3^- > I^- > Cl^- > OH^- > NO_2^-$ for anionic ligands [27]. A comparison of equilibrium constants corresponding to ground-state energy differences of Pt(II)-DMSO complexes showed the trans influence to vary as NH_3 > DMSO ~ C_2H_4 > Br^- ~ Cl^- ~ H_2O with factors relative to H_2O of 10(NH_3):4(DMSO):3(C_2H_4):1(Br^-, Cl^-) [28]. In the case of σ-bonding ligands, there appears to be a correlation between the trans influence and the kinetic trans effect discussed below [7,27]. For π-bonding ligands such as ethylene, however, there is no correlation between these two phenomena, presumably owing to differences in their origin.

There is some disagreement over the relative importance of the cis influence in square planar complexes. Molecular orbital calculations [27,29] indicate the cis influence to be of comparable magnitude to the trans influence. For a series of Pt(II) iodide complexes it was calculated that the cis influence is highest for ligands having empty, low-lying d orbitals [27]. The order of cis influence was found to be PH_3 > H_2S > C_2H_4 > NH_3 > H_2O > CO for $PtLI_3^-$, when L is neutral, with the same order for trans-PtI_2L_2, except that CO > H_2O. When L is anionic the order decreases as SiH_3^- > CH_3^- > OH^- > I^- > CN^- > H^- > Cl^- > NO_2^- for $PtLI_3^{2-}$ and SiH_3^- > CH_3^- > I^- > H^- > OH^- > Cl^- > NO_2^- for trans-$PtI_2L_2^{2-}$. The experimental evidence for the cis influence is much less than that for the trans influence [7], however. The cis influences of Pt(II)-DMSO compounds were found to be NH_3 > H_2O ~ Cl^- ~ Br^- > C_2H_4 ~ DMSO with factors relative to H_2O of 1-2 (NH_3):1(Br^-, Cl^-):0.3(C_2H_4):0.1(DMSO) [28]. These values are all lower than the corresponding trans influence factors.

2.5. Kinetics of Substitution Reactions

Square platinum(II) complexes are, comparatively speaking, kinetically inert. Their substitution reaction kinetics have been widely studied and these studies have been extensively reviewed [1,10,30-34]. The complexes are susceptible both to nucleophilic attack, owing to the positive charge on the metal center, and to electrophilic attack, owing to considerable d-electron density of the metal. Substitution reactions, such as that shown in Eq. (4), where Y is the entering group and X the leaving group, occur with

$$PtL_3X^n + Y^y \longrightarrow PtL_3Y^{n'} + X^x \tag{4}$$

retention of configuration (cis or trans). The reactions generally follow the rate law given in Eq. (5) where k_1 is the first-order

$$\text{rate} = (k_1 + k_2[Y])[\text{complex}] \tag{5}$$

rate constant reflecting a solvolytic pathway and k_2 is the second-order rate constant appropriate for a bimolecular substitution process. In general, $k_2 >> k_1$.

Although the evidence is indirect, the favored reaction mechanism for substitution is associative, involving a five-coordinate intermediate as shown in Fig. 3. In the figure, S represents the solvent. The associative mechanism is supported by kinetic data in which the complex charge, solvent, and bulk of the side groups have been varied. The evidence is briefly summarized below. A more thorough review of the literature is available elsewhere [1,10,30-35]. For a series of reactions in which the entering and leaving groups are the same, a variation in charge on the complex has a very small effect on the first-order rate constant, even though the entering group may be charged. For example, for the platinum(II) ammine chloride complexes $[PtCl_4]^{2-}$ to $[Pt(NH_3)_3Cl]^+$ in aqueous media, k_1 only varied from 0.62×10^{-5} to 9.8×10^{-5} sec^{-1} [1]. The value of k_1 changes with changes in solvent. For the reaction shown in Eq. (6), k_1 changed from 3.5×10^{-5} to 38×10^{-5} sec^{-1} upon going from H_2O

$$\text{trans-Pt(py)}_2Cl_2 + 2R_4N^{36}Cl \longrightarrow \text{trans-Pt(py)}_2{}^{36}Cl_2 \tag{6}$$

to DMSO [1]. A sterically hindered system results in a sharp decrease in reaction rate; 2-substituted pyridines are much less reactive toward square planar complexes than are 3- and 4-substituted derivatives [34]. The value of k_2 depends on the nature of the entering ligand, Y. For the reaction shown in Eq. (7), k_2 changed

$$Pt(dien)Br^{+} + Y \longrightarrow Pt(dien)Y^{+2} + Br^{-} \tag{7}$$

from 0 to 4300 × 10^4 for Y = OH^- and SCN^-, respectively [1], in aqueous solution at 25°C. The volumes of activation for these reactions are very negative. Negative values for volumes of activation are indicative of bond formation in the transition state.

The intermediate of the associative reaction mechanism is thought to be a trigonal bipyramid [Fig. 3(b)], which is consistent

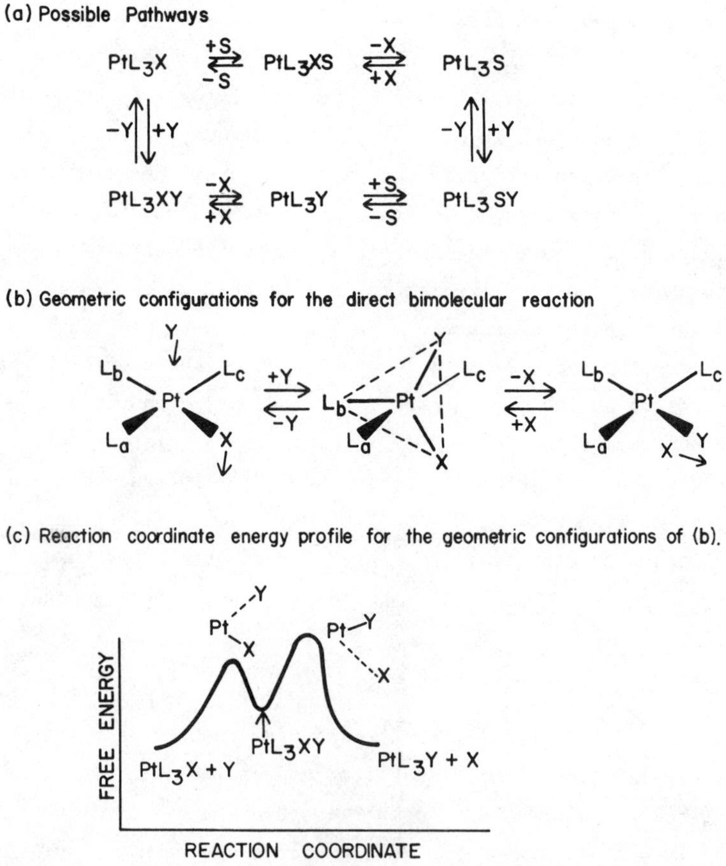

(a) Possible Pathways

$$PtL_3X \underset{-S}{\overset{+S}{\rightleftharpoons}} PtL_3XS \underset{+X}{\overset{-X}{\rightleftharpoons}} PtL_3S$$

$$PtL_3X \underset{-Y}{\overset{+Y}{\rightleftharpoons}} PtL_3XY \qquad PtL_3S \underset{-Y}{\overset{+Y}{\rightleftharpoons}} PtL_3SY$$

$$PtL_3XY \underset{+X}{\overset{-X}{\rightleftharpoons}} PtL_3Y \underset{-S}{\overset{+S}{\rightleftharpoons}} PtL_3SY$$

(b) Geometric configurations for the direct bimolecular reaction

L_b, L_c, L_a, Pt, X, Y $\underset{-Y}{\overset{+Y}{\rightleftharpoons}}$ trigonal bipyramid (L_b, L_c, L_a, Pt, X, Y) $\underset{+X}{\overset{-X}{\rightleftharpoons}}$ L_b, L_c, L_a, Pt, Y, X

(c) Reaction coordinate energy profile for the geometric configurations of (b).

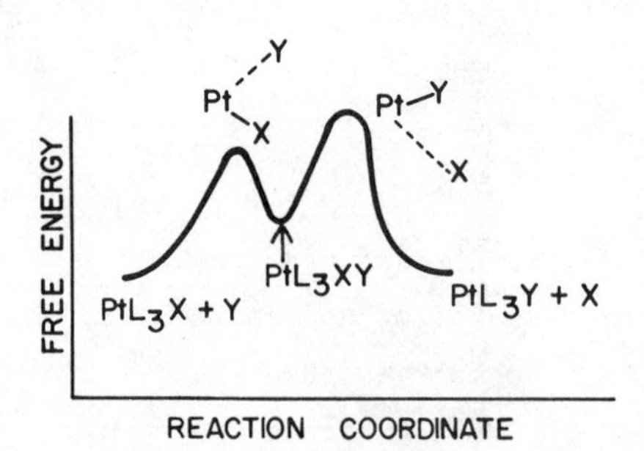

FIG. 3. Associative reaction mechanism for Pt(II) substitution reactions. (Terms are defined in text.)

with the observed retention of configuration. Moreover, all known five-coordinate platinum(II) complexes are trigonal bipyramids, at least in the solid state [1,3-5], and a five-coordinate intermediate in an Au(III) substitution reaction was actually isolated and shown to have this geometry [36]. Au(III) complexes are also square planar. Theoretical calculations using both ligand field [37,38] and molecular orbital [8] theory support an associative mechanism and a trigonal bipyramidal (tbp) transition state geometry. The molecular orbital analysis used a tbp intermediate and generated a double-humped potential energy curve as shown in Fig. 3(c).

The dependence of the bimolecular reaction rate on the nature of the entering ligand has led to the definition of a nucleophilicity scale for platinum(II) complexes [39]. The reactivity of various ligands is related to their polarizability and to the amount of ionic character and σ versus π interactions in their binding to the metal center. The nucleophilicity scale is based on the reactivity of a ligand toward the complex trans-$[Pt(py)_2Cl_2]$ in methanol. The nucleophilicity of a ligand is given the symbol n^o_{Pt}, defined in Eq. (8), where k_y and k^o_s are the second-order rate constants for the

$$n^o_{Pt} = \log \frac{k_y}{k^o_s} \tag{8}$$

bimolecular and for the solvolytic reaction pathways, respectively ($k^o_s \equiv k_1$, [MeOH]). Table 5 contains values of n^o_{Pt} for a variety of ligands. These data can be extended to other ligands if the nucleophilic discrimination factor s, defined by Eq. (9), is known. The

$$sn^o_{Pt} = \log \frac{k^y_2}{k^o_s} \tag{9}$$

nucleophilic discrimination factor for $[Pt(en)Cl_2]$ in water is 0.64; the value given to trans-$[Pt(py)_2Cl_2]$ in MeOH is 1.00 [39(a)]. The n^o_{Pt} values are calculated from kinetic measurements; they have not been meaningfully correlated to any other properties of the ligand.

TABLE 5

Nucleophilic Reactivity Constants, n^{o}_{Pt} Values[a]

Nucleophile	n^{o}_{Pt}	Nucleophile	n^{o}_{Pt}
CH_3OH	0.00	C_6H_5SH	4.15
CH_3COO^-	<2.0	Br^-	4.18
F^-	<2.2	$S(4\text{-}NH_2C_6H_4)_2$	4.27
α-Picoline	2.2	$S(C_2H_5)_2$	4.52
CH_3O^-	<2.4	$P(N(C_2H_5)_2)_3$	4.54
$S(4\text{-}NO_2C_6H_4)_2$	2.4	$S(CH_3)_2$	4.87
Cl^-	3.04	$PO(OCH_3)_2^-$	5.01
NH_3	3.07	$S(CH_2)_5$	5.02
Piperidine	3.13	$S(CH_2)_4$	5.14
Aniline	3.16	I^-	5.46
Pyridine	3.19	$Se(CH_2C_6H_5)_2$	5.53
$S(4\text{-}ClC_6H_4)_2$	3.21	$Se(CH_3)_2$	5.70
NO_2^-	3.22	SCN^-	5.75
$S(C_6H_5)_2$	3.22	SO_3^{2-}	5.79
$S(C_6H_5)(4\text{-}ClC_6H_4)$	3.25	$C_6H_{11}NC$	6.34
$S(4\text{-}FC_6H_4)_2$	3.30	$Sb(C_6H_5)_3$	6.79
$S(4\text{-}NH_2C_6H_4)(4\text{-}NO_2C_6H_4)$	3.31	$As(C_6H_5)_3$	6.89
$S(CH_2C_6H_5)_2$	3.43	$SeCN^-$	7.11
Imidazole	3.44	CN^-	7.14
N_3^-	3.58	$C_6H_5S^-$	7.17
$S(C_6H_5)(4\text{-}CH_3OC_6H_4)$	3.64	$SC(NH_2)_2$	7.17
$S(4\text{-}CH_3C_6H_4)_2$	3.68	$P(OCH_3)_3$	7.23
$S(4\text{-}CH_3OC_6H_4)_2$	3.73	$S_2O_3^{2-}$	7.34
NH_2OH	3.85	$As(C_2H_5)_3$	7.68
$S(4\text{-}OHC_6H_4)_2$	3.85	$P(C_6H_5)_3$	8.93
NH_2NH_2	3.86	$P(C_2H_5)_3$	8.99

[a]Data are taken from Refs. 1 and 34.

The order given in Table 5, however, is similar to the order of the kinetic trans effect, which depends upon the nonlabile ligands in a complex. The kinetic trans effect is discussed in the following section.

The rate of a substitution reaction also depends upon the composition of the metal complex, in which both labile and nonlabile ligands are influential. The leaving group, or labile ligand, has been found to affect the bimolecular rate constant k_2 in the order $NO_3^- > H_2O > Cl^- \geq Br^- \geq I^- > N_3^- > SCN^- > NO_2^- > CN^-$ for a variety of complexes [10,34]. Moreover, for any given class of leaving groups, the more basic the ligand, the slower the reaction rate [40]. Also of importance is the fact that the nonlabile ligands can influence not only the rate of a substitution reaction, but also which labile ligand is the leaving group. The ligand of greatest importance is the one trans to a potential leaving group.

2.6. Kinetic Trans and Cis Effects

The kinetic trans effect, sometimes referred to as the labilizing trans effect, is measured by the effect on the reaction rate of the ligand T trans to the leaving group. The kinetic trans effect differs from the thermodynamic trans influence. The latter reflects the ground state of a complex, while the former involves a reaction transition state or intermediate, assumed to have a trigonal bipyramidal structure. The trans influence helps determine which ligand will be the leaving group; the trans effect is of importance in determining how quickly the designated ligand will depart. These phenomena may or may not be related. A compilation of kinetic data from numerous sources has produced the following order of trans labilizing ability: olefins $\simeq$ NO $\simeq$ CO $\simeq$ $CN^- > R_3Sb > R_3P \geq R_3As \simeq H^- \simeq SC(NH_2)_2 > CH_3^- > C_6H_5^- > SCN^- > NO_2^- > I^- > Br^- > Cl^- >$ py $\geq NH_3 > OH^- > H_2O$. DMSO has also been found to have a large trans effect [17,28]; in one report the labilizing ability was found [28] to be

$H_2O < NH_3 < Cl^- < Br^- < DMSO < C_2H_4$ as $1 < 200 < 330 < 3000 < 2 \times 10^6 < 10^{11}$.

There is also a kinetic cis effect which is generally of lesser magnitude than the corresponding trans phenomenon [28,34]. The cis effect ligand order is $C_2H_4 < Br^- \sim Cl^- < NH_3 \sim H_2O < DMSO$ as $0.05 < 0.3 \sim 0.4 < 1 \sim 1 < 5$ [28]. Interestingly, it was found that the less basic the cis ligand, the larger the cis effect, that is, the faster the reaction rate [41]. For a series of amines cis to an entering amine ligand in a substitution reaction of cis-$[Pt(DMSO)(am)Cl_2]$ in methanol, the basicity of the entering amine did not affect the reaction rate but the basicity of the cis ligand was found to affect k_2 in the manner shown in Eq. (10), where C is

$$\log k_2 = -0.4pK_a + C \tag{10}$$

a constant. It should be noted, however, that the order assigned to ligands for either the cis or trans effect depends in part upon the nucleophilicity of the entering group, n^o_{Pt} [34].

There have been several attempts to elucidate the factors contributing to square planar substitution reaction kinetics. Kinetic parameters in close agreement with experimental values for reactions of the $Pt(NH_3)_x(H_2O)_yCl_{4-x-y}$ system have recently been generated [8]. The bond-making process was found to be most often rate-determining. The rate of reaction increases, owing to a lowering of the initial barrier to reaction intermediate formation, with increasing σ-donor strength of the entering ligand, π acceptor orbitals on the entering ligand, good interactions with (n + 1)s,p orbitals on the metal by the entering ligand, entering ligand softness, decreasing σ-donor strength of the trans ligand, decreasing σ-donor strength of the leaving ligand, and increasing σ-donor strength of the cis ligands. The trans effect of a ligand was also found to increase with decreasing σ-donor strength of the ligand T, increasing π-acceptor strength of T, good interaction of T with (n + 1)s,p orbitals on the metal, and increasing class b character of T.

2.7. Kinetics Related to Chelation

Also of interest to the substitution reaction kinetics of platinum(II) are the kinetics of chelation. Studies have been made of the kinetics of ring closure of trans-$[Pt(enH)_2Cl_2]^{2+}$ [42], both ring-closing and -opening for a series of cis-$[Pt(DMSO)(amH)Cl_2]^{2+}$ complexes where amH is a protonated diamine of varying length [43], and rates of ring closure involving a second chelate ring using triamines [44]. In all cases the ring-closing rate is extremely fast compared to the addition of a monodentate amine. For example, the rate of ring closure for ethylenediamine in the DMSO complex is 1.5×10^4 sec^{-1}, compared with 3.79 sec^{-1} for the addition of a second cyclohexylamine ligand [43(a)]. The rate of ring opening of ethylenediamine is only slightly less than substitution of a cyclohexylamine species, 1.0×10^{-4} sec^{-1} versus 1.5×10^{-4} sec^{-1} [43(a)]. These data suggest that the chelate effect is a combination of a larger than expected ring-closing rate together with a smaller than expected ring-opening rate.

Complexes containing aromatic chelate ring systems such as $[Pt(bipy)Cl_2]$ and $[Pt(terpy)Cl]^+$ react much faster, respectively, than $[Pt(py)_2Cl_2]$ or $[Pt(dien)Cl]^+$ [45]. This result is ascribed to the high trans labilizing effect of ligands such as bipy and terpy [10,46].

Bridged binuclear complexes also can react like chelated systems. For monatomic bridges such as halides, the reaction rates are from 10^2 to 10^3 times faster than corresponding monomeric systems, reflecting the strain expected in a four-membered ring [10]. The rate law usually follows the same form as mononuclear complexes, however, with a nucleophile needed to open the bridge, regardless of the strain in the parent compound. Additionally, ligand exchange sometimes proceeds through a dimeric intermediate. Halide exchange in cis-$[(NH_3)_2PtX_2]$ systems is catalyzed by the presence of $[PtX_4]^{2-}$, X = Cl^- or Br^-. A doubly bridged intermediate is proposed [47].

2.8. Additional Kinetic Factors

Although the majority of substitution reactions are associative, in some cases a dissociative mechanism may be of importance. Anomalous reactions, not obeying the rate law of Eq. (5), and dissociative reactions have been summarized [35,48]. There is considerable evidence for a dissociative mechanism, especially when the metal is sterically hindered from axial coordination [34,35,49] or undergoes photosubstitution [50]. Moreover, platinum(II) complexes catalyze substitution in platinum(IV) systems [51]; this topic has been recently reviewed [52].

In addition to substitution reactions, both free-ligand catalyzed and spontaneous cis-trans isomerizations occur for platinum(II) complexes. The mechanism of isomerization, long believed to be associative, is now thought to depend upon both the solvent system and the complex; it may be either associative or dissociative [34,35,53].

Reactions involving coordinated ligands of platinum(II) complexes also occur. For example, chelated products form in the reaction of amines with monodentate isocyanide complexes [54], and water is postulated to add to coordinated aromatic chelate rings [55]. The latter reaction will be discussed in Sec. 3.5.

3. HYDROLYSIS AND AQUEOUS SUBSTITUTION CHEMISTRY

3.1. Effects of Buffers and Salts on Complex Stabilities

Many buffer-salt systems, both in vitro and in vivo, contain potential ligands, e.g., Cl^-, phosphate, HCO_3^-, citrate, acetate, $C_2O_4^{2-}$, OH^-, or ammonia, the concentrations of which can affect the nature of a platinum(II) complex in solution. In aqueous media, of course, water is present in highest concentration. Although water is a potential ligand, it is also an excellent leaving group and hydrolysis products can be suppressed in the presence of better nucleophiles in sufficiently high concentration. The chloride ion is especially effective in inhibiting hydrolysis reactions. Water is ~70 times faster a leaving group than Cl^- in some amine systems [56] and ~40 times faster in a DMSO complex [43(a)].

In a chloride medium comparable to blood plasma ($[Cl^-] \sim 103$ mM) and at physiological pH and temperature (pH = 7.4, T ~ 37°C), the species present in a solution containing $[(en)PtCl_2]$ are $[(en)Pt(H_2O)_2]^{2+}$: $[(en)Pt(OH)_2]$: $[(en)Pt(H_2O)(OH)]^+$: $[(en)Pt(H_2O)(Cl)]^+$: $[(en)Pt(OH)(Cl)]$: $[(en)PtCl_2]$ in the ratio 0.00134:0.033:0.053:1:1: 37.3 [56]. In a low-chloride-ion medium, comparable to the cytoplasm of a cell ($[Cl^-]$ = 4 mM, pH = 7.4), the same species have the ratio 0.0345:0.086:1.38:1:1:1.45. These results suggest [56] that $[(en)PtCl_2]$ would be about 10 times as reactive in the cell than in the plasma. Using data found in Tables 6 and 8, and assuming the pK_a of the aquochloro complex of the cis-diammine is the same as that attributed to the ethylenediamine compound, 7.4, the species derived from cis-$[(NH_3)_2PtCl_2]$ would be diaquo:dihydroxo:aquohydroxo:aquochloro:hydroxochloro:dichloro in the ratio 0.0178:1.41: 1.12:1:1:24 in the plasma and 0.46:3.63:2.90:1:1:0.915 in the cytoplasm. Comparable data are not available for complex formation with buffers or salts composed of other potentially coordinating ligands. It may be noted, however, that acetate ion is a poorer nucleophile toward platinum(II) than is chloride (see Table 5) and thus is a poorer entering group. On the other hand, ammonia is a slightly better nucleophile than chloride.

Perchlorate and sulfate ions are considered to be noncoordinating for platinum(II) and nitrate is a better leaving group than water [10,34]. Buffer systems composed of these anions are noncoordinating and may be used when the integrity of a complex containing a labile ligand is desired. Hydrolysis reactions and kinetic studies of platinum(II) complexes are often carried out in acidic perchlorate media.

TABLE 6

Equilibrium Constants for Hydrolysis Reactions[a]
(Reactant + $H_2O \rightleftharpoons$ product + X)

X	Product	K_{eq} (mM^{-1})	T (°C)	Ref.
Cl^-	$[(en)Pt(H_2O)Cl]^+$	2.19	25	57
Cl^-	$[(en)Pt(H_2O)Cl]^+$	2.76	35	57
Cl^-	$[(en)Pt(H_2O)_2]^{2+}$	0.143	25	57
Cl^-	$[(en)Pt(H_2O)_2]^{2+}$	0.138	35	57
Cl^-	cis-$[(NH_3)_2Pt(H_2O)Cl]^+$	3.63	25	47(a)
Cl^-	cis-$[(NH_3)_2Pt(H_2O)Cl]^+$	4.37	35	47(a)
Cl^-	cis-$[(NH_3)_2Pt(H_2O)_2]^{2+}$	0.111	25	47(a)
Cl^-	cis-$[(NH_3)_2Pt(H_2O)_2]^{2+}$	1.88	35	47(a)
Br^-	cis-$[(NH_3)_2Pt(H_2O)Br]^+$	1.13	25	47(b)
Br^-	cis-$[(NH_3)_2Pt(H_2O)_2]^{2+}$	0.042	25	47(b)
Cl^-	trans-$[(NH_3)PtCl_2(H_2O)]$	14	25	28
Cl^-	cis-$[(NH_3)PtCl_2(H_2O)]$	<0.25	25	28
Cl^-	$[PtCl_3(H_2O)]^-$	12.6	25	58
Cl^-	$[PtCl_3(H_2O)]^-$	8.0	25	16
Cl^-	$[Pt(Cl)(H_2O)_3]^+$	0.10	25	16
Cl^-	$[Pt(H_2O)_4]^{2+}$	0.011	25	16
Cl^-	trans-$[PtCl_2(H_2O)_2]$	1.2	25	28
Cl^-	cis-$[PtCl_2(H_2O)_2]$	0.80	25	28
Cl^-	trans-$[(DMSO)PtCl_2(H_2O)]$	0.013	25	59
Cl^-	trans-$[(DMSO)PtCl_2(H_2O)]$	5.2	25	28
Cl^-	cis-$[(DMSO)PtCl_2(H_2O)]$	0.115	25	28
Cl^-	cis-$[(C_2H_4)PtCl_2(H_2O)]$	<0.25	25	28
Cl^-	trans-$[(C_2H_4)PtCl_2(H_2O)]$	3	25	28

[a] K_{eq} values correspond to the monohydrolysis reaction.

TABLE 7

Rate Constants for Hydrolysis Reactions, k, and for the Reversible Halide Anations, k_{-1}, at 25°C

Leaving group	Product	k (sec^{-1})	k_{-1} (sec^{-1} M^{-1})	Ref.
Cl^-	$[(en)PtCl(H_2O)]^+$	3.4×10^{-5}	1.5×10^{-2}	57
Cl^-	$[(en)Pt(H_2O)_2]^{2+}$	4.4×10^{-5}	3.1×10^{-1}	57
Cl^-	cis-$[(NH_3)_2PtCl(H_2O)]^+$	2.5×10^{-5}		8
Cl^-	cis-$[(NH_3)_2Pt(H_2O)_2]^{2+}$	3.3×10^{-5}		8
Cl^-	trans-$[(NH_3)_2Pt(H_2O)Cl]^+$	9.8×10^{-5}		8
Cl^-	trans-$[(NH_3)PtCl_2(H_2O)]$	4×10^{-6}	3×10^{-4}	8
Cl^-	cis-$[(NH_3)PtCl_2(H_2O)]$	3×10^{-5}	2.7×10^{-4}	8
Cl^-	cis-$[(NH_3)PtCl(H_2O)_2]^+$	5.9×10^{-5} [a]		8
Cl^-	trans-$[Pt(H_2O)_2Cl_2]$	2.8×10^{-8}	2.3×10^{-5}	28
Br^-	trans-$[Pt(H_2O)_2Br_2]$	1.4×10^{-8}	9×10^{-5}	28
Cl^-	cis-$[Pt(H_2O)_2Cl_2]$	3×10^{-5}	3.8×10^{-2}	28
Cl^-	trans-$[(DMSO)PtCl_2(H_2O)]$	0.10	18.2	28
Br^-	trans-$[(DMSO)PtBr_2(H_2O)]$	0.10	54	28
Cl^-	cis-$[(DMSO)PtCl_2(H_2O)]$	2×10^{-4}	1.7	28
Cl^-	cis-$[(DMSO)PtCl(H_2O)_2]^+$	0.34 [b]	390	28
Br^-	cis-$[(DMSO)PtBr(H_2O)_2]^+$	1.5 [b]	6000	28
Cl^-	$[(DMSO)Pt(H_2O)_3]^{2+}$	0.47 [b]	3900	28
Cl^-	$[(DMSO)Pt(H_2O)_3]^{2+}$	1.3×10^{-6} [c]	0.7	28

[a] Leaving group is trans to NH_3.
[b] Leaving group is trans to DMSO.
[c] Leaving group is cis to DMSO.

3.2. Hydrolysis of Platinum(II) Complexes: Substitution Reactions

The absence of better nucleophiles or poorer leaving groups than water in aqueous systems facilitates the investigation of hydrolysis reactions of platinum(II) complexes. Table 6 contains equilibrium constants for a number of hydrolysis reactions and Table 7 gives some typical rate constants. Although Table 6 reveals that cis-$[(NH_3)_2PtCl_2]$ is less stable to hydrolysis than $[(en)PtCl_2]$, Table 7 shows that the latter compound reacts ~35% faster than the former. That cis-$[(NH_3)_2PtBr_2]$ is more stable to hydrolysis than the chloride compound (Table 6) is expected for a class b metal. The rates of reaction for trans-DMSO complexes (Table 7) reflect the kinetic trans effect discussed earlier while the rates of halide anation of Cl^- < Br^- (Table 7) are expected from the n^o_{Pt} values of Table 5.

The hydrolysis of $[PtCl_4]^{2-}$ to form $[PtCl_3(H_2O)]^-$ proceeds faster in the presence of Zeise's anion, $[PtCl_3(C_2H_4)]^-$, than in its absence [60]. The catalyst is thought to be trans-$[Pt(C_2H_4)Cl_2$-$(H_2O)]$, since the increase in the rate of hydrolysis of $[PtCl_4]^{2-}$

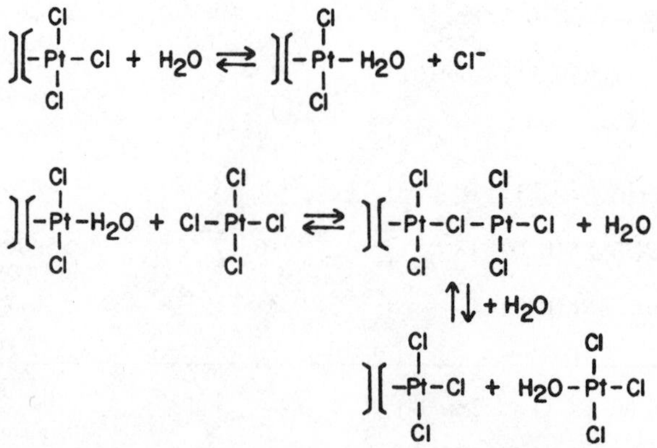

$$\rangle\!\!\big[\!-\!PtCl_2\!-\!Cl + H_2O \rightleftarrows \rangle\!\!\big[\!-\!PtCl_2\!-\!H_2O + Cl^-$$

$$\rangle\!\!\big[\!-\!PtCl_2\!-\!H_2O + Cl\!-\!PtCl_2\!-\!Cl \rightleftarrows \rangle\!\!\big[\!-\!PtCl_2\!-\!Cl\!-\!PtCl_2\!-\!Cl + H_2O$$

$$\big\updownarrow + H_2O$$

$$\rangle\!\!\big[\!-\!PtCl_2\!-\!Cl + H_2O\!-\!PtCl_2\!-\!Cl$$

FIG. 4. Mechanism proposed for the catalysis of $[PtCl_4]^{2-}$ hydrolysis by Zeise's anion, omitting charges on platinum-containing species. (Adapted from Ref. 60.)

parallels an increase in the concentration of the trans-[Pt(C_2H_4)-Cl_2(H_2O)]. The mechanism proposed [60] involves the chloride-bridged intermediate shown in Fig. 4. The suggestion that decomposition of the bridge results in reformation of Zeise's anion is interesting in view of the much larger trans-labilizing ability of C_2H_4 compared with that of chloride ion.

3.3. Acidity of Coordinated Water

Platinum aquo complexes are weak acids with reported pK_a values ranging from >2.5 for $[Pt(H_2O)_4]^{2+}$ to 7.6 for $[(en)Pt(H_2O)(OH)]^+$. The pK_a values of some platinum hydrolysis products are given in Table 8. The greater the trans influence of the ligand trans to the water molecule undergoing deprotonation, the less tightly bound is that water molecule and the higher the expected pK_a of the complex. For the complexes in Table 8, the pK_a values increase with H_2O < amines < Cl^- in the trans position as predicted from the trans influence series.

TABLE 8

pK_a Values of Platinum(II) Hydrolysis Products

Complex	pK_a	T (°C)	Ref.
$[Pt(en)(H_2O)_2]^{2+}$	5.8	25	56
$[Pt(en)(H_2O)(OH)]^+$	7.6	25	56
cis-$[Pt(NH_3)_2(H_2O)_2]^{2+}$	5.6	20	56
cis-$[Pt(NH_3)_2(H_2O)(OH)]^+$	7.3	20	56
$[Pt(dien)(H_2O)]^{2+}$	6.13	25	61
$[Pt(H_2O)_4]^{2+}$	>2.5	25	62
$[PtCl_3(H_2O)]^-$	~7	25	63

3.4. The Hydroxide Ion as Ligand

The result of deprotonating coordinated water is the formation of a hydroxo complex. Whereas the water molecule is an excellent leaving group, hydroxide ion is not. The affinity of hydroxide for platinum(II) is approximately that of ammonia. Substitution of coordinated hydroxide occurs after protonation to form water. Moreover, a platinum(II) bound hydroxide ion is still a good nucleophile, and it may act as a bridging ligand. Bridged hydroxide complexes are stable toward protonation and thus exceedingly inert to substitution.

Although $[(en)Pt(OH)_2]$ appears to be indefinitely stable at high pH, the addition of one equivalent of hydroxide to a 6 mM solution of $[(en)Pt(H_2O)_2]^{2+}$ when left at ambient temperature for a few days produces a stable, unreactive product with no titratable group, assumed to be the dimeric $[(en)Pt(OH)]_2^{2+}$ [56]. A variety of hydroxo-bridged diammine complexes have been isolated as crystalline solids [64]. The first of these was $[(NH_3)_2Pt(OH)]_2(NO_3)_2$, an air-stable compound having nearly perfect D_{2h} symmetry obtained from a solution of pH = 6.44 [64(a)]. The infrared stretching band at 1040 cm^{-1} shifts upon deuteration and is assigned to the bridging hydroxo ligand. A crystalline carbonate analog with four formula units per unit cell is also known [64(b)]. The stacking of this compound is believed to be stabilized by hydrogen-bonding interactions between formula units rather than metal-metal interactions. Both complexes are stable over a large pH range, from $\sim$ pH 2-11. Cyclic trimeric species with formulae $[(NH_3)_2Pt(OH)]_3(SO_4)_3 \cdot 6H_2O$ [64(c)] and $[(NH_3)_2Pt(OH)]_3(NO_3)_6$ [64(d)] have also been characterized by x-ray crystallography. Crystalline cis-$[(NH_3)_2Pt(NO_3)_2]$ has also been isolated from a cis-$[(NH_3)_2Pt(H_2O)_2]^{2+}$ nitrate solution at pH 2 [65]. Between pH 4 and pH 7, the predominant species in a concentrated solution of cis-$[(NH_3)_2Pt(H_2O)_2]^{2+}$ are thought to be the dimeric and trimeric hydroxo-bridged complexes.

3.5. Proposed Hydrolysis of Coordinated Ligands

Water not only reacts as a nucleophile toward platinum(II), but also toward aromatic carbon atoms α to coordinated nitrogen. A covalently hydrated complex $[(bipy)Pt(CN)_2]\cdot H_2O$ with the structure shown in Fig. 5(a), has been proposed [55(a)]. The anhydrous material, a red compound, has an nmr spectrum at 100°C that is comparable to that of $[(bipy)PtCl(H_2O)]$. At 25°C, the complex is a yellow material with an nmr spectrum quite different from that of the red species, showing two separate signals assigned to the protons on the two differing α carbons on the basis of ^{195}Pt (I = 1/2) coupling.

The hydroxide ion may also interact with coordinated pyridine rings as shown in Fig. 5(b). The complex $[(bipy)_2Pt]^{2+}$ and its

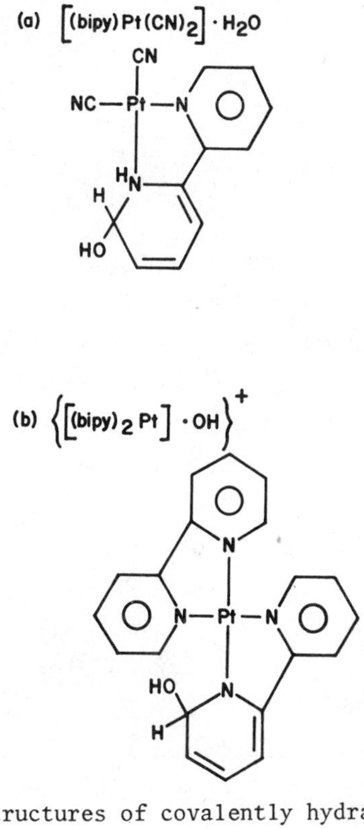

FIG. 5. Proposed structures of covalently hydrated complexes. (From Ref. 55.)

5,5'-dimethyl analog undergo reversible electronic and nmr spectral changes as a function of pH [55(b),55(c)]. The pK_a of the unsubstituted bipyridyl complex is 9.0 while that of the 5,5'-dimethyl analog is 9.5. The complex ions $[(bipy)Pt(NH_3)_2]^{2+}$, $[(bipy)Pt(py)_2]^{2+}$, and $[(py)_4Pt]^{2+}$ do not exhibit these effects nor does the spectrum of $[(bipy)Pt(OH)_2]$ correspond to those of the bis-bipyridyl compounds. The complex $[(o\text{-}phen)PtCl_2]$ also exhibits reversible spectral changes with pH adjustments [55(b)]. These changes are believed to be a result of addition of hydroxide ion to the α-carbon atom of the aromatic amine. Hydroxide substitution and five-coordinate complexes have been ruled out on spectral grounds, as has formation of a conjugate base by proton abstraction from the coordinated ligand, since there is no exchange in alkaline D_2O.

Note added in proof: The above interpretations have recently been challenged. The attack of hydroxide ion appears to be on the metal center after all [135].

4. BINDING SITES IN BIOLOGICAL SYSTEMS

Biological systems have many donor atoms and binding sites for metal ions and platinum(II) is no exception. Because of the complexity of in vivo processes, however, most information about metal binding comes from studies of simpler, in vitro systems. The potential and known and/or suspected binding sites for platinum(II) with nucleic acids and proteins are separate topics and will be treated as such.

The identification of platinum(II) as a class b metal suggests that biomolecules containing atoms of sulfur, carbon, and/or nitrogen that can serve as metal donors will form the most stable complexes with platinum(II). Metal binding will be kinetically controlled, at least initially, and the nucleophilicity of a potential ligand (Table 5), together with its leaving-group ability, are important considerations. The degree of structural accessibility of a potential ligand and its relative concentration are also important factors.

4.1. Nucleic Acids: Potential Binding Sites

The monomer units of polynucleotides consist of a phosphate diester, a pentose ring, and an organic base. Each portion of this nucleotide contains a binding site for platinum(II). The negative charge of the phosphate moiety is a source of electrostatic attraction to a positive platinum(II) ion. Moreover, the terminal oxygen atoms of the phosphate group are potential metal-ion donor ligands. Although the affinity of the class b platinum(II) for class a phosphate oxygen is quite low, the phosphate residue is geometrically situated so as to allow metal coordination without major disruption of the structure of the biopolymer. Additionally, these readily accessible, negatively charged species are present in a relatively high concentration. In DNA there are two phosphate residues every 3.4 Å along the double helix [66]. Phosphate oxygen, however, is expected to be about as good a leaving group as nitrate oxygen, which has been shown to be a better leaving group than water [10,34].

The sugar residue, either ribose or deoxyribose, is a neutral entity containing a ring oxygen atom, which theoretically is a potential ligand. Ether oxygens, however, are known to be poor donor atoms in general and, in particular, have little or no affinity for platinum(II) [1]. Apart from the ring oxygen, the ribose (but not deoxyribose) ring carries a 2'-hydroxyl group, which may also coordinate to a metal. The affinity of alcoholic oxygen for platinum(II), however, is also quite low [1]. Although sugar residues are present in concentrations equivalent to that of phosphates, nucleic acid tertiary structure is such that in polymeric duplexes the pentose oxygens are generally oriented away from the hydrophilic solvent, and thus 2'-hydroxyl groups are not as geometrically accessible as are the phosphate oxygens to metal ions in solution. The 3'- and 5'-oxygen atoms of the phosphodiester linkage are also potential ligands but are unlikely to bind strongly to platinum(II) both for steric and electronic reasons.

The organic bases of the nucleotides offer the best potential donor atoms for platinum(II) coordination. The common bases and

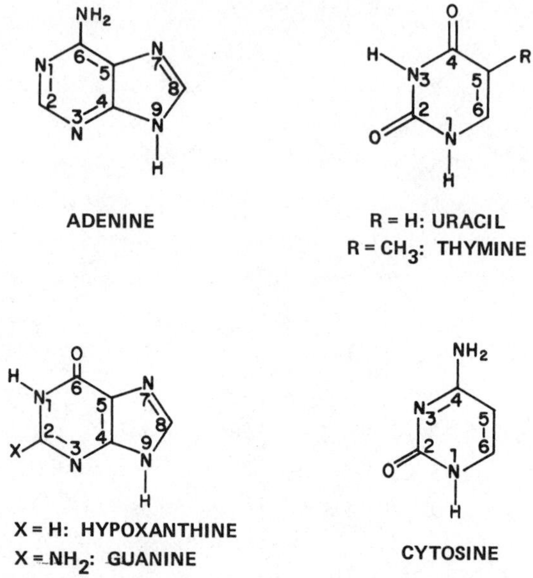

ADENINE

R = H: URACIL
R = CH_3: THYMINE

X = H: HYPOXANTHINE
X = NH_2: GUANINE

CYTOSINE

FIG. 6. Nucleic acid bases. The purine bases are joined to the pentose ring at N-9, the pyrimidine bases at N-1, to form nucleosides.

their respective numbering schemes are shown in Fig. 6. DNAs are generally composed of adenosine (A), cytidine (C), guanosine (G), and thymidine (T) monophosphates while RNAs contain uridine (U) instead of thymidine monophosphate. The adenine and guanine purine bases and cytosine each contain heterocyclic nitrogen atoms that can serve as metal ligands. These pyridine-like systems are expected to have a reasonably high nucleophilicity for platinum(II) (see Table 5) and be relatively poor leaving groups. Although thymine and uracil, in their keto forms, have no pyridine-like nitrogen atoms at neutral pH, they do exist in the enolic forms of these nucleotides. In addition to the ring nitrogens, the purines and cytosine have exocyclic amine groups which, like that of aniline, could serve as a metal donor. It should be noted, however, that an exocyclic amine in a nitrogen heterocycle, especially when attached to an α carbon atom, is a poor nucleophile and thus less attractive as a ligand than the amine of aniline (Table 5). All bases except adenine also have exocyclic oxygen atoms. In their keto forms these oxygen atoms, although not class b ligands, could act as metal donors for platinum(II). Although the affinity for platinum should drop drastically once the oxygens are in their enolic form, the deprotonated enolic oxygen would be a good ligand. The potential for a heterocyclic nitrogen and an exocyclic group to chelate a metal should also be noted. Less common bases with sulfur substituted for exocyclic oxygen, such as 4-thiouracil and 6-thioguanine, are expected to be excellent ligands for platinum(II).

Although the organic bases in nucleic acids contain the best potential ligands for platinum(II), the tertiary structure of the biopolymer does not always render these binding sites geometrically accessible to the metal. The bases are buried in the hydrophobic portion of the biomolecule, and metal coordination will be accompanied by changes in polymer tertiary structure. The concentration of any given base is also less than that of the phosphate or sugar moieties, although this factor is probably not too important.

4.2. Binding Sites on Nucleic Acid Constituents

Nucleic acid binding of heavy metals in general [67], and of platinum(II) with DNA constituents in particular [68], have been recently reviewed. The most definitive information is available for platinum(II) binding to nucleic acid constituents, i.e., the bases, nucleosides, and nucleotides, rather than to DNA, RNA, or even oligonucleotides. Platinum(II) is known to bind covalently to each of the bases in Fig. 6. The binding sites observed under conditions of neutral or near-neutral pH are summarized in Table 9.

As may be seen in Table 9, only N-1 and N-7 of adenosine bind platinum(II), although the exocyclic amino group on C-6 was proposed to bind the metal when its pK_a is lowered through methylation at N-1. The N-7 atom is the favored site on guanosine in neutral (or acid) solutions and no evidence for chelation through O-6 has been found at any pH, either in the reported crystal structures or in Raman or proton

TABLE 9

Known Platinum(II) Binding Sites on Nucleic Acid Constituents in the pH Range 6-8.5[a]

Complex	Coordination site	Type of evidence
$[Pt(adenosine)_4]^{2+}$	N-7	^1H nmr
cis-$[Pt(NH_3)_2(adenosine)_2]^{2+}$	N-7 and/or N-1	Ultraviolet spectroscopy; pH titrations
$[\{Pt(dien)\}_2(adenosine)]^{4+}$	N-7 and N-1	^1H nmr
$[Pt(en)(guanosine)_2]^{2+}$	N-7	X-ray crystallography
cis-$[Pt(NH_3)_2(guanosine)_2]^{2+}$	N-7	X-ray crystallography
trans-$[\{Pt(NH_3)_2(OH)\}_2(5'\text{-GMP})]$	N-7 and N-1	Raman spectroscopy[b]
$[Pt(en)(5'\text{-CMP})]_2 \cdot 2H_2O$	N-3 and O(phos)	X-ray crystallography
$[Pt(dien)(uridinate)]^+$	N-3	^1H nmr[c]
$[Pt(dien)(thymidinate)]^+$	N-3	^1H nmr[c]
$[Pt(en)(uridinate)_2]$	N-3	^1H nmr[c]

[a]Data are from Refs. 67 and 68.

[b]Data are from Ref. 69.

[c]Data are from Ref. 70.

nuclear magnetic resonance spectroscopic studies of cis- and trans-diammineplatinum(II) with 5'-GMP and polyguanylate [69]. Platinum(II) also binds to the deprotonated N-1 position of 5'-GMP under slightly alkaline conditions, pH = 8.3, and Pt-GMP ratios > 1. In cytosine, the ring nitrogen N-3 is the preferred binding site. No data exist to support chelation through neighboring exocyclic keto or amino groups. Uridine and thymidine also bind platinum(II) through N-3, the metal substituting for the hydrogen normally found at that position. Although uracil and thymine form polymeric, blue compounds which presumably arise from bridged chelation with cis-diammineplatinum (see Sec. 5), neither uridine nor thymidine appears to form mononuclear chelated compounds with platinum(II) at neutral pH [70].

The relative order of nucleophilicity of the ribonucleotides toward cis- and trans-diammineplatinum(II) was determined to be GMP > AMP >> CMP >> UMP at 25°C and pH 7 [71]. No differences between the ribo- and deoxyribonucleotides are expected and thymidine monophosphate should have a nucleophilicity comparable to UMP. The kinetics of the reaction shown in Eq. (10), where L is a nucleoside,

$$[^{14}C]\text{-Pt(en)Cl}_2 + L \xrightarrow{k_2} [^{14}C]\text{-Pt(en)LCl}^+ + Cl^- \tag{10}$$

were studied at 37°C. The k_2 values for guanosine, 7-methylguanosine, and adenosine were found to be 0.106, 0.024, and 0.006 M^{-1} sec^{-1}, respectively [72]. These data indicate that guanosine derivatives are more reactive than adenosine toward platinum(II), even when the N-7 position is blocked. Moreover, the difference between the rates of substitution of guanosine and its 7-methyl derivative clearly point to N-7 as the kinetically preferred binding site for platinum(II). The trans effect of guanosine has been reported to be less than that of chloride or bromide since cis-$[Pt(guanosine)_2X_2]$ (with X^- = Cl^-, Br^-) has been isolated, but the trans isomer has not [73]. In contrast to the observed kinetic differences, the stability constants of the various nucleic acid constituents with platinum(II) are all reported to be of comparable magnitude [74].

4.3. Binding to Polynucleotides

Many investigations of platinum(II) binding to DNA and RNA have been undertaken but in the majority of these studies the binding site of the metal has not been determined. One major exception involves platinum binding to crystals of yeast $tRNA^{Phe}$; tRNA crystals soaked in a $[PtCl_4]^{2-}$ solution bind platinum in four specific sites, three of which were used along with other heavy atom binding sites to determine phases in the crystal structure of the tRNA [75]. Although the specific donor atoms on the tRNA have not yet been identified, they presumably could be, with further refinement of the structure. Additionally, trans-$[(NH_3)_2PtCl_2]$ has been shown to bind specifically to N-7 of the guanosine-34 residue of yeast $tRNA^{Phe}$ [76]. This residue is in the anticodon loop of the tRNA. Cis-$[(NH_3)_2PtCl_2]$ does not appear to bind similarly to tRNA.

No evidence of platinum(II) binding to the ribose ring, either to the ring oxygen or to the exocyclic hydroxyl groups, has been found in any investigation involving nucleic acids or their constituents. Neither has any evidence for covalent interactions with phosphate moieties of natural DNAs or RNAs been reported. However, a covalent platinum(II)-phosphate oxygen link has been identified by x-ray crystallography in solid [Pt(en)(5'-CMP)]$\cdot 2H_2O$ (see Table 9) [77]. Quantitative binding of $[^3H]$[Pt(terpy)Cl]$^+$ to the sulfur atom of $[^{35}S]$poly($_s$AU), a polynucleotide prepared from adenosine 5'-O-(1-thiotriphosphate) and UTP, has also been reported [78]. Apart from these two cases, reported platinum(II)-nucleic acid interactions have been either electrostatic in nature or involved binding to the bases. There does not appear to be a unique base-binding site, however.

The extent of platinum(II) nucleic acid binding is a function of the composition of the metal complex employed, the base composition of the polynucleotide, the ionic strength and composition of the medium, and the time allowed for incubation. It has been shown through potentiometric titrations and ethidium fluorescence inhibition studies that whereas [Pt(dien)Cl]$^+$ and $[Pt(NH_3)_3Cl]^+$, each with only one labile ligand, bind DNA covalently at only one site, complexes with two or more labile ligands, e.g., cis- and trans-$[Pt(NH_3)_2Cl_2]$ and $[PtCl_4]^{2-}$, bind more than one site on DNA [79-81]. The latter complexes also cross-link DNA [81]. Kinetically inert compounds such as $[Pt(terpy)(HET)]^+$, HET = 2-hydroxyethanethiolate, and $[Pt(NH_3)_4]^{2+}$ do not bind DNA or RNA covalently [79,80,82]. The nonaromatic, nonplanar kinetically inert compounds, e.g., $[Pt(NH_3)_4]^{2+}$ or $[Pt(py)_2(en)]^{2+}$, exhibit only electrostatic interactions, while the planar, aromatic compounds, e.g., $[Pt(terpy)HET)]^+$ and $[Pt(bipy)(en)]^{2+}$, bind to double-helical nucleic acids by intercalation [82,83].

The buoyant density of DNA containing bound cis-$[Pt(NH_3)_2Cl_2]$ is proportional to the G•C content of the biopolymer and depends upon its sequence [84]. The buoyant densities of platinum(II)-bound natural DNAs increase with G•C base pair content, whereas the buoyant density of platinum bound to poly dG•poly dC is much larger than that of poly d(G•C) after incubation with equal quantities of reagent. Other workers, utilizing techniques to quantitate the bound platinum, have confirmed the dependence of covalent DNA-platinum(II) binding on G•C content [85]. Platinum(II)-DNA intercalative binding is also a function of the G•C content of the biopolymer [86].

Variations occur in platinum(II)-DNA binding with varying ionic strength or incubation times. At low (~1 mM) salt and high Pt-DNA phosphate ratios (r_f), $[Pt(en)Cl_2]$ and *E. coli* DNA show no further reaction after 12 hr at 20°C [87]. In contrast, at r_f of 0.8 in 0.2 M NaCl and ambient temperatures with calf thymus DNA, differences were observed in the binding of $[(en)PtCl_2]$ for reaction times of 66 and 139 hr [82(b)]. Generally, low concentrations of a potentially competing ligand such as Cl^-, low total ionic strength, and long incubation times produce greater covalent platinum binding, provided that conditions are such as to prevent precipitation.

Although cis-$[Pt(NH_3)_2Cl_2]$ and $[Pt(en)Cl_2]$ have at least a kinetic preference for the base guanine (see Ref. 67 for a summary of these data), platinum(II) binds to all the DNA bases. It has been reported [88] that at saturation with $[PtCl_3DMSO]^-$ both homopolymers and natural DNAs bind 2 platinums/adenine, 1 platinum/cytosine, 2 platinums/guanine, and 0.6-0.8 platinum/thymine. The platinum product is trans-$[PtCl_2(DMSO)L]$, owing to the large kinetic trans effect of DMSO, and the DNAs are denatured by the binding. The binding sites are assumed to be the same as those reported in Table 9 for nucleosides. After 3 days incubation at 37°C in 5 mM phosphate, cis-$[Pt(NH_3)_2Cl_2]$ was found [71] either to bind primarily to guanine (r_f = 0.2) or guanine and adenine (r_f = 0.4), whereas cis- and trans-$[Pt(NH_3)_2(H_2O)_2]^{2+}$ bound to all bases (r_f = 0.2). The binding sites in this latter study were also assigned to be the ring nitrogens designated in Table 9. X-ray photoelectron spectroscopic data have been interpreted to indicate a guanine O-6, N-7 chelate with cis-$[Pt(NH_3)_2Cl_2]$ [79]. This conclusion is not supported by other work, however; Refs. 67 and 69 contain a good review of this subject.

Note added in proof: Evidence for the binding of cis-$[Pt(NH_3)_2Cl_2]$ to a $d(G)_4(C)_4$ sequence in a naturally occurring DNA has recently been obtained [136].

4.4. Proteins: Potential Binding Sites

Every protein or polypeptide chain has at least as many potential metal binding sites as there are peptide linkages. The carboxylate (C-terminal) and amino (N-terminal) ends of the chain are also potential ligands. At neutral pH, both terminal groups are charged and interact electrostatically with any charged platinum species present. Both carboxylates and amines can bind platinum(II). Carboxylates have fairly low nucleophilic reactivity constants, however, and are reasonably good leaving groups. Conversely, the affinity of amines for platinum(II) is high and these ligands are known to be very poor leaving groups, unless trans-labilized. The N-terminus is thus a better ligand for divalent platinum than the C-terminus. The amide nitrogens and the carbonyl oxygen atoms of the peptide linkages have a low affinity for platinum(II). These

TABLE 10

Amino Acids Containing Donor Atoms for Platinum(II)[a]

Amino acid	pK_a	Residue serving as binding site	Affinity for Pt(II)
Cysteine	8.3	Sulfur atom	Very high
Methionine		Sulfur atom	Very high
Histidine	6.00	Imidazole nitrogen	High
Arginine	12.48	Amine	Moderate
Lysine	10.53	Amine	Moderate-low
Tryptophan		Cyclic nitrogen atom	Moderate-low
Aspartic acid	3.86	Carboxylate	Low
Glutamic acid	4.25	Carboxylate	Low
Asparagine		Amide	Low
Glutamine		Amide	Low
Tyrosine	10.07	Phenolate	Low
Serine		Alcohol	Very low
Threonine		Alcohol	Very low

[a]Amino acids are listed in approximate order of affinity for platinum(II) at neutral pH.

residues are present in much larger concentrations than any of the individual side chains in the majority of polymers. At least some of these amide groups are accessible on any protein or polypeptide, regardless of the molecule's tertiary structure. Binding to peptide nitrogen and oxygen atoms, especially if deprotonated, should therefore be considered as a possibility.

Selected amino acids containing potential donors for platinum(II) on their side chains and corresponding affinities for the metal at neutral pH are summarized in Table 10. The sulfur-containing species cysteine and methionine have the highest affinity for platinum(II). Nucleophilicities of sulfur atoms are high (n^o_{Pt} values increase as $RSH < R_2S < RS^-$) and the Pt-S bond is kinetically inert. The pyridine-type imidazole nitrogen atom of histidine also has a high nucleophilicity for platinum(II) and is a poor leaving group. Arginine, although positively charged, is similar to imidazole, having one tertiary nitrogen atom that would have a reasonably high affinity for the metal. The primary amine lysine is positively charged at neutral pH and will bind to platinum(II) with loss of a hydrogen ion. The cyclic nitrogen atom of tryptophan is nonaromatic, but binding to platinum would be enhanced by loss of the proton. The remaining potential donors consist of carboxylates, amides, and alcohols, all of which have low or very low affinities for platinum(II) and are good leaving groups. Platinum(II) is thus expected to bind to sulfur-containing residues and to imidazole rings whenever the tertiary structure of a protein or polypeptide renders these moieties available. Another possible binding site in proteins is the disulfide linkages formed by two cysteine residues. Platinum(II) will bind to disulfides almost as readily as to cysteine or methionine.

4.5. Platinum(II)-Amino Acid Binding

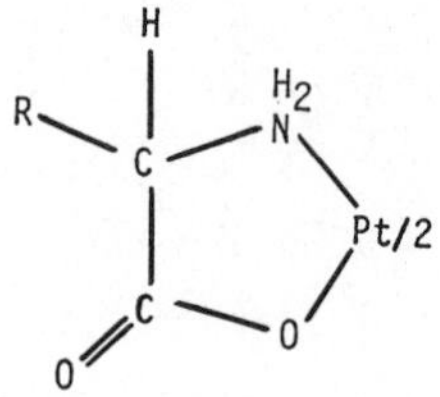

The binding of platinum(II) to amino acids generally, and to sulfur-containing units in particular, was reviewed in 1972 [89,90]. More current information is included in a recent review on platinum-protein binding [91]. Amino acids with no potential donor atoms in their side chains, e.g., glycine and alanine, bind platinum(II) through their terminal amino and carboxyl groups. At neutral pH stable, five-membered chelate rings form, whereas under acidic or basic conditions the amino acid is monodentate. Under acidic conditions, the proton competes successfully with the metal for the carboxyl group and under basic conditions and sufficiently high amino acid concentrations the greater affinity of the amine group becomes evident.

There is some evidence based on spectroscopic assignments that the cis isomers of bis(amino acid)platinum(II) chelates are the kinetically preferred products at neutral pH when chloride is the leaving group [89,91]. It is suggested that a carboxylate oxygen atom binds the metal first, followed by ring closure with amine coordination, since both chloride ions and amines are better trans labilizers than oxygen. Under basic conditions, however, where the amino group is expected to bind first, trans isomers of [Pt(L-proline)(L-alanine)] and [Pt(L-proline)(L-valine)] have been found [92]. The kinetic product, under basic conditions, in the reaction of $[Pt(DMSO)Cl_3]^-$ with glycine to form [Pt(DMSO)Cl(gly)] has the sulfur and nitrogen atoms trans to one another, whereas the thermodynamically more stable product, obtained from heating the complex, is one in which oxygen is trans to sulfur [93]. In this latter complex, the donor atom with the lowest affinity for platinum(II) is trans to the donor with the highest affinity for the metal.

Dipeptides containing no potential ligands in their side chains bind to platinum(II) readily under basic conditions. Both peptide and terminal groups serve as ligands. Reaction of Zeise's salt, $[Pt(C_2H_4)Cl_3]^-$, with $(L\text{-valine})_2$, L-leucine-L-valine, and L-valine-L-leucine gave bridged binuclear products with the amino terminal nitrogen and peptide oxygen atoms of one amino acid bound to one cis-$[Pt(C_2H_4)Cl]^+$ entity and the peptide nitrogen and terminal carboxylate oxygen atoms of the second amino acid bound to another platinum(II) unit [94].

Amino acids containing potential donor atoms in their side chains may bind to platinum(II) through these residues and/or through their terminal groups. The binding of platinum(II) to methionine occurs through the sulfur atom and, with chelate ring formation, through both terminal groups [89-91]. The reaction of any platinum(II) complex containing a labile ligand with the methionine sulfur is rapid. The trans labilizing effect of the bound sulfur atom often aids the addition of a second methionine. For example, the reaction under basic conditions of cis-[$Pt(NH_3)_2$-Cl_2] with methionine yields [$Pt(NH_3)(met)_2$], in which one methioine is bidentate and the other monodentate. The sulfur atoms are trans to one another. The reaction of trans-[$Pt(NH_3)_2Cl_2$] with methionine, however, yields trans-[$Pt(NH_3)_2(met)_2$], in which both amino acids are monodentate. This latter reaction product is stable to all ligands except those having higher nucleophilicities than sulfur for platinum(II). Reaction of trans-[$Pt(NH_3)_2(met)_2$] with iodide yields trans-[$Pt(NH_3)_2I_2$] [89,90]. Under all conditions, the highest methionine-platinum(II) ratio observed is 2:1. Less work has been reported involving cysteine and platinum(II), although the metallointercalation reagent $[Pt(terpy)cys]^+$ has been synthesized [82(b)]. Cysteine coordinates through the sulfur atom, which loses a proton upon binding. Cysteine is a monodentate ligand under neutral conditions, but at pH $\geq$ 8 the complex decomposes, presumably with loss of terpyridine resulting from attack by the terminal amine of cysteine.

Histidine reacts very slowly with platinum(II) under basic conditions to form a bis-chelated complex [89,91]. Coordination is reported to be through the imidazole and the α-amino nitrogen atoms. Reactions of $[PtX_4]^{2-}$ (X = halide) or $[Pt(C_2O_4)_2]^{2-}$ with N-methyl imidazole (MeIm) yield mixtures of cis- and trans-[$Pt(MeIm)_2X_2$], the cis compound exclusively, or $[Pt(MeIm)_4]^{2+}$ depending upon X^- and the conditions [95,96]. Trans-$[Pt(NH_3)_2(MeIm)_2]^{2+}$ is also known [97].

Information about platinum binding to the other amino acids listed in Table 10 is scarce or nonexistent. Platinum(II) binds to both L-lysine and L-aspartate [98]. No evidence for binding to tyrosine, tryptophan, serine, or threonine has been reported.

4.6. Binding to Proteins and Polypeptides

Many platinum(II)-protein binding sites have been determined by x-ray crystallography, since platinum(II) is one of several metals employed as a heavy atom label for proteins. An extensive review of the use of platinum(II) complexes to help phase protein crystal structures is available [99]. A less comprehensive but more encompassing review on complexes of platinum with proteins has also recently appeared [91].

As with nucleic acids, the extent and type of platinum(II)-protein binding is a function of the composition of the platinum complex, the composition and tertiary structure of the protein, the ionic strength and composition of the medium, and the time allowed for incubation. Complexes such as $[Pt(CN)_4]^{2-}$ with no labile ligands bind proteins electrostatically, while complexes such as $[PtCl_4]^{2-}$ generally form covalent protein linkages. Charged complexes such as $[Pt(CN)_4]^{2-}$ and $[Pt(NH_3)_4]^{2+}$ are less likely to penetrate into the hydrophobic interior of proteins than are neutral species such as $[Pt(en)Cl_2]$.

The composition and tertiary structure of a protein or polypeptide dictate the number of available binding sites. The binding of several cis-amine platinum(II) nitrate complexes with poly(L-glutamate), poly(L-aspartate), and poly(L-lysine) at pH 6.4 in the absence of halide ion has been investigated [100]. With the molar concentration of the platinum only 5% that of amino acid monomers, the metal binds bifunctionally to the carboxylate side chains of poly(L-glutamate), monofunctionally to the carboxylate side chains of poly(L-aspartate), and either bi- or monofunctionally to the amino side chains of poly(L-lysine). The leaving group was the nitrate ion. The difference in binding observed between the glutamate and aspartate polymers is attributed to the difference in length of the side chains of the two amino acids. Geometry prohibits bifunctional binding to nearest neighbors on the shorter aspartate side chains. Although the lysine residue is long enough to allow nearest-neighbor, bifunctional binding, the two possible binding modes could not be distinguished from each other. The polypeptides were all assumed to be in a random coil conformation.

The importance of the ionic strength and composition of the medium is illustrated by the following observations. The reaction between α-chymotrypsin crystals and $[PtCl_4]^{2-}$ is slow and nonreproducible in an ammonium sulfate medium. This reaction proceeds both more rapidly and with a higher degree of reproducibility in a phosphate medium [101]. The differences in reactivity are attributed to the formation of $[Pt(NH_3)_xCl_{4-x}]^{-2+x}$ species in the $(NH_4)_2SO_4$ medium. Ammonia is a very poor leaving group unless trans labilized. The $[PtCl_4]^{2-}$ ion does not bind to cysteine residues of triosephosphate isomerase in phosphate buffer, but does bind to these residues within 2 days in ammonium sulphate [102]. Both solutions were at pH 7. Additionally, ribonuclease S binding to $[PtCl_4]^{2-}$ is limited to a methionine residue at pH 5.5 while at pH 7, binding includes a histidine.

The platinum(II) binding sites on various proteins as determined by x-ray crystallography are listed in Table 11. In each case the designated platinum reagent was added to a protein crystal. Both covalent and ionic binding sites are given. A large number of the covalent binding sites are methionines, especially when the reagent is K_2PtCl_4. Indeed, $[PtCl_4]^{2-}$ is considered to be methionine-specific under appropriate reaction conditions and short incubation times [103]. It has also been proposed that the platinum, at least in some cases, is oxidized to platinum(IV) [103]. This proposal has met with criticism from a number of other workers, however [89, 91,99]. The ionic associations of $[Pt(CN)_4]^{2-}$ contrast quite sharply with the covalent binding sites of $[PtCl_4]^{2-}$. The former complex is more likely to bind charged amine residues than methionine, cysteine, or disulfide bonds. Identification of $[Pt(CN)_4]^{2-}$ binding sites, either by x-ray crystallography or nmr spectroscopy, is instructive in determining possible in vivo anion binding sites of proteins.

In recent years, various workers have studied the effect of platinum(II) complexes on enzymes such as leucine aminopeptidase

TABLE 11

Platinum(II) Binding Sites on Various Proteins[a]

Protein	Reagent	Reagent concentration (mM)	Buffer[b]	pH	Time of soak	Site no.	Z	Binding site
Concanavalin A	K_2PtCl_4	0.5	2.1 M phosphate	6	3 days	1	61	Met 129, His 127
						2	23	Met 129
Chironomus hemoglobin	K_2PtCl_4		3.75 M phosphate	7		1	80	Met H17
						2	55	His G2, C-terminus
						3	7	His G19
Chironomus hemoglobin	$Pt(NO_2)_2(NH_3)_2$		3.75 M phosphate	7		1	45	His G2
						2	74	His G19
Ribonuclease S	$Pt(en)Cl_2$	2	3.2 M AS	8		1		His 119
Ribonuclease S	$Pt(en)Cl_2$	2	3.2 M AS	5.5	30-50 hr (fresh sol'n every 10 hr)	1	64	Met 29
Lactate dehydrogenase	$Pt(en)Cl_2$	2.5				1	32	Cys(SH)
						2	81	Cys(SH)
Concanavalin	K_2PtCl_4	1			2 days	1	75	His 127, Met 129
						2	11	Met 42
Horse ferricytochrome c	K_2PtCl_4		4.6 M phosphate	6.2		1	35	Met 65 (close together)
						2	40	Met 65 (close together)
						3	8	His 33 (close together)
Tuna Ferrocytochrome c	K_2PtCl_4	0.1	95% AS	6	2 days	1	24	Met 65
Cytochrome c_{550}	K_2PtCl_4	1.3			7 days	1	69	

TABLE 11 (continued)

Protein	Reagent	Reagent concentration (mM)	Buffer[b]	pH	Time of soak	Site no.	Z	Binding site
α-Chymotrypsin	K_2PtBr_4, K_2PtCl_4 or K_2PtI_4		3.5 M phosphate 2-4% dioxane	4.2		1,2	95	N-terminus and S-S of Cys 1-127
						3,4	55	Met 192
Subtilisin BPN'	K_2PtCl_4	0.65	2.1 M AS 0.05 M acetate	5.9	10-40 days	1	78	Met 50
						2	14	His 64
Subtilisin novo	K_2PtCl_4					1		Met 50
						2		Trp 241, His 238, Trp 106
						3		Ala 1 (N-terminus)
Thermolysin	K_2PtCl_4	6	5% DMSO 0.01 $CaAc_2$ 0.01 M tris/acetate	5.8	10 days	1		His 250
						2		His 216
Carboxypeptidase A	K_2PtCl_4		0.2 M LiCl 0.02 M tris	7.5	42 days	1	74	Cys 161 (-S-S-)
						2	45	Met 103
						3	68	N-terminus: Ala 1
						4	27	His 303
High potential Iron protein (HiPIP)	$K_2Pt(NO_2)_4$	10	3.2 M AS	6.5	7 days	1		Met 49
						2		(major site)
Adenyl kinase	$K_2Pt(NO_2)_4$	2			68 days	1	40	His 36
						2	17	(major site)
Carbonic anhydrase	MMTGA $+K_2Pt(CN)_4$		2.3 M AS	8.5		1		Zn, Thr 197, X139

Ribonuclease S	$K_2Pt(CN)_4$	5	3 M AS 0.1 M acetate	5.5		1	24	
						2	28	
						3	16	
						4	12	
						5	8	
Tuna ferrocytochrome c	$K_2Pt(CN)_4$	6	95% AS	6	1 day	1	30	No near neighbor, Lys 53, Ala 4, Lys 7
						2	15	Ser 100, Val 3, Glu 44, Gln 70, Lys 72, Lys 73
						3	9	Lys 99, Lys 99', Ser 103
						4	5	Ser 103', Glu 21, Lys 7, Lys 25
						5	6	Ile 269 (main chain)
Liver alcohol dehydrogenase	$K_2Pt(CN)_4$	1	0.05 M tris/HCl	8.4	Cocryst.	1		Asp 223, Lys 228, Arg 47, Arg 369
Adenyl kinase	$K_2Pt(SCN)_4$	2			8 days	1	90	
						2	38	Major site near His 36
						3	27	
						4	19	

[a] Data are taken from Ref. 99.

[b] AS = ammonium sulfate.

and several dehydrogenases. The results of these experiments are summarized in a recent review [91]. The complexes $[PtX_4]^{2-}$ (X = halide ion), cis- and trans-$[Pt(NH_3)_2Cl_2]$, and $[Pt(en)Cl_2]$ have been studied most extensively and, although the enzyme binding sites have not been identified, the evidence points to sulfur-containing residues as the site of covalent interaction.

4.7. Other Biological Molecules

In addition to binding proteins and nucleic acids, platinum(II) has been shown to interact with smaller molecules of biological importance. The binding of platinum(II) to thiamine (vitamin B_1) and its phosphate esters has been reported [104]. The metal binds, not to the sulfur of the five-membered ring, but to the cyclic nitrogen of the six-membered ring, para to the exocyclic amino group. Platinum(II) porphyrins are also known. They are generally very stable and their solutions are strongly phosphorescent [105].

5. THE PLATINUM BLUES: A CASE STUDY

5.1. Statement of the Problem

Because many platinum(II) complexes are yellow or red, the occurrence of a blue product in the reaction of a platinum(II) salt has attracted particular attention. The first report of such a compound was issued at the turn of the century [106], but it was not until quite recently that the molecular and electronic structures of any platinum blues were elucidated. The original *Platinblau* was obtained in a reaction between $[(CH_3CN)_2PtCl_2]$ and silver salts; it was formulated as a mononuclear platinum(II) acetamide complex, $[(CH_3CONH)_2Pt]\cdot H_2O$. The acetamide ligand arose through the hydrolysis of acetonitrile during the course of the reaction. Unfortunately, the product was not crystalline, so the postulated structure could not be verified crystallographically after the subsequent advent of x-ray diffraction methods.

Various workers investigated or at least thought about the unusual blue platinum complexes in later years [107-110]. A blue, crystalline material was isolated from the reaction of trimethylacetamide with $[(CH_3CN)_2PtCl_2]$ and it appeared that the structure would finally be elucidated by x-ray crystallography [108]. The crystals proved to be a 7:2:1 mixture of three components, the first two of which were yellow, crystalline products isomorphous with the blue crystals. The blue component was an amorphous solid, however, formulated as a mononuclear platinum(IV) compound $[(t\text{-}C_4H_9CONH)_2\text{-}PtCl_2]$ on the basis of extensive spectroscopic analyses. By analogy, the original *Platinblau* was assigned the formula $[(CH_3CONH)_2Pt(OH)_2]$.

Blue platinum compounds also arise when divalent platinum complexes, such as the antitumor drug cis-$[PtCl_2(NH_3)_2]$ [111], are allowed to react with pyrimidine bases or with DNA or RNA [112]. These bases, such as uracil and thymine (Fig. 6), are cyclic amides that are presumed to serve the same ligand function as acetamide. Unfortunately, these blue compounds are also noncrystalline. Various investigations [110,113-115] revealed them to be paramagnetic oligomers of differing chain length, which accounts for the difficulty in obtaining crystals. Platinum pyrimidine and amide blues have good antitumor drug activity with less nephrotoxicity than the parent cis-dichlorodiammineplatinum(II) [116].

In order to achieve a full understanding of the platinum blues, it was necessary to prepare a crystalline derivative, determine its structure, and compare its properties with other, noncrystalline members of the class. The fulfillment of this objective is described in the following sections together with the DNA-binding properties of platinum uracil blue.

5.2. Molecular Structure of Platinum Blues

A key to getting crystals of a blue platinum complex was to limit the oligomerization reaction. Since it seemed likely that the chain of platinum atoms is propagated through bridging amide ligands and hydrogen bonding interactions [110], a ligand with minimal hydrogen-bonding potential, α-pyridone, was employed. From a solution of

hydrolysis products [56] of cis-diammineplatinum(II), α-pyridone, and sodium nitrate, kept at low pH and temperature to stabilize the blue color, dark blue parallelepipeds formed over a 12-hr period [117, 118]. Analytical and x-ray crystallographic data revealed the formula to be $[Pt_2(NH_3)_4(C_5H_4ON)_2]_2(NO_3)_5 \cdot H_2O$. The structure of the tetranuclear cation is shown in Fig. 7 [117,119]. Two cis-diammineplatinum units are bridged by two deprotonated α-pyridone ligands.

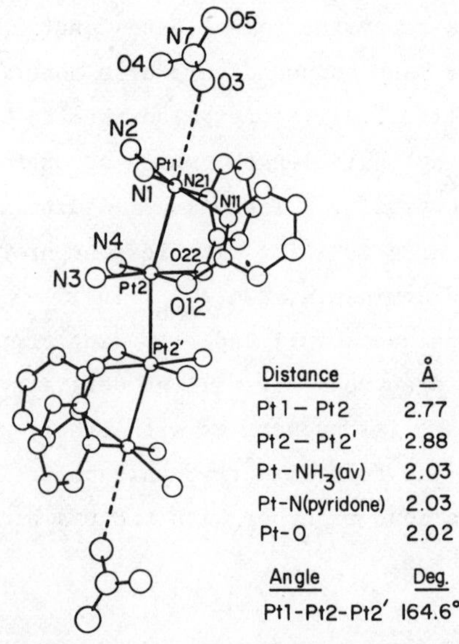

FIG. 7. Structure of the tetranuclear cation in $[Pt_2(NH_3)_4(C_5H_4ON)_2]_2(NO_3)_5 \cdot H_2O$, showing the 40% probability thermal ellipsoids and omitting nonassociated nitrate ions and hydrogen atoms.

The platinum atom at the chain end is bonded to two pyridine nitrogen atoms, while the inner platinum atom coordinates to the exocyclic oxygen atoms. The presence of these Pt-O bonds further stresses the point made earlier that, under the appropriate circumstances, the class b platinum metal can coordinate strongly to oxygen. As shown in Fig. 7, two α-pyridonate bridged diplatinum units are further linked across a crystallographically required center of symmetry. This linkage is supported by partial platinum-platinum bonding, discussed in the following section, and by four hydrogen bonds between the N-H protons of ammine ligands on one platinum atom and the acceptor oxygen atom on the adjacent platinum atom in the chain. The two center platinum coordination planes are strictly parallel and eclipsed whereas the outer planes are canted by 27.4° with respect to one another and twisted by 22° about the Pt-Pt bond axis. The canting and twisting minimize nonbonded steric repulsions between the ammine ligands on adjacent planes and are features also observed in the closely related polymeric structure of cis-$[(NH_3)_4Pt_2P_2O_7]_n$ [120].

The aromatic ring hydrogens of α-pyridone in the above structure preclude additional hydrogen-bonding interactions with adjacent tetranuclear cations and doubly bridged polymers do not form. Uracil and thymine (Fig. 6) have additional exocyclic oxygen atoms that could, and probably do, promote polymerization. Various possible related structures for the platinum pyrimidine and acetamide blues have been discussed [121,122].

5.3. Electronic Structure of Platinum Blues

From the charge on the $[Pt_2(NH_3)_4(C_5H_4ON)_2]^{5+}$ cation it is apparent that platinum has a nonintegral oxidation state of +2.25. Formally, then, the tetranuclear chain consists of three platinum(II) ions and one platinum(III) ion, accounting for the paramagnetism of the platinum blues. A temperature-dependent magnetic susceptibility study of cis-diammineplatinum α-pyridone blue showed it to be a simple Curie paramagnet with a magnetic moment of 1.81 μ_B, consistent with the presence of one unpaired electron per tetranuclear unit [119]. The electron is delocalized over the four platinum atoms. Single-crystal electron spin resonance measurements revealed a nearly axial spectrum, with the unpaired spin residing in a molecular orbital comprised mainly of platinum d_{z^2} orbitals and directed along the chain axis. Although ^{195}Pt hyperfine interactions were not observed in the solid state esr spectrum, they do appear in solution and reflect the delocalization of the unpaired electron along the platinum chain [122]. Similarly, an x-ray photoelectron spectral study of cis-diammineplatinum α-pyridone blue failed to detect any difference between the platinum-4f binding energies of the structurally distinguishable inner and outer pairs of platinum atoms (Fig. 7) [118].

There are several lines of evidence that together establish that the molecular and electronic structures of cis-diammineplatinum α-pyridone blue are representative of the entire class of platinum blues. The solution optical, redox, and esr spectral properties of the compounds are very similar to one another [113,119]. The x-ray photoelectron spectra of the original *Platinblau*, platinum uracil blue, and the α-pyridone blue are virtually identical [118]. Moreover, an EXAFS (extended x-ray absorption fine structure) investigation of cis-diammineplatinum uridine blue showed it to have a 2.9 Å Pt-Pt bond length and other features consistent with the structure of the α-pyridone blue [123]. It therefore appears that the platinum blues all share the features of bridging amidate ligands, mixed valency, and oligomerization typified by the tetranuclear cis-diammineplatinum α-pyridone blue.

The one remaining task is to define precisely the electronic transition responsible for the blue color. It is clear that the band involves a transition either into or from the nearly filled molecular orbital housing the unpaired electron, since loss of color and paramagnetism have been experimentally linked with one another [122]. Single-crystal optical studies would provide a definitive answer.

5.4. Solution and Redox Chemistry

The color of solutions of the platinum blues bleaches with time [110,122]. The absorbance of the blue chromophore depends on the anion present and is also sensitive to temperature. Chloride ion discharges the blue color, while addition of nitrate ion or lowering of the temperature stabilizes it. Kinetic studies reveal the decomposition of cis-diammineplatinum α-pyridone blue, followed by the disappearance with time of either the blue chromophore at 680 nm or the esr signal intensity, to be first-order in platinum at high concentration but higher-order at low concentration [122]. This result indicates a multistep decomposition. For example, two tetramers may decompose to form a diamagnetic platinum(III) dimer, two platinum(II) dimers, and two platinum(II) monomers.

Oxidative titrations of cis-diammineplatinum α-pyridone blue with ceric ion show a linear decrease in esr signal intensity and A_{260} upon addition of three equivalents of oxidant [122]. The product is presumably cis-$[Pt_2(NH_3)_4(C_5H_4NO)_2]^{4+}$, the platinum(III) dimer. A similar result was obtained by potentiometric titration, following which oxidation to platinum(IV) was observed [124]. Reductive titrations with excess ferrous sulfate and back titration with permanganate give formal oxidation states of 2.27 ± 0.10 for cis-diammineplatinum α-pyridone blue, 2.08 ± 0.15 for platinum uracil blue, and 2.28 ± 0.17 for a green hypoxanthine analog [122]. These results and the similarity in optical and esr spectral properties of the platinum blues in solution further support the conclusion that they are mixed-valent oligomers. Gel electrophoretic studies of the cis-diammineplatinum blues indicate the length of the oligomers to increase along the ligand series α-pyridone < hypoxanthine < uracil, assuming an identical charge per monomer unit [122].

5.5. Binding to DNA

The DNA-binding properties of platinum uracil blue (PUB) have been extensively studied [125]. This compound has antitumor drug activity and has been used as a cytological stain [126,127]; both functions are likely to involve platinum-DNA interactions. At low salt concentration, 0.5 M or less, a precipitate forms between PUB and closed circular DNAs which redissolves by increasing the ionic strength to greater than 1.0 M. The buoyant density of the DNA in CsCl increases cooperatively with the concentration of PUB. Using [^{14}C]-radiolabeled uracil, it was shown that platinum binds to the DNA and releases its uracil ligand since no label is transferred to the DNA [125]. The reaction can largely be reversed with cyanide to form the very stable $[Pt(CN)_4]^{2-}$ ion. Nothing is known about the binding sites of platinum on the DNA, however, and future work will have to be directed at elucidating this feature.

6. DETECTING PLATINUM IN BIOLOGICAL SYSTEMS

The growing interest in studying the interaction of platinum(II) complexes with biological molecules has led to the development of analytical techniques for detecting platinum. While the effects of platinum on the properties (activity, electrophoretic mobility, spectrum, etc.) of a biopolymer are readily monitored it is usually desirable, upon separation of the biopolymer from a platinum reagent, to ascertain whether any platinum remains with the polymer and, if so, where and how much. A brief review of the analytical techniques employed to detect platinum in the presence of a biopolymer and to elucidate the nature of its binding are therefore provided in this section.

6.1. Methods of Detection

Platinum attached to a biopolymer can be determined either directly or indirectly. Indirect methods usually involve monitoring a radioactively labeled, kinetically inert ligand, or studying platinum loss in a solution from which the biopolymer has been removed. Direct determinations are preferable and will be stressed here. Unfortunately, such methods are not always feasible.

The direct detection of platinum can be accomplished by several methods. Perhaps the least expensive and most versatile of these is flameless atomic absorption spectroscopy. Platinum can be detected by this method, even in the presence of a biopolymer, in amounts as low as 1-20 μg when other metals are absent. Atomic absorption has been employed by a variety of workers for studying platinum bound to DNA [79,80,128]. There is no indication that the DNA, buffer systems, or composition of the platinum reagent employed interfere with platinum detection by this method.

Another method for directly determining platinum, applicable over a wide concentration range, is the use of the radioactive isotope, ^{195m}Pt. This metastable isotope was used [85] to determine the quantity of cis-$[Pt(NH_3)_2Cl_2]$ bound to DNA. However, its short half-life (4.1 days) and limited availability limit the utility of this approach.

Platinum-195, the stable isotope, is 33.8% abundant and has a nuclear spin, I, of one-half. Detection of this species is, therefore, possible by nmr spectroscopy. ^{195}Pt nmr has not, as yet, been employed as a means of detecting platinum in biological systems. The conditions for detection are concentrations in the millimolar range and, even with a ^{195}Pt-enriched sample [129], many systems are too dilute for this method to become practical. ^{13}C and ^{1}H nmr spectroscopy have been used to determine platinum binding to both nucleosides [69,71,130] and amino acids [131]. Platinum-195 coupling occurs with both ^{13}C and ^{1}H nuclei and in solutions of ~1 mM platinum concentration this coupling can be readily observed at natural abundance. Concentration is again the limiting factor in applying this technique for platinum detection to biopolymers, however. Enrichment of platinum reagents with ^{195}Pt would lower the concentration requirements, but only by a factor of 3.

The presence of platinum in a biopolymer may also be determined by x-ray diffraction. The high electron density of platinum facilitates its identification in the presence of the lighter elements of which biopolymers are normally composed. Detection of platinum by this method is easiest when the system is highly ordered. Platinum binding to proteins [99] and to tRNA [75,76] has been elucidated by x-ray crystallography and, in the case of the metallointercalation reagents, platinum binding to DNA has been ascertained by fiber x-ray diffraction analysis [83(a)].

Other methods of directly detecting platinum are also available. Instrumental techniques such as EXAFS and mass spectrometry have been used to measure distances between platinum and its ligands on DNA [132] and binding to nucleosides, respectively [133]. Colorimetric tests may also be used. Here the platinum-biopolymer complex is degraded, the platinum reduced to Pt(0), removed from any potential ligands, and then reoxidized to Pt(II) or Pt(IV). One such colorimetric test uses p-nitrosodimethylaniline as the reagent and is optimal over a range of 3.6×10^{-6} to 1.2×10^{-5} M in platinum [134].

Finally, the presence of platinum in a biopolymer may generate a new spectral band characteristic of a platinum-ligand bond. The presence of platinum on poly- and mononucleotides containing phosphorothioate groups was monitored [78] by the appearance of an electronic absorption band specific to sulfur-bonded platinum terpyridine systems. Platinum binding to polypeptides was ascertained [100] by charge transfer bands indicative of carboxylates or amines bound to platinum bipyridine and platinum o-phenanthroline systems.

6.2. Nature and Amount of Bound Platinum

Once platinum has been detected in a biochemical system, it is often of interest to learn where and how much of the material is bound. Covalent binding can be differentiated from electrostatic or other associative interactions by electrophoresis or sometimes by exhaustive dialysis of the platinum-biopolymer complex. Methods for determining intercalative binding to both linear and circular DNAs and for differentiating between this type of binding and either covalent or electrostatic interactions have been thoroughly described [82(b)]. Here ethidium bromide fluorescence inhibition studies and sedimentation and electrophoretic studies on closed circular DNAs were employed.

The question of where the platinum is bound on the biopolymer is not always easy to answer and has been the topic of discussion in Sec. 4. How much of the platinum is bound can often be determined through the direct detection of platinum or through the various indirect methods mentioned above. Atomic absorption spectroscopy and radioactive isotopic labeling are two techniques that are excellent for quantitating the amount of bound platinum, but they give no information about where or how the metal is bound. It is rare that these techniques are employed merely to detect platinum without quantitation; the same is true of colorimetric analyses.

Conversely, nmr studies yield better information about where the platinum is bound than about how much. This result is also true of Raman spectroscopy [69,71,130], EXAFS [132], x-ray photoelectron spectroscopy [79], and mass spectrometry [132].

X-ray crystallography provides definitive information about where and how much platinum is bound. This technique is limited, however, to fairly small nucleic acids or oligonucleotides and to proteins. Fiber diffraction analyses of larger, polycrystalline nucleic acids can be instrumental in determining where the platinum is, but reveal little about how much metal is bound.

Systems where platinum binding gives rise to a spectroscopic change, such as charge transfer bands in the ultraviolet or visible portion of the spectrum, are generally excellent for determining both where and how much of the metal is bound [78].

ABBREVIATIONS

am	amine
bipy	2,2'-bipyridine
dien	diethylenetriamine
DMSO	dimethylsulfoxide
en	ethylenediamine
EXAFS	extended x-ray absorption fine structure
HET	2-hydroxyethanethiolate
MeIm	N-methyl imidazole
o-phen	o-phenanthroline
PUB	platinum uracil blue
py	pyridine
tbp	trigonal bipyramidal
terpy	2,2',2"-terpyridine

REFERENCES

1. F. R. Hartley, The Chemistry of Platinum and Palladium, John Wiley and Sons, New York, 1973.
2. F. A. Cotton and G. W. Wilkinson, Advanced Inorganic Chemistry (3rd ed.), Interscience, New York, 1972.
3. E. A. H. Ebsworth, J. M. Edward, E. J. S. Reed, and J. D. Whitelock, *J. Chem. Soc. Dalton,* 1161 (1978).
4. (a) G. Natile, L. Maresca, L. Cattalini, U. Belluco, P. Uguagliati, and U. Croatto, *Inorg. Chim. Acta, 20,* 49 (1976);
   (b) L. Maresca, G. Natile, and L. Cattalini, *Inorg. Chim. Acta, 14,* 79 (1975).
5. (a) P. Meakin and J. P. Jesson, *J. Amer. Chem. Soc., 96,* 5751 (1974);
   (b) J. P. Jesson and P. Meakin, *J. Amer. Chem. Soc., 96,* 5760 (1974).

6. Ref. 10, Table 2.4, p. 69.

7. T. G. Appelton, H. C. Clark, and L. E. Manzer, *Coord. Chem. Rev.*, *10*, 335 (1973).

8. J. K. Burdett, *Inorg. Chem.*, *16*, 3013 (1977), and references cited therein.

9. S. Ahrland, J. Chatt, and N. R. Davies, *Quart. Rev. (London)*, *12*, 265 (1958).

10. F. Basolo and R. G. Pearson, Mechanisms of Inorganic Reactions (2d ed.), John Wiley and Sons, New York, 1967, and references cited therein.

11. J. Chatt and D. M. P. Mingos, *J. Chem. Soc.*, *A*, 1770 (1969).

12. D. S. Martin, Jr., R. A. Jacobson, L. D. Hunter, and J. E. Benson, *Inorg. Chem.*, *9*, 1276 (1970).

13. J. K. Barton and S. J. Lippard, *Ann. N.Y. Acad. Sci.*, *313*, 686 (1978).

14. J. S. Miller and A. J. Epstein, in Progress in Inorganic Chemistry, Vol. 20, S. J. Lippard, ed., Interscience, New York, 1976, p. 1.

15. K. W. Nordquest, D. W. Phelps, W. F. Little, and D. J. Hodgson, *J. Amer. Chem. Soc.*, *98*, 1104 (1976).

16. L. I. Elding, *Inorg. Chim. Acta*, *28*, 255 (1978).

17. B. P. Kennedy, R. Gosling, and M. L. Tobe, *Inorg. Chem.*, *16*, 1744 (1977).

18. L. I. Elding and L. Gustafson, *Inorg. Chim. Acta*, *19*, 165 (1976).

19. T. D. Harrigan and R. C. Johnson, *Inorg. Chem.*, *16*, 1741 (1977).

20. I. Mochida, J. A. Mattern, and J. C. Bailer, Jr., *J. Amer. Chem. Soc.*, *97*, 3021 (1975).

21. A. Pidcock, R. E. Richards, and L. M. Venanzi, *J. Chem. Soc.*, *A*, 1707 (1966).

22. L. J. Manojlović-Muir and K. W. Muir, *Inorg. Chim. Acta*, *10*, 47 (1974).

23. P. B. Hitchcock, B. Jacobsen, and A. Pidcock, *J. C. S. Dalton*, 2043 (1977).

24. M. H. Chisholm, H. C. Clark, L. E. Manzer, J. B. Stothers, and J. E. H. Ward, *J. Amer. Chem. Soc.*, *95*, 83 (1973).

25. M. A. M. Meester, D. J. Stufkens, and K. Vrieze, *Inorg. Chim. Acta*, *21*, 251 (1977).

26. G. M. Bancroft and K. D. Butler, *J. Amer. Chem. Soc.*, *96*, 7208 (1974).

27. D. R. Armstrong, R. Fortune, P. G. Perkins, R. J. Dickinson, and R. V. Parish, *Inorg. Chim. Acta*, *17*, 73 (1976).

28. L. I. Elding and O. Gröning, *Inorg. Chem.*, *17*, 1872 (1978).

29. S. S. Zumdahl and R. S. Drago, *J. Amer. Chem. Soc.*, *90*, 6669 (1968).

30. C. H. Langford and H. B. Gray, Ligand Substitution Processes, W. A. Benjamin, New York, 1965.

31. R. G. Wilkins, The Study of Kinetics and Mechanisms of Reactions of Transition Metal Complexes, Allyn and Bacon, Boston, 1974.

32. M. L. Tobe, Inorganic Reaction Mechanisms, Nelson, London, 1972.

33. L. Cattalini, in Reaction Mechanisms in Inorganic Chemistry, M. L. Tobe, ed., Butterworths, London, 1972.

34. A. Peloso, *Coord. Chem. Rev.*, *10*, 123 (1973), and references cited therein.

35. R. J. Mureinik, *Coord. Chem. Rev.*, *25*, 1 (1978).

36. A. J. Hall and D. P. N. Satchell, *J. C. S. Chem. Commun.*, 163 (1976).

37. J. Vranckx and L. G. Vanquickenbirne, *Inorg. Chim. Acta*, *11*, 159 (1974).

38. L. G. Vanquickenbirne, J. Vranckx, and C. Görller-Walrand, *J. Amer. Chem. Soc.*, *96*, 4121 (1974).

39. (a) U. Belluco, L. Cattalini, F. Basolo, R. G. Pearson, and A. Turo, Jr., *J. Amer. Chem. Soc.*, *87*, 241 (1965);
(b) R. G. Pearson, *Chem. Brit.*, *3*, 103 (1967);
(c) R. G. Pearson, H. Sobel, and J. Songstad, *J. Amer. Chem. Soc.*, *90*, 319 (1968).

40. L. Cattalini, G. Maragini, G. Degetto, and M. Brunelli, *Inorg. Chem.*, *10*, 1545 (1971).

41. (a) P. D. Braddock, R. Romeo, and M. L. Tobe, *Inorg. Chem.*, *13*, 1170 (1974);
(b) R. Romeo and M. L. Tobe, *Inorg. Chem.*, *13*, 1991 (1974).

42. M. G. Carter and J. K. Beattie, *Inorg. Chem.*, *9*, 1233 (1970).

43. (a) R. Romeo, S. Lanza, and M. L. Tobe, *Inorg. Chem.*, *16*, 785 (1977);
(b) R. Romeo, S. Lanza, D. Minniti, and M. L. Tobe, *Inorg. Chem.*, *17*, 2436 (1978).

44. G. Natile, G. Albertin, E. Bordignon, and A. A. Orio, *J. C. S. Dalton*, 626 (1976).

45. (a) F. Basolo, H. B. Gray, and R. G. Pearson, *J. Amer. Chem. Soc.*, *82*, 4200 (1960);
(b) P. Haake and R. A. Cronin, *Inorg. Chem.*, *2*, 879 (1963).

46. G. Annibale, L. Cattalini, L. Maresca, G. Michelon, and G. Natile, *Inorg. Chim. Acta*, *10*, 211 (1974).

47. (a) K. W. Lee and D. S. Martin, Jr., *Inorg. Chim. Acta*, *17*, 105 (1976);
(b) G. F. Vandegrift, III and D. S. Martin, Jr., *Inorg. Chim. Acta*, *12*, 179 (1975).

48. D. S. Martin, Jr., *Inorg. Chim. Acta Rev.*, *1*, 87 (1967), and references therein.

49. C. J. May and J. Powell, *Inorg. Chim. Acta*, *26*, L21 (1978).

50. C. Bartocci, A. Ferri, V. Carassiti, and F. Scandola, *Inorg. Chim. Acta*, *24*, 251 (1977).

51. W. R. Mason, *Coord. Chem. Rev.*, *7*, 241 (1975).

52. L. I. Elding and L. Gustafson, *Inorg. Chim. Acta*, *19*, 31 (1976).

53. (a) W. J. Louw, *Inorg. Chem.*, *16*, 2147 (1977), and references therein;
(b) R. Romeo, D. Minniti, and M. Trozzi, *Inorg. Chim. Acta*, *14*, L15 (1975).

54. A. L. Balch and J. E. Parks, *J. Amer. Chem. Soc.*, *96*, 4114 (1974).

55. (a) R. D. Gillard, L. A. P. Kane-Maguire, and P. A. Williams, *Trans. Met. Chem.*, *1*, 247 (1976);
(b) E. Bielli, R. D. Gillard, and D. W. James, *J. C. S. Dalton*, 1837 (1976);
(c) R. D. Gillard and J. R. Lyons, *J. C. S. Chem. Commun.*, 585 (1973).

56. M. C. Lim and R. B. Martin, *J. Inorg. Nucl. Chem.*, *38*, 1911 (1976).

57. R. F. Coley and D. S. Martin, Jr., *Inorg. Chim. Acta*, *7*, 573 (1973).

58. L. I. Elding, *Acta Chem. Scand.*, *24*, 1331 (1970).

59. Y. N. Kukushkin, *Inorg. Chim. Acta*, *9*, 117 (1974).

60. M. Green and M. G. Swanwick, *J. C. S. Dalton*, 158 (1978).

61. R. M. Alcock, F. R. Hartley, and D. E. Rogers, *J. C. S. Dalton*, 1070 (1973).

62. L. I. Elding, *Inorg. Chim. Acta*, *20*, 65 (1976).

63. A. K. Johnson and J. D. Miller, *Inorg. Chim. Acta*, *16*, 93 (1976).

64. (a) R. Faggiani, B. Lippert, C. J. L. Lock, and B. Rosenberg, *J. Amer. Chem. Soc.*, *99*, 777 (1977);
(b) B. Lippert, C. J. L. Lock, B. Rosenberg, and M. Zvagulis, *Inorg. Chem.*, *17*, 2971 (1978);

65. B. Lippert, C. J. L. Lock, B. Rosenberg, and M. Zvagulis, *Inorg. Chem.*, *16*, 1525 (1977).

66. V. A. Bloomfield, D. M. Crothers, and I. Tinoco, Jr., Physical Chemistry of Nucleic Acids, Harper and Row, New York, 1974.

67. (a) J. K. Barton and S. J. Lippard, in Metal-Nucleic Acid Interactions (T. Spiro, ed.), John Wiley and Sons, New York, 1980, Vol. 1, pp. 32-113.
(b) D. J. Hodgson, in Progress in Inorganic Chemistry, Vol. 23 (S. J. Lippard, ed.), Interscience, New York, 1977, p. 211;
(c) L. G. Marzilli, in Progress in Inorganic Chemistry, Vol. 23 (S. J. Lippard, ed.), Interscience, New York, 1977, p. 255.

68. R. C. Harrison and C. A. McAuliffe, *Inorg. Persp. Biol. Med.*, *1*, 267 (1978).

69. G. Y-H. Chu, S. Mansy, R. E. Duncan, and R. S. Tobias, *J. Amer. Chem. Soc.*, *100*, 593 (1978).

70. M. C. Lim and R. B. Martin, *J. Inorg. Nucl. Chem.*, *38*, 1915 (1976).

71. S. Mansy, G. Y-H. Chu, R. E. Duncan, and R. S. Tobias, *J. Amer. Chem. Soc.*, *100*, 607 (1978).

72. A. B. Robins, *Chem. Biol. Inter.*, *6*, 35 (1973).

73. N. Hadjiliadis and T. Theophanides, *Inorg. Chim. Acta*, *16*, 77 (1976).

74. W. M. Scovell and T. O'Connor, *J. Amer. Chem. Soc.*, *99*, 120 (1977).

75. S. H. Kim, G. Quigley, F. L. Suddath, A. McPherson, D. Sneden, J. J. Kim, J. Weinzierl, P. Blattmann, and A. Rich, *Proc. Nat. Acad. Sci. U.S.*, *69*, 3746 (1972).

76. A. Jack, J. E. Ladner, D. Rhodes, R. S. Brown, and A. Klug, *J. Mol. Biol.*, *111*, 315 (1977).

77. S. Louie and R. Bau, *J. Amer. Chem. Soc.*, *99*, 3874 (1977).

78. K. G. Strothkamp and S. J. Lippard, *Proc. Nat. Acad. Sci. U.S.A.*, *73*, 2536 (1976).

79. (a) J.-P. Macquet and T. Theophanides, *Biopolymers*, *14*, 781 (1975);
(b) M. M. Millard, J.-P. Macquet, and T. Theophanides, *Biochim. Biophys. Acta*, *402*, 166 (1975);
(c) J.-P. Macquet and T. Theophanides, *Biochim. Biophys. Acta*, *442*, 142 (1976).

80. J.-L. Butour and J.-P. Macquet, *Eur. J. Biochem.*, *78*, 455 (1977).

81. J. M. Pascoe and J. J. Roberts, *Biochem. Pharmacol.*, *23*, 1345 (1974).

82. (a) K. W. Jennette, G. A. Vassiliadis, S. J. Lippard, and W. R. Bauer, *Proc. Nat. Acad. Sci. U.S.*, *71*, 3839 (1974);
(b) M. Howe-Grant, K. C. Wu, W. R. Bauer, and S. J. Lippard, *Biochem.*, *15*, 4339 (1976);
(c) J. K. Barton and S. J. Lippard, *Biochem.*, *18*, 2661 (1979).

83. (a) S. J. Lippard, P. J. Bond, K. C. Wu, and W. R. Bauer, *Science*, *194*, 726 (1976);
(b) B. Nordén, *Inorg. Chim. Acta*, *31*, 83 (1978).

84. (a) P. J. Stone, A. D. Kelman, and F. M. Sinex, *Nature*, *251*, 736 (1974);
(b) P. J. Stone, A. D. Kelman, F. M. Sinex, M. M. Bhargava, and H. O. Halvorson, *J. Mol. Biol.*, *104*, 793 (1976).

85. (a) L. L. Munchausen and R. O. Rahn, *Cancer Chemother. Rep.*, Part 1, *59*, 643 (1975);
(b) L. L. Munchausen and R. O. Rahn, *Biochim. Biophys. Acta*, *414*, 242 (1975);
(c) J.-P. Macquet and T. Theophanides, *Inorg. Chim. Acta*, *18*, 189 (1976);
(d) V. Guantieri, L. DeNardo, and A. M. Tamburro, *Inorg. Chim. Acta*, *30*, 155 (1978).

86. M. Howe-Grant and S. J. Lippard, *Biochem.*, *18*, 5762 (1979).

87. I. A. G. Roos, *Chem.-Biol. Interact.*, *16*, 39 (1977).

88. R. F. Whiting and F. P. Ottensmeyer, *Biochim. Biophys. Acta*, *474*, 334 (1977).

89. A. J. Thomson, R. J. P. Williams, and S. Reslova, in Structure and Bonding, Vol. 11, Biochemistry, Springer-Verlag, New York, 1972, p. 1.

90. C. A. McAuliffe and S. G. Murray, *Inorg. Chim. Acta Rev.*, *6*, 103 (1972).

91. P. Melius and M. E. Friedman, *Inorg. Perspect. Biol. Med.*, *1*, 1 (1978).

92. R. D. Gillard and O. P. Slyudkin, *J. C. S. Dalton*, 152 (1978).

93. L. E. Erickson and W. F. Hahne, *Inorg. Chem.*, *15*, 2941 (1976).

94. L. E. Nance and H. G. Frye, *J. Inorg. Nucl. Chem.*, *38*, 637 (1976).

95. B. J. Graves, D. J. Hodgson, C. G. van Kralingen, and J. Reedijk, *Inorg. Chem.*, *17*, 3007 (1978).

96. C. G. van Kralingen and J. Reedijk, *Inorg. Chim. Acta*, *30*, 171 (1978).

97. J. W. Carmichael, N. Chan, A. W. Cordes, C. K. Fair, and D. A. Johnson, *Inorg. Chem.*, *11*, 1117 (1972).

98. T. N. Bhat and M. Vijayan, *Acta Crystallogr.*, *B32*, 891 (1976).

99. T. L. Blundell and L. N. Johnson, Protein Crystallography, Academic Press, New York, 1976, Chap. 8.

100. Y.-Y. H. Chao, A. Holtzer, and S. H. Mastin, *J. Amer. Chem. Soc.*, *99*, 8024 (1977).

101. P. B. Sigler and D. M. Blow, *J. Mol. Biol.*, *12*, 17 (1965).

102. G. A. Petsko, Doctoral dissertation, Oxford University, Oxford, England, 1973.

103. R. E. Dickerson, D. E. Eisenberg, J. Varnum, and M. L. Kopka, *J. Mol. Biol.*, *45*, 77 (1969).

104. N. Hadjiliadis, J. Markopoulous, G. Pneumatikakis, D. Katakis, and T. Theophanides, *Inorg. Chim. Acta*, *25*, 21 (1977).

105. (a) J. A. Mercer-Smith and D. G. Whitten, *J. Amer. Chem. Soc.* *100*, 2620 (1978);
(b) F. R. Hopf and D. G. Whitten, in Porphyrins and Metalloporphyrins (K. M. Smith, ed.), Elsevier, Amsterdam, 1975, p. 667.

106. K. A. Hofmann and G. Bugge, *Berichte*, *41*, 312 (1908).

107. R. D. Gillard and G. Wilkinson, *J. Chem. Soc.*, 2835 (1964).

108. D. B. Brown, R. D. Burbank, and M. B. Robin, *J. Amer. Chem. Soc.*, *91*, 2895 (1969), and references cited therein.

109. D. Cohen, Doctoral dissertation, Northwestern University, 1973.

110. E. I. Lerner, Doctoral dissertation, Columbia University, 1976.

111. B. Rosenberg, L. Van Camp, J. E. Trosko, and V. H. Mansour, *Nature*, *222*, 385 (1969).

112. J. P. Davidson, P. J. Faber, R. G. Fischer, Jr., S. Mansy, H. J. Peresie, B. Rosenberg, and L. Van Camp, *Cancer Chemother. Rep.*, *59*, 287 (1975).

113. B. Lippert, *J. Clin. Hematol. Oncol.*, *7*, 26 (1977).

114. R. D. Macfarlane and D. F. Torgerson, *Science*, *191*, 920 (1976).

115. E. I. Lerner, W. R. Bauer, and S. J. Lippard, unpublished results, 1976.

116. (a) R. J. Speer, H. Ridgway, L. M. Hall, D. P. Stewart, K. E. Howe, D. Z. Lieberman, A. D. Newman, and J. M. Hill, *Cancer Chemother. Rep.*, *59*, 629 (1975);
(b) J. M. Hill, E. Loeb, A. MacLellan, N. O. Hill, A. Khan, and J. J. King, *Cancer Chemother. Rep.*, *59*, 647 (1975).

117. J. K. Barton, H. N. Rabinowitz, D. J. Szalda, and S. J. Lippard, *J. Amer. Chem. Soc.*, *99*, 2827 (1977).

118. J. K. Barton, S. A. Best, S. J. Lippard, and R. A. Walton, *J. Amer. Chem. Soc.*, *100*, 3785 (1978).

119. J. K. Barton, D. J. Szalda, H. N. Rabinowitz, J. V. Waszczak, and S. J. Lippard, *J. Amer. Chem. Soc.*, *101*, 1434 (1979).

120. (a) J. A. Stanko, results quoted by M. J. Cleare, in Platinum Coordination Complexes in Chemotherapy (T. A. Connors and J. J. Roberts, eds.), Springer-Verlag, New York, 1974, pp. 24-26; (b) J. A. Stanko, private communication, 1976.

121. J. K. Barton and S. J. Lippard, *Ann. N.Y. Acad. Sci.*, *313*, 686 (1978).

122. J. K. Barton, C. Caravana, and S. J. Lippard, *J. Amer. Chem. Soc.*, *101*, 7269 (1979).

123. B. K. Teo, K. Kijima, and R. Bau, *J. Amer. Chem. Soc.*, *100*, 621 (1978).

124. M. Laurent, private communication, 1978.

125. W. R. Bauer, S. L. Gonias, S. K. Kim, K. C. Wu, and S. J. Lippard, *Biochem.*, *17*, 1060 (1978).

126. R. W. Wagner, B. Rosenberg, and S. Aggarwal, *Fed. Amer. Soc. Exp. Biol.*, *33*, 1385 (1974).

127. P. K. McAllister, B. Rosenberg, S. K. Aggarwal, and R. W. Wagner, *J. Clin. Hematol. Oncol.*, *7*, 717 (1977).

128. G. L. Cohen, W. R. Bauer, J. K. Barton, and S. J. Lippard, *Science*, *203*, 1014 (1979).

129. J. Pesek and W. R. Mason, *J. Mag. Res.*, *25*, 519 (1977).

130. G. Y.-H. Chu, R. E. Duncan, and R. S. Tobias, *Inorg. Chem.*, *16*, 2625 (1977).

131. L. E. Erickson, J. W. McDonald, J. K. Howie, and R. P. Chow, *J. Amer. Chem. Soc.*, *90*, 6371 (1968).

132. B. K. Teo, P. Eisenberger, J. Reed, J. K. Barton, and S. J. Lippard, *J. Amer. Chem. Soc.*, *100*, 3225 (1978).

133. I. A. G. Roos, A. J. Thomson, and J. Eagles, *Chem.-Biol. Inter.*, *8*, 421 (1974).

134. J. J. Kirkland and J. H. Yoe, *Anal. Chem.*, *26*, 1340 (1954).

135. O. Farver, O. Mønsted, and G. Nord, *J. Amer. Chem. Soc.*, *101*, 6118 (1979).

136. G. L. Cohen, J. A. Ledner, W. R. Bauer, H. M. Ushay, C. Caravana, and S. J. Lippard, *J. Amer. Chem. Soc.*, *102*, 2487 (1980).

Chapter 3

CLINICAL ASPECTS OF PLATINUM ANTICANCER DRUGS

Barnett Rosenberg
Department of Biophysics
Michigan State University
East Lansing, Michigan

1. INTRODUCTION 128
   1.1. The History of Platinum Drugs 128
   1.2. Utility of Chemotherapy in Cancer 132
2. ANTICANCER ACTIVITY OF PLATINUM DRUGS IN ANIMALS 134
   2.1. Review of Results Accumulated to Date 134
   2.2. Schedule, Dose, and Route Dependencies 139
3. TOXIC SIDE EFFECTS IN ANIMALS 141
   3.1. Dose-limiting and Minor Side Effects 141
4. DRUG FATE IN ANIMALS 143
   4.1. Excretion Rates 143
   4.2. Organ Distribution 146
   4.3. Transport 150
   4.4. Biochemical Reactions 152
5. COMBINATION CHEMOTHERAPY IN ANIMALS 153
   5.1. Additivity and Synergism 153
   5.2. Combinations with Radiation 158
6. DEVELOPMENT OF RESISTANCE TO THE PLATINUM DRUGS IN ANIMALS 160
   6.1. Tissue Culture: Whole Animal Studies 160
7. MECHANISMS OF ACTION 163
   7.1. Host Activities for Responses 163

7.2. Molecular Mechanisms of Action 166
8. MUTAGENESIS AND CARCINOGENESIS OF PLATINUM DRUGS 172
8.1. Mutagenesis in Bacteria and Mammalian Cells 172
8.2. Carcinogenesis of Platinum Drugs 174
9. ANTICANCER ACTIVITY OF CISPLATIN IN HUMANS 175
9.1. Early Trials of Cisplatin as a Single Agent 175
9.2. Nephrotoxicity in Humans and Its Amelioration . . . 178
10. ACTIVE COMBINATION CHEMOTHERAPIES OF HUMAN CANCERS WITH CISPLATIN 179
10.1. Testicular Cancers 179
10.2. Ovarian Cancer 183
10.3. Head and Neck Cancers 185
10.4. Squamous Cell Carcinoma of the Cervix 186
10.5. Bladder Cancers 187
10.6. Oat Cell Lung Cancers 188
10.7. Other Cancers 188
11. GENERAL CONCLUSIONS 189
REFERENCES 190

1. INTRODUCTION

1.1. The History of Platinum Drugs

It is now convincingly clear that platinum coordination complexes play a significant role in the treatment of human cancers. This brief review is undertaken with two intentions. The first is to present the accumulated evidence of seven years of clinical experiences in animals and humans with platinum drugs; and the second is to describe some of the chemistry and biological effects of platinum complexes that may be relevant to their anticancer activity.

An unexpected and interesting biologic activity of platinum complexes was first noted in the author's laboratory in a study that was designed to test the effects of electric fields on growing cells. A certain effect was found when *E. Coli* were grown in a continuous culture apparatus containing platinum electrodes. When the electric field was applied, bacterial growth continued, but cell division was inhibited, and the bacterial rods grew into very long filaments. It was found eventually that this effect was due not to the action of the electric fields on the bacteria but to a small amount (~10 ppm) of an electrolysis product of the platinum electrode formed in the presence of the ammonium chloride in the nutrient medium. This stable chemical species was responsible for the filamentation effect [1]. Further analysis showed [2] that this chemical was the classic Peyrone's chloride, cis-dichlorodiammineplatinum(II), or its higher oxidation state equivalent, cis-tetrachlorodiammineplatinum(IV). The Peyrone's chloride will be hereafter referred to by its generic drug name, cisplatin.

Further studies of this unique bacterial action suggested that an entire class of platinum complexes, and indeed other metal complexes as well, was capable of causing this cell division inhibition leading to filamentation. Some other interesting bacteriologic effects were discovered during these studies: bactericidal activity and the ability to induce lysis in lysogenic strains of bacteria [3]. The class of metal complexes which are charged ions in solution exhibit potent bactericidal effects, while the class of complexes that are neutral in solution cause both the filamentation and the lytic phenomena.

These studies continued over a period of years and led to a better, but never complete, understanding of the bacterial actions of this broad class of chemicals. It was not until 1968, however, that the purely intuitive jump was made to test these interesting complexes for anticancer activity in a mouse tumor model system. The standard protocols of the National Cancer Institute (NCI) for determining any anticancer activity of new drugs called for the injection of a certain number of transplantable tumor cells into the mouse on day 0. The drug is injected into the peritoneal cavity on day 1, and the progressive development of the tumor is followed with some specific endpoint determination (i.e., death of the animal or size of the tumor after a given number of days). Cisplatin completely inhibited the development of the tumor. Other complexes were tested and some of these were also found to be active. Eventually, four of these drugs were submitted to the National Cancer Institute for testing in a different screening system. Their results confirmed a significant degree of activity [4].

This unique new class of anticancer chemicals elicited an ambivalent response. Scientists were enthusiastic over the discovery of a new class of chemicals with anticancer activity, but their enthusiasm was blunted by the discomfiture of introducing heavy metal complexes back into medicine. Nevertheless, after further extensive animal tests, cisplatin was chosen to enter on Phase I clinical trials in humans. To do this required suitable preparation techniques for larger-scale production of the drug in high purity; methods for characterizing the purity; new techniques for preparing the drug in vial form for intravenous administration; some knowledge of the pharmacologic and toxicologic reactions that might be anticipated in humans; and finally, some estimate of the minimally toxic dose to start with in humans. These were all achieved with support from the National Cancer Institute.

The early results of the clinical trials with terminal patients were optimistic, in that some notable tumor regressions were observed. The first report of clear anticancer activity in human patients was presented by Dr. J. Hill and his associates at the Wadley Institutes of Molecular Medicine at the Seventh International Chemotherapy Congress in Prague (1971) [5]. Their work was done quite independently of the National Cancer Institute.

Within the next year, additional reports of clinical trials by various oncology study groups supported by the NCI appeared in the literature. By 1973, it had become quite clear that cisplatin had interesting activity against cancers in patients considered terminal and unresponsive to the best prior therapies. A particularly high degree of activity against testicular cancers by Wallace and coworkers [6] and ovarian cancers by Wiltshaw [7] was noted. Again, however, the enthusiasm was muted by the observation of a marked kidney toxicity. Thus a prevailing sentiment was expressed by Dr. Hill--"Cis-platinum(II) diamminechloride appears to be too good a therapeutic agent to abandon, yet too toxic for general use . . ." [101]. While all earlier cancer chemotherapy drugs exhibited a wide variety of toxicities, none of them were so toxic to the kidneys. Clinicians, therefore, had little previous experience to fall back on in handling this new toxicity.

A significant breakthrough occurred when Cvitkovic, Krakoff, and their associates at the Sloan-Kettering Institute for Cancer Research discovered that the simple pharmacologic trick of hydrating the patient markedly diminished the kidney toxicity of cisplatin, with little or no loss of anticancer activity [8]. The drug could then be given at doses up to three times the previous limit without compromising the patient's kidney function.

A further advance was made by Merrin [9] at the Roswell Park Memorial Institute when he found that administration of the drug as a slow infusion over 6 to 8 hr also ameliorated the kidney toxicity without loss of anticancer function. With these results confirmed, the major hurdle to the acceptability of the cisplatin drug was overcome; further clinical trials have increased at an exponential rate (with a doubling time of about one year) thereafter.

Studies in animals by Venditti and his coworkers [10] during this same time showed that cisplatin would act either additively or synergistically with a number of other anticancer drugs, yielding a substantial improvement in the treatment of animal tumors. Their work, confirmed by many others, stimulated the testing of a variety of combinations involving the cisplatin drug with other known anticancer agents in human clinical trials. Substantial successes were established and confirmed in the treatments of testicular and ovarian cancers. In the case of the former, a combination of bleomycin and vinblastine had been shown by Samuels [11] at the M. D. Anderson Hospital to be effective in producing a significant number of complete remissions. Unfortunately, the duration of these remissions was relatively short. Einhorn and coworkers [12] at Indiana University added the cisplatin drug to the Samuels regimen and reported a significant increase, both in the number of complete remissions, and in their duration. Golbey and his colleagues [13] at the Memorial Sloan-Kettering Cancer Center also included cisplatin in their combination chemotherapy for testicular cancers with similar results. In the ovarian case, Wiltshaw [14] at the Royal Marsden Hospital extended her studies to include combination chemotherapy with chlorambucil, and eventually to the high-dose therapy with cisplatin. Here again, the results were a significant improvement over prior therapies. A new combination of cisplatin and adriamycin was introduced by Bruckner and his coworkers [15] at Mount Sinai Hospital for advanced metastatic ovarian cancer, again with a very significant increase of activity both in response rate and duration of remissions.

With the firm establishment of cisplatin as a drug of high potency in the treatment of these two human malignancies, a concerted effort began to study other combination chemotherapies involving cisplatin against a variety of cancers.

1.2. Utility of Chemotherapy in Cancer

Historically, the first successful method of treating cancers was by surgical removal of the tumor masses and some surrounding tissues. This works well when the tumor is a localized mass, as may occur in an early stage of growth. However, most cancers shed actively growing malignant cells, which can spread throughout the body via the lymph or blood circulation. These cells lodge in many locations distant from the initial tumor site and start proliferative growth. Surgery, alone, can no longer succeed in treating such a condition of advanced metastatic cancer.

The next modality of treatment to evolve was radiation therapy. Again, if the tumor or the metastases are restricted to a reasonably local portion of the body, ionizing radiation can cause relatively selective death of cancer cells, with an acceptable amount of damage to the normal cells. Again, however, widespread metastases cannot be treated with radiation.

The modern era of cancer chemotherapy began in the late 1940s with the introduction of the nitrogen mustards for the treatment of leukemias. The obvious advantages of chemotherapy for metastatic cancers stimulated the search for other chemicals with more selective activity against a broad spectrum of cancers, with less toxicity and greater efficacy. After 30 years of testing (mostly using the "Edisonian" approach), a stable of about 30 workhorses is currently in use, falling into eight broad categories of chemicals: the alkylating agents (e.g., cyclophosphamide); natural products (e.g., bleomycin); antibiotics (e.g., adriamycin); modified nucleosides (e.g., 5-fluorouracil); antifolates (e.g., methotrexate); mitotic inhibitors (e.g., vinblastine); steroids (e.g., prednisone); and the most recent, and the subject of this review, metal coordination complexes (e.g., cisplatin). The author should like to emphasize here that despite furious activity, the mechanism of selective cancer cell destruction cannot yet be specified in molecular detail for any of these substances. The crucial underpinnings for any such theories still elude researchers.

Most clinicians would probably agree with the statement that any single drug therapy against cancer is, with rare exception, of limited value. It is the development of the concept of combination chemotherapy for cancers in the last two decades that produced the present optimism in regard to the eventual cure of cancers. Two or more drugs, given in carefully tested and continually evolving schedules (protocols), have been shown in animal studies to be additive or synergistic in curing cancers. Similar, but not as quantifiable, improvements in therapeutic responses occur in human patients, as well. One of the more striking characteristics of the platinum coordination complexes, as will be described later, is their ability to act additively or synergistically with a wide variety, if not all, of the other anticancer chemicals. They can, therefore, be added to previously developed therapies for given types of cancers, with a reasonable expectation of improved responses.

TABLE 1

Metastatic Cancers Curable With Chemotherapy

Tumor	Cure rate (%)
Choriocarcinoma	75
Rhabdomyosarcoma	50
Advanced Hodgkin disease	58
Burkitt's lymphoma	45-55
Wilm's tumor	58
Acute lymphocytic leukemia (in children)	50
Histiocytic lymphoma	42
Ewing's tumor	31
Embryonal testicular cancer[a]	32-70

[a]Longer observation time needed.

Source: Adapted from Ref. 16.

Zubrod [16] has listed those cancers where good evidence exists for a substantial cure rate by chemotherapy (Table 1). Additions to this table of many other types of cancers very likely will be made in the near future.

2. ANTICANCER ACTIVITY OF PLATINUM DRUGS IN ANIMALS

2.1. Review of Results Accumulated to Date

To expedite the testing of new compounds for anticancer activity, the National Cancer Institute has established a set of standard test systems [17]. They consist mainly of particular mouse strains with transplantable tumors adapted to them. The tumors may be leukemias/lymphomas (disseminated blood or lymph cancers), or solid tumors (carcinomas and sarcomas), and are implanted in the animal at specific sites (peritoneal cavity, under the skin, etc.) on day 0. The tumors are allowed 24 hr to establish themselves and the chemical to be tested is administered on day 1, usually by injection at various dose levels in the peritoneal cavity. Either the tumor sizes or the animals day of death are noted for a treated group, as well as for a control group. A comparison of the results between these two groups then gives a measure of the efficacy and toxicity of the chemical against these standard cancers. There are approximately 100 standard cancers available for tests in about 80 animal hosts. If a chemical meets the defined criteria for one screening system, it is usually tested against a number of others, as well, to determine the spectrum of tumors that respond to it. A good drug should show activity in a number of such systems. However, most laboratories dedicated to research with anticancer drugs carry no more than one or two tumor-host systems, and cooperation between laboratories is essential to evaluate the breadth of the spectrum of tumors responding.

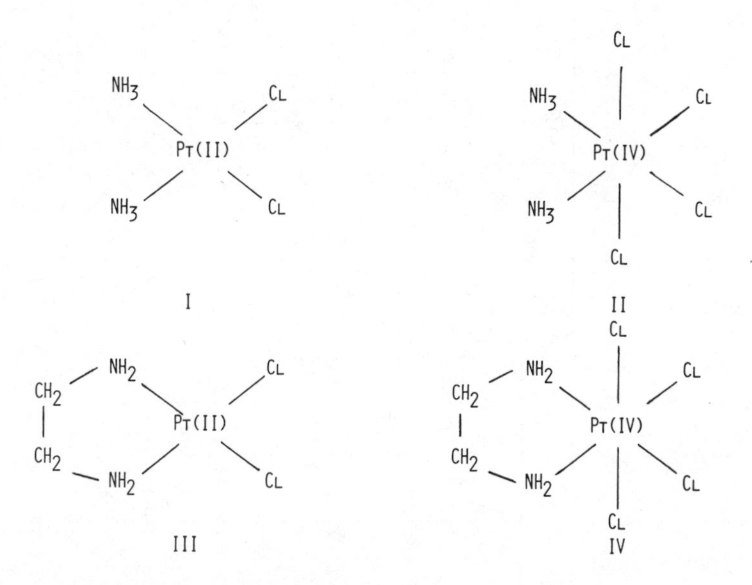

FIG. 1. Chemical structures of the first four platinum coordination complexes found to have anticancer activity: (I) cis-dichlorodiammineplatinum(II); (II) cis-tetrachlorodiammineplatinum(IV); (III) dichloroethylenediamineplatinum(II); (IV) tetrachloroethylenediamineplatinum(IV).

TABLE 2

Best Results of the Anticancer Activity of Cisplatin in Animal Systems

Tumor	Host	Best results
Sarcoma 180 solid	Swiss white mice	T/C = 2-10%[a]
Sarcoma 180 solid (advanced)	Swiss white mice	100% cures
Sarcoma 180 ascites	Swiss white mice	100% cures
Leukemia L1210	BDF_1 mice	%ILS = 379%; 4/10 cures[b]
Primary Lewis lung carcinoma	BDF_1 mice	100% inhibition
Ehrlich ascites	BALB/c mice	%ILS = 300%
Walker 256 carcinosarcoma (advanced)	Fisher 344 rats	100% cures; T.I. > 50[c]
Dunning leukemia (advanced)	Fisher 344 rats	100% cures
P388 lymphocytic leukemia	BDF_1 mice	%ILS = 534%;[b] 7/10 cures
Reticulum cell sarcoma	C+ mice	%ILS = 141%[b]
B-16 melanocarcinoma	BDF_1 mice	%ILS = 268%;[b] 8/10 cures
ADJ/PC6	BALB/c mice	100% cures; T.I. = 8[c]
AK leukemia (lymphoma)	$AKR/_{Lw}$ mice	%ILS = 225%;[b] 3/10 cures
Ependymoblastoma	C57BL/6 mice	%ILS = 169%;[b] 1/6 cures
Rous sarcoma (advanced)	15-I chickens	65% cures

DMBA-induced mammary carcinoma	Sprague-Dawley rats	77% total regressions[d]
ICI 42, 464-induced myeloid and lymphatic leukemias	Alderly Park rats	%ILS = 400%[c]
AK leukemia	$AKR/_{Lw}$ mice	%ILS = 163%; 3/10 cures
$CD8F_1$ mammary	$CD8F_1$ mice	T/C = 0%; 10/10 cures
Colon 26 (C6)	CDF_1 mice	%ILS = 257%; 9/10 cures
Colon 38 (C8)	BDF_1 mice	T/C = 11%; %ILS = 299; 5/10 cures
M5076 ovarian carcinoma	BDF_1 mice	T/C = 0%; %ILS = 321; 6/10 cures
Ridgeway osteogenic sarcoma	AKD_2F_1 mice	%ILS = 297%; 2/10 cures
LX_1 lung xenograft	Nude mice	T/C = 0%
Colon 06-A	CDF_1 mice	%ILS = 150; 2/10 cures

[a] 3/9 free of all tumors.

[a] $T/C = \frac{\text{tumor mass in treated animals}}{\text{tumor mass in control animals}} \times 100.$

[b] %ILS = % increase in lifespan of treated over control animals.

[c] T.I. = therapeutic index (LD_{50}/ED_{90}), ED_{90} = effective dose to inhibit tumors by 90%.

[d] 3/9 free of all tumors.

The first tests for antitumor activity of the platinum complexes were performed in the author's laboratory in 1968, using the solid Sarcoma 180 in the ICR strain of mice [4]. The structures of the first four active complexes are shown in Fig. 1. These were sent to the National Cancer Institute for testing against a second system, L1210, in BDF_1 mice. Their confirmation of activity justified publication of the results. Continued interest in the prime candidate for clinical trials, cis-dichlorodiammineplatinum(II), has produced a spate of test results against more than 50 tumor-host systems. The best results from a selection of these systems are compiled in Table 2. Since the number of animals in most of these tests is small, no adequate statistical analyses are possible. The best results are presented merely to give an impression of the activity.

Some of the more significant generalities that may be derived from this myriad of tests are described below. Foremost is the fact that the anticancer activity of cisplatin covers a broad spectrum in animals, a fact which is now becoming obvious in human cancers, as well (see below). The spectrum includes disseminated leukemias/lymphomas (e.g., L1210, P388), as well as solid cancers (e.g., B16, sarcoma 180). Good results are obtained against normally drug-resistant tumors (e.g., L1210), as well as drug-sensitive ones (e.g., Walker 256 carcinosarcoma). Activity occurs in slow-growing (e.g., ADJ/PC6, Lewis lung), as well as rapidly growing, tumors. The level of activity for most tumors is high rather than marginal. Cisplatin is active against tumors that are transplantable, chemically induced (e.g., DMBA mammary carcinoma), or virally induced (e.g., Rous sarcoma). Even some tumors of human origin transplanted to immunologically deprived mice (xenografts) respond well (e.g., LX_1). Cisplatin works well in a variety of animal hosts such as mice, rats, and chickens. Finally, it is of some significance that the drug may be given to animals whose tumors are well advanced and which would produce death within a few days, and still cause complete regressions of these advanced tumors (e.g., ADJ/PC6,

Dunning leukemia, Walker 256 carcinosarcoma). Workers are justified in concluding from these results that cisplatin is a potent, broad-spectrum anticancer drug in animals.

2.2. Schedule, Dose, and Route Dependencies

When a drug is found to be active in these screens, it is obligatory to determine the optimum dose levels, schedules, and routes of administration for various tumors. Usually the first tests are done with a single intraperitoneal injection at different dose levels for ease of manipulation. There is, at this stage, little or no information on the kinetics of distribution of the chemical out of the peritoneal cavity; its distribution in the organs; its metabolism; and its excretion rate and route (urine/feces). The toxicities, and the efficacy against the tumor, however, are controlled in the main by these variables.

For cisplatin, it was established early by the National Cancer Institute screening tests [18] that the drug is most effective when administered either as an intraperitoneal or intravenous injection. Subcutaneous and intramuscular injections were less effective, and oral administration was not effective at all (Table 3).

TABLE 3

Route-of-Administration Dependency of Cisplatin Activity Against the L1210 Tumor in Mice

Schedule (days)	Route	Best % ILS
1-5	Intraperitoneal	217
1-5	Subcutaneous	143
1-5	Intravenous	122
1-5	Oral	$<120^a$

[a]% ILS $\leq$ 120 is considered inactive.

Source: Adapted from Ref. 18.

TABLE 4

Schedule Dependency of Cisplatin Activity Against the L1210 Tumor

Schedule	Route	Best % ILS
Day 1	Intraperitoneal	378
Days 1-5	Intraperitoneal	217
Days 1-9	Intraperitoneal	207
Every 3 hr × 8 on days 1, 9	Intraperitoneal	196
Every 3 hr × 8 on days 1, 5, 9	Intraperitoneal	206
Every fourth day for 3 times	Intraperitoneal	381

Source: Adapted from Ref. 18.

In human clinical trials, the preferred method of administration is intravenous. Other techniques are usually developed after the drug is of proven significance, with oral administration as the preferred choice.

As a single intraperitoneal injection, the administered dose which kills 50% of the animals (LD_{50}) is 13-14 mg of the cisplatin per kilogram of animal body weight. The therapeutic dose is 7 mg kg^{-1}, which is less than the LD_{10}.

A comparison of the various schedules and doses for intraperitoneal injections of cisplatin against the L1210 tumor is shown in Table 4. There does not seem to be a clearly preferable schedule. However, Davidson et al. [19] reported that multiple injections at 1 mg kg^{-1} each, given every 3 hr on day 1 produced a marked increase in the efficacy (5/6 cures versus 1/12 cures for one injection at 7 mg kg^{-1}) and a marked decrease in the kidney toxicity. This suggests that a slow infusion of the drug over 24 hr may be preferable to a fast-push injection.

Slow release techniques of the drug cisplatin have also been attempted with some degree of success. Yolles et al. [20] have tested a composite of polylactic acid (a biodegradable polymer), incorporating various amounts of cisplatin, and injected it as a slurry into the peritoneal cavity of ICR mice containing the ascites Sarcoma 180 tumor. The release rate was such that 50% of the dose was released over a 92-day test period. The results were a decrease of the toxicity by at least a factor of 2 and an increase in the efficacy by a factor of 2 to 3. Thus, slow-release methods may eventually provide a distinct advantage in the administration of cisplatin.

3. TOXIC SIDE EFFECTS IN ANIMALS

3.1. Dose-limiting and Minor Side Effects

After convincing evidence of anticancer activity has been obtained from the animal screening systems, further developments require some information on the nature of the toxic side effects occurring upon drug administration. In the mouse, the first such study was done by Toth-Allen [21], and in rats, by Kociba and Sleight [22]. Further toxicologic studies using the dog and the monkey were reported by Schaeppi et al. [23]. The results for all four animals have been reviewed recently by Guarino et al. [24]. These are described in Table 5. The dose-limiting side effect is due to damage to the kidney function, in particular, to the proximal convoluted tubules. This causes a decrease in the filtering capacity of the kidney, with a consequent elevation of the blood-urea nitrogen and a decrease in the creatinine clearance. It also, as will be shown later, constitutes the major toxic side effect in man. There are at present many ongoing studies attempting to understand the nature of the damage caused to the tubules and to develop effective techniques to circumvent it [25,26].

In the mouse and rat, damage to the lining of the intestine is extensive, leading to loss of appetite (anorexia) and eventual starvation. Nausea and vomiting do not appear in mice and rats because they are incapable of vomiting. In the dog and the monkey, however, vomiting is severe. While the platinum drug does produce

TABLE 5

Major Clinical Toxicities of Cisplatin in Various Animal Species

Type of toxicity	Species			
	Mouse	Rat	Dog	Monkey
Gastrointestinal				
Emesis	-	-	+	+
Anorexia	+	+	+	+
Renal				
BUN increase	+	+	+	+
Creatinine increase	ND[a]	+	+	-
Polyurea	ND	+	+	+
Tubular necrosis	+	+	+	+
Hematologic				
WBC decrease	ND	+	+	+
Platelet decrease	ND	+	+	-
Bone marrow hypocellularity	+	+	+	+
Otologic	ND	+	ND	+

[a]ND = not determined.

Source: Adapted from Ref. 24.

mild damage to the blood-forming cells [27], this is generally not considered to be dose-limiting and is much less severe than for most other anticancer drugs. Hair loss, considered a significant indication of toxicity to rapidly growing cells, does not occur with cisplatin either in animals or man. One unusual toxicity for cisplatin was first described in man and then later detected in animals, as well. It is the destruction of the hair cells of the organ of Corti, leading to a loss in high-frequency hearing and, in some cases, total deafness [28].

The information garnered from the preclinical pharmacology and toxicology using dogs and monkeys is used to predict the expected toxicities in man. It also provides an indication of the

TABLE 6

Toxic Dose Levels of Cisplatin in Animals and Man (dose levels in mg M^{-2}, iv)

	Beagle dog		Rhesus monkey
	Single dose	Daily × 5	Daily × 5
Lethal dose (LD)	100	30	30
Toxic dose high (TDH)	50	15	15
Toxic dose low (TDL)	25	7.5	3.5
Highest nontoxic dose (HNTD)	12.5	3.5	1.7
	Rat		Mouse
LD_{90}	49		48
LD_{50}	44	13	39
LD_{10}	41	11	33
	Man		
Typical clinical schedules	75-100	20	

Source: Adapted from Ref. 24.

lowest dose that can be given with safety for the initial clinical trials in human patients. This is called the "highest nontoxic dose." The tabulated results of the various dose-level categories are shown in Table 6. Some suggestions for the sensitivity of the various animals organs to the drug will be described in Sec. 4.

4. DRUG FATE IN ANIMALS

4.1. Excretion Rates

The beginning of a study of the pharmacologic distribution of a drug requires some knowledge of the excretion rates. This measures the body retention time of the cisplatin, the modes of excretion, and the changes in the drug during its residence in the animal.

The radioisotope ^{195m}Pt was created at Oak Ridge National Laboratories by Poggenberg from enriched isotopes of platinum.

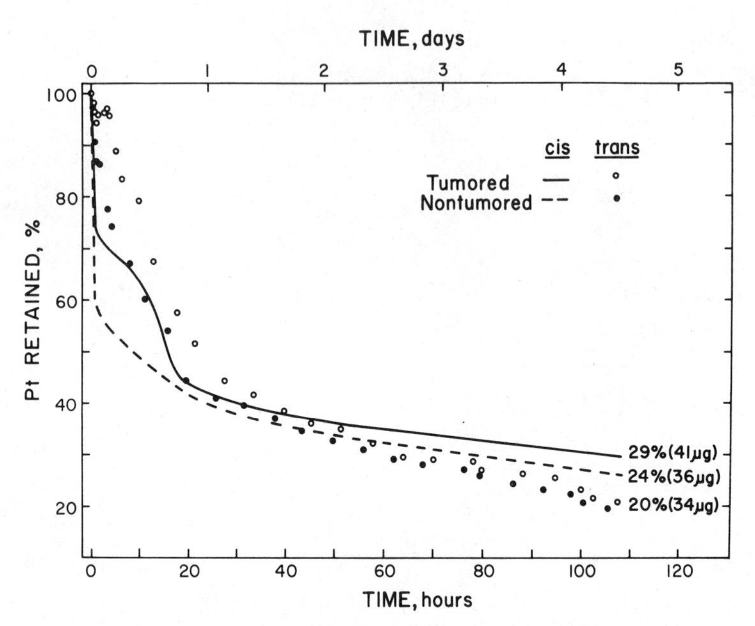

FIG. 2. Rate of loss of a pulse-injected dose of cisplatin and its trans isomer, both containing radioactive ^{195m}Pt in mice. Measured with a small animal, whole body scintillation counter. There is an early rapid loss phase, followed by a slower second phase.

It was then diluted with appropriate amounts of cold platinum to achieve the desired specific activity. The drug cisplatin is synthesized from this in a 'hot lab.' The isotope has a half-life of only about 4 days and decays by isomeric transitions emitting three γ rays of energies--0.031, 0.099, and 0.130 MeV. The γ emissions make it particularly suitable for whole body studies using small animal, whole body counters. The cisplatin is injected as a pulse dose, intraperitoneally, and the radioactivity remaining in the animals is monitored periodically. Hoeschele and Van Camp [29] first reported the decay of the drug in mice, both with and without tumors, for both cis- and trans-dichlorodiammineplatinum(II). Their results are shown in Fig. 2.

The excretion of these drugs follows a biphasic pattern. This is usually described by a "two-compartment model," and is given mathematically by

$$\text{Rate of loss} = Ae^{-\alpha t} + Be^{-\beta t} \quad (1)$$

and

$$T_{1/2}\alpha = \frac{0.693}{\alpha} \qquad T_{1/2}\beta = \frac{0.693}{\beta} \quad (2)$$

This excretion pattern has been corroborated in a number of other studies [30-32]. The initial rapid phase α has a half-life of $T_{1/2}\alpha = 1.5$ hr; while the second slow phase β has a half-life of $T_{1/2}\beta \simeq 20$ hr. More detailed studies by Wolf and Manaka [30], and later by Litterst et al. [31] suggest shorter $T_{1/2}\alpha$ times. These are listed for some different animals in Table 7. This subject was reviewed recently by Litterst et al. [33].

The principal mode of excretion in mice appears to be primarily through the urine, with about 90% of the dose excreted within 5 days after injection. Little or no excretion in mice occurs via the feces. The urinary platinum excreted shortly after injection

TABLE 7

Half-Lives for Elimination of Cisplatin in Various Animals

	$T_{1/2}$		
Species	α	β	Technique
Mouse	1.5 hr	20 hr	Whole-body scintillation counting
Rat	5.9 min	15.6 hr	Dynamic blood loop
	8.6 min	44.9 days	Serial sacrifice
	16.1 min	6.6 days	Tumored; serial sacrifice
	9.7 min	35.7 hr	Control group; serial sacrifice
	12.0 min	22.9 hr	Hydrated group; serial sacrifice
Dog	22 min	5 days	Cannulated
Rat	43 min	2 days	Serial sacrifice
Shark	71 min	17.5 days	Serial sacrifice

Source: Adapted from Ref. 33.

was analyzed by thin-layer chromatography. It moves with the same R_F value as cisplatin, but about 5% of the platinum remained at the spotting point. This latter was presumed to be protein-bound platinum. Later studies [34] suggest that a much higher concentration of protein-bound platinum occurs at longer times in the blood serum of human patients.

The trans complex, which has no antitumor activity, does not appear to have a markedly different excretion profile than the antitumor-active cis complex. Thus, it may be assumed that the differences in antitumor activity cannot be accounted for by different excretion rates.

An alternate method to the use of radioactive isotopes for determining platinum drug concentration is the use of atomic absorption spectrophotometry. While this technique is not subject to the time constraints of a rapidly decaying radioactive isotope, it does have the disadvantage of requiring an elaborate processing of the biologic samples to isolate the platinum free from interferences. The detectability of good instruments is low, about 50 ng ml^{-1}, which is well below the range required for most experiments with cisplatin.

4.2. Organ Distribution

A pulse injection of cisplatin into the peritoneal cavity of mice and rats disappears in 10-15 min due to uptake by the blood and lymph system. It is then rapidly distributed in the extracellular fluid of the body.

Cisplatin reacts primarily by exchange of the labile chlorides for water or hydroxyl ions. The kinetics of this exchange in water was reported many years ago by Reishus and Martin [35], and more recently by Cleare et al. [36]. Below is a sequential reaction:

$$\text{cis-Pt(NH}_3)_2\text{Cl}_2 \underset{H_2O}{\rightleftharpoons} \text{cis-[Pt(NH}_3)_2\ \text{Cl(H}_2\text{O)]}^+ + \text{Cl}^-$$

$$\underset{H_2O}{\rightleftharpoons} \text{cis-Pt[(NH}_3)_2\text{(H}_2\text{O)}_2]^{2+} + \text{Cl}^-$$

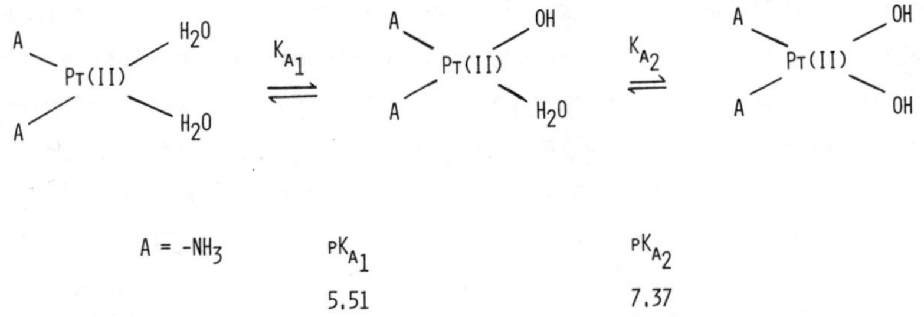

$$A_2Pt(II)(H_2O)_2 \underset{}{\overset{K_{A1}}{\rightleftharpoons}} A_2Pt(II)(OH)(H_2O) \underset{}{\overset{K_{A2}}{\rightleftharpoons}} A_2Pt(II)(OH)_2$$

$A = -NH_3$ $\qquad pK_{A1}$ 5.51 $\qquad pK_{A2}$ 7.37

FIG. 3. Three possible fully aquated species of cisplatin. Other possible species include the dichloro, the monochloro-monoaquo, the monochloro-monohydroxy, and the hydroxo-bridged dimers and trimers.

with the final diaquo species acting as a weak acid, leading to the acid-base equilibrium shown in Fig. 3. The activation enthalpy for the aquation reaction rate is 18 kcal mol^{-1}. Drobnik and Horacek have shown that the rate of reaction of cisplatin and DNA [37] and the rate of virus inactivation by cisplatin [38] both have about the same activation enthalpies as does the aquation reaction. This suggests the hypothesis that the rate-limiting step in the biologic reactions of the drug is aquation.

The presence of high chloride ion concentrations (>5 meq of Cl^- $liter^{-1}$) suppresses the aquation reaction products. Thus, in the presence of the 100 meq $liter^{-1}$ of Cl^- in the extracellular fluid, almost all of the cisplatin is in the unreacted form. The cisplatin passively permeates the cell membranes from the extracellular fluids (no evidence exists for a carrier molecule as will be shown below). In most cells of the body, the intracellular Cl^- concentration is much lower than the extracellular fluid, and the aquation products are formed in high concentrations. This provides the only known activation process required for cisplatin to react with molecules in the cell. Metabolic activation is not required. However, the Cl^- concentrations of cells may vary from the low value of about 10 meq $liter^{-1}$ in muscle cells to a high value of about 160 meq $liter^{-1}$ in stomach parietal cells; for a given type of cell, Cl^- concentration may vary from one animal species to another or even from one specimen to another. For an ion such as Cl^- of such physiologic import, there is a surprising dearth of data on its intracellular concentrations. It is now believed to play the controlling role in determining the cellular reactions of cisplatin. Therefore, the fundamental information necessary to interpret the excretion rates, the organ distributions, the cellular uptakes, the toxicities, and the anticancer activities of the platinum complexes is lacking.

Subject to this restriction, however, the way cisplatin is distributed throughout the various organs of the body can still be studied. The first report by Toth-Allen [39] used the technique of neutron activation of platinum to determine the levels in body organs of mice as a function of time after a pulse dose. These preliminary results have been extended and confirmed by Hoeschele and Van Camp [29], Wolf and Manaka [30], and Litterst et al. [31], and have been reviewed recently by Litterst et al. [33].

Table 8, taken from the data of Wolf and Manaka [30], lists the percent of the injected cisplatin, containing the radioisotope ^{195m}Pt, per gram of wet organ tissue for a variety of organs of the rat. These values do change with time, and this set was obtained 3 hr after injection. The organs of highest uptake in the rat are the kidneys, liver, adrenals, lung, bone, skin, and tail. The Walker 256 carcinosarcoma solid tumor exhibited no particularly significant uptake compared to normal tissues, and the brain showed quite a low uptake. Such results have been fairly consistently found in rodents. A few pertinent comments concerning these results need to be made. Impairment of kidney function is dose-limiting for cisplatin. This is consistent with its high uptake in that organ. However, there are relatively little toxicities that can be attributed to the roughly similar high uptake in some of the other organs. For example, damage to the liver in animals and humans is so rare that when one case surfaced recently of a liver toxicity (manifest as jaundice) in a bladder cancer patient being treated with cisplatin, it was published [40].

There is no evidence for a central nervous system impairment due to the drug. In higher animals, the severe nausea and vomiting

TABLE 8

Organ Distribution of Cisplatin in Walker-256-Carcinosarcoma-Bearing Rats (3 hr After Pulse Injection)

Organ	% of injected dose per gram of tissue
Blood	0.378
Skin	0.495
Thyroid	0.119
Liver	0.755
Spleen	0.334
Pancreas	0.306
Stomach	0.110
Small intestine	0.411
Large intestine	0.351
Kidneys	1.747
Adrenals	0.614
Heart	0.166
Lung	0.569
Brain	0.035
Muscle	0.002
Tumor	0.226
Bone	0.584
Marrow	0.234
Bladder	0.520
Tendon	0.675

Source: Adapted from Ref. 30.

(the latter of which can cause gastrointestinal damage as a secondary effect) is most likely due to a reaction of the cisplatin in the blood with the chemoreceptor sites of the brain that are known to control nausea. In this area of the brain, the blood-brain barrier is necessarily less exclusive.

It should be emphasized here that organs are masses of sometimes quite heterogeneous cell types. Each cell type may have its own unique uptake and toxicity. In the kidney, for example, the primary damage occurs in the cells of the proximal convoluted tubules of the glomerulus in all species tested [41]. There is a generalized, hydropic degeneration in these cells in animals given frankly toxic doses of cisplatin, while the other cell types of this organ are apparently not harmed.

4.3. Transport

The platinum drug either passively diffuses through the cell membrane to enter the cell, or it may require a carrier molecule to transport it, as do some of the nitrogen mustards [42]. A standard test to distinguish these two cases is based upon Michaelis-Menten kinetics for the formation of a carrier-substrate intermediate complex and the Lineweaver-Burk equation for analysis of the rates. Such a study was done by Gale et al. [43]. The results are given in Fig. 4, where a double reciprocal plot is shown of the uptake of the drug by the cells (Ehrlich ascites tumor cells) against the external drug concentration for two different times. The lines, within experimental error, converge at the origin, implying that the Michaelis-Menten constant is approximately infinite. Thus, either the rate of formation of a complex is essentially 0, or the rate of breakup of the complex is infinitely slow. In either case, the results strongly mitigate against a carrier transport mechanism, and it is presumed at present that the platinum drug passively diffuses through the plasma membrane of the cell.

A second test supporting this conclusion emerged from the effects of a variety of substances on the rate of uptake. None slowed it down, but those chemicals known to compromise the integrity of the cellular membrane increased the uptake rate. A few caveats are necessary here--the platinum drug used in these tests was not cisplatin, but the close analog cis-dichloro(dipyridine)platinum(II)

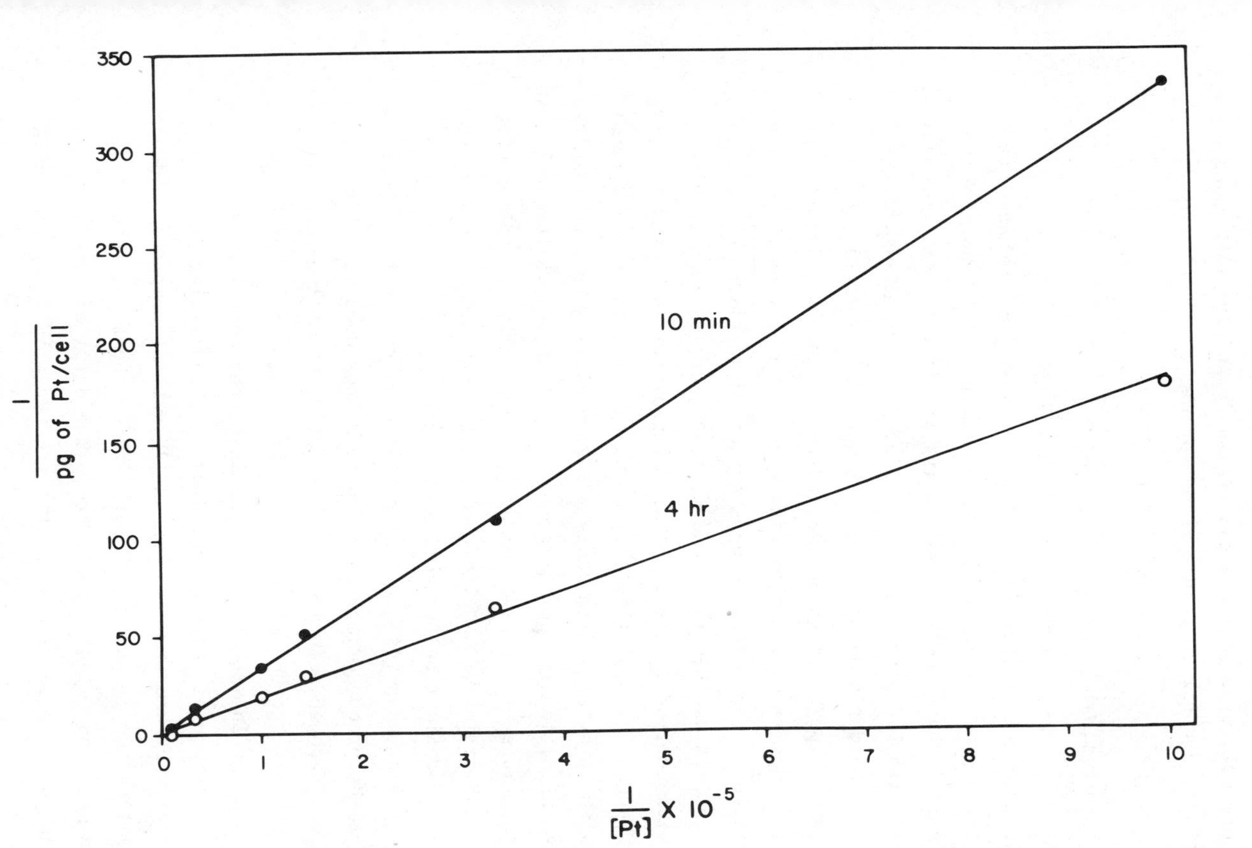

FIG. 4. Double reciprocal (Lineweaver-Burk) plot of the intracellular concentration of cis-dichloro-(dipyridine)platinum(II) as a function of the extracellular concentration for Ehrlich ascites tumor cells in tissue culture at two different exposure times. The lines converge on the origin of the axis, indicating that no carrier molecule is necessary for membrane transport [43].

with a tritium label on the pyridines; these tests were done with only one cell type (a mouse cancer line), and in tissue culture. Therefore, general relevance is not proven.

4.4. Biochemical Reactions

Cisplatin, in less than frankly toxic doses, has profound effects upon cells grown in tissue culture. The nature of some of these biochemical changes was probed, using the incorporation of radioactively labeled precursors of such biopolymers as protein, RNA, and DNA. Harder and Rosenberg [44] and simultaneously, Howle and Gale [45], reported a selective and persistent inhibition of new DNA synthesis without concomitant inhibition of transcription and translation. These studies were performed in tissue culture (AV_3 [44], Ehrlich ascites [45], and HeLa [46] cells), and in vivo in mice with Ehrlich ascites [45], in lymphocytes [47], and in intestinal mucosa. The similar actions of the drug in such a diversity of systems provide a firm basis for generalizing that new DNA synthesis inhibition is the major biochemical lesion in the cell. At concentrations of the platinum drugs higher than the therapeutic value, inhibition of RNA and then protein synthesis progressively increases, until finally, cell death occurs.

It was shown [44] that cisplatin does not inhibit the formation of the precursor molecules for DNA, nor does it inhibit their transport through membranes. And finally, the polymerases necessary to synthesize the DNA are not inhibited in the cell (although they are inhibited in vitro). This leaves as the most likely reaction a direct attack of the platinum drug on the cellular DNA. An excellent review by Roberts and Thomson [49] of the complete evidence implicating the cisplatin-DNA reaction as the significant cell reaction has been recently published. The detailed nature of the reactions possible between these two reactants is a topic of high current interest (see below).

While these lines of evidence pick out DNA as the cellular target for the antitumor actions, we may not ignore the possibility that cisplatin could react with special proteins to produce at least some of the cellular toxicities. Friedman and Teggins [50,51] have demonstrated enzyme inhibition with platinum drugs, although it occurs at drug concentrations far above those usually found in cells.

Guarino et al. [24] have reviewed the evidence regarding inhibition by platinum drugs of p-aminohippuric acid transport in kidney tubules and suggest that this reflects the platinum drug toxicity to the kidneys. They attribute it to a direct inhibition of Na and K ATPases by the platinum drugs. They also speculate that ototoxicity, peripheral neuropathy, and gastrointestinal toxicities may have this same mechanism as their cause. This needs to be looked at more closely since, for the first time, it directly implicates a cisplatin-protein interaction as a toxic cellular reaction.

5. COMBINATION CHEMOTHERAPY IN ANIMALS

5.1. Additivity and Synergism

While the search for new, more effective anticancer drugs goes on, a major advance in successful therapy over the past two decades was the introduction of combination chemotherapy. The basic concept is simple; two or more drugs are administered on an optimum schedule, whose actions against the cancers are additive or synergistic, but whose toxicities to normal cells are dissimilar and less than additive. At the present time, it is rare to find a cancer that does not respond better to combinations than to single drugs. While it is relatively easy to document additivity or synergism quantitatively in animal model systems, it is difficult to do so in humans. An air of controversy still hangs over the use of these two words in clinical studies and no particular definitions are in general acceptance. This has led to a proliferation of euphemisms such as "superadditivity," "potentiation of therapeutic potency," "therapeutic advantage," etc.

TABLE 9

Effect of Combination Chemotherapy with Cisplatin on Survival of Mice with L1210 Tumors

	Alone		In combination with cisplatin	
	%ILS	60-day survivors	%ILS	60-day survivors
Cisplatin	55	0/8		
Isophosphamide	100	0/8	>469	4/8
Cyclophosphamide	140	0/8	>495	4/8
ICRF-159	122	0/8	300	2/8
Emetine	61	0/8	122	0/8
Cytosine arabinoside	133	0/8	194	0/8
5-Fluorouracil	80	0/8	120	0/8
BIC	78	0/8	117	0/8
Methotrexate	78	0/8	111	0/8
Phosphoramide mustard	158	0/8	205	0/8
5-Azacytidine	144	0/8	189	1/8
Camptothecin	144	0/8	178	0/8
5-HP	83	0/8	95	0/8
BCNU	>560	6/8	>395	4/8

Source: Adapted from Ref. 10.

When it was clear that cisplatin produced significantly less suppression of blood-forming elements (myelosuppression) than other drugs, and that these others were more sparing of kidney damage, it became interesting to look for additive/synergistic combinations with cisplatin in animals. The first such activity was reported by Woodman et al. [10,52], where 13 other antileukemic drugs were tested in combinations with cisplatin against the L1210 tumor in BDF_1 mice. Table 9 presents the results in terms of percent increase in life span (%ILS) and cures. The indications of synergism were clearly found for combinations with isophosphamide, cyclophosphamide, and ICRF-159. Additivity occurred with the other drugs listed.

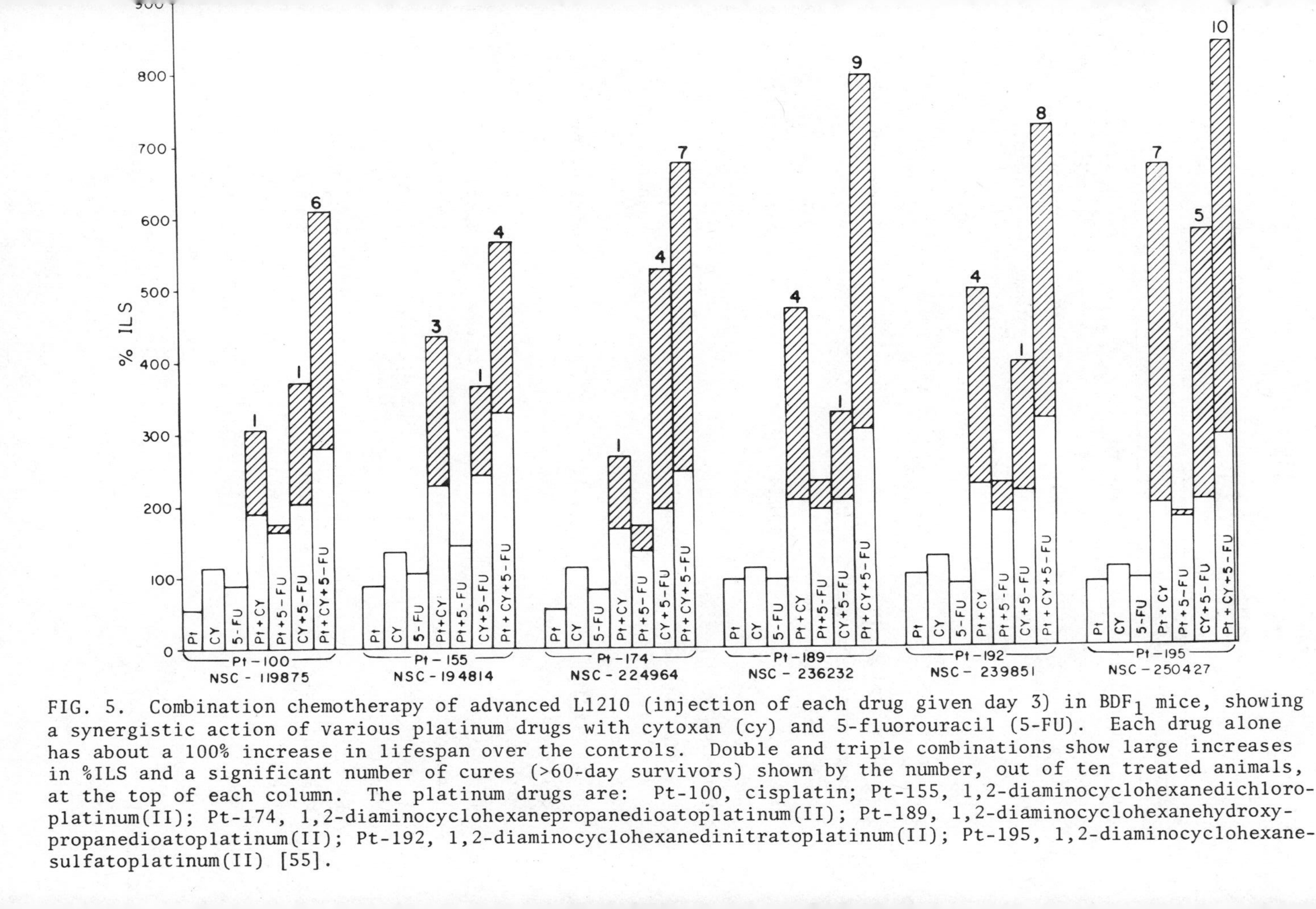

FIG. 5. Combination chemotherapy of advanced L1210 (injection of each drug given day 3) in BDF_1 mice, showing a synergistic action of various platinum drugs with cytoxan (cy) and 5-fluorouracil (5-FU). Each drug alone has about a 100% increase in lifespan over the controls. Double and triple combinations show large increases in %ILS and a significant number of cures (>60-day survivors) shown by the number, out of ten treated animals, at the top of each column. The platinum drugs are: Pt-100, cisplatin; Pt-155, 1,2-diaminocyclohexanedichloroplatinum(II); Pt-174, 1,2-diaminocyclohexanepropanedioatoplatinum(II); Pt-189, 1,2-diaminocyclohexanehydroxypropanedioatoplatinum(II); Pt-192, 1,2-diaminocyclohexanedinitratoplatinum(II); Pt-195, 1,2-diaminocyclohexanesulfatoplatinum(II) [55].

TABLE 10

Additivity or Synergism of Combination Chemotherapies with Cisplatin

Drug(s) added to cisplatin	Tumor	Comments
Cyclophosphamide	Sarcoma-180	Additive if not synergistic
	L1210	Synergistic
	Reticulum cell sarcoma	Synergistic
	FANFT-induced bladder carcinoma	Synergistic
	DMBA-induced mammary tumor	Synergistic
Isophosphamide	L1210	Synergistic
Phosphoramide mustard	L1210	Synergistic
Yoshi 864	L1210	Synergistic
BIC	L1210	Synergistic
DIC	L1210	Synergistic
Cytosine arabinoside	L1210	Synergistic
Hydroxyurea	L1210	Slightly more than additive
5-Fluorouracil	L1210	Synergistic
5-Azacytidine	L1210	Synergistic
6-Thioguanine	L1210	Synergistic
Methotrexate	L1210	Synergistic
Bleomycin	B-16 melanocarcinoma	Synergistic
Adriamycin	Sarcoma-180	Synergistic
Vincristine	L1210	Synergistic
Vinblastine	L1210	Synergistic
Emetine	L1210	Synergistic
Camptothecin	L1210	Synergistic
ICRF-159	L1210	Synergistic
SHP	L1210	Synergistic

TABLE 10 (continued)

Drug(s) added to cisplatin	Tumor	Comments
In three-drug combinations		
Cyclophosphamide		
+ Adriamycin	L1210	Synergistic
+ Cytosine arabinoside	L1210	Synergistic
+ 5-Fluorouracil	L1210	Synergistic
+ Hydroxyurea	L1210	Synergistic
+ Methotrexate	L1210	Synergistic
+ Yoshi 864	L1210	Synergistic

Source: Adapted from Ref. 60.

The advantages of cisplatin in combination chemotherapy were soon verified by Speer et al. [53], again against L1210 in BDF_1 mice, and by Van Camp and Rosenberg [54] against the solid sarcoma 180 in ICR mice. But perhaps the prettiest piece of work in this area is due to Gale and his associates [55-57] using three drug combinations. A sample of one set of results with a variety of antitumor-active platinum analogs in combination with cytoxan and hydroxyurea against the L1210 in BDF_1 mice is shown in Fig. 5. Synergism and a high cure rate are clearly evident.

The advantages of combination chemotherapy with cisplatin in other animal systems modeling human cancers have been shown now for a chemically induced bladder cancer in mice [58] and myeloma in mice [59]. Table 10, taken from a recent review by Schabel et al. [60], compiles the results of many studies of combinations, with the individual authors' evaluations.

At this time, there are neither good experimental evidence nor usable concepts of the molecular mechanisms for synergism. It remains an area in urgent need of further development.

5.2. Combinations With Radiation

Radiation therapy combined concomitantly with chemotherapy is a more recent development. Wodinsky and coworkers [61,62] first reported the enhanced response (increase in %ILS) in mice with intraperitoneally or intravenously inoculated P388 leukemia cells treated with cisplatin combined with γ irradiation from a ^{60}Co source. Table 11 gives the results for cisplatin injected 4 hr before irradiation. It is obvious that there is an approximate additivity of the results. Similar tables occurred when the cisplatin was given just prior to or 4 hr after irradiation. One point noted by the authors was that in the irradiation treatment given 4 hr before cisplatin, there was a perceptible decrease in %ILS at the higher doses compared to the other schedules. They also point out that since the extrapolated dose of γ rays to cause a 50% ILS is about 1000 rad, and since this value can be achieved at 100 rad with cisplatin given at 4 mg kg^{-1}, that cisplatin increases the efficiency of radiation by a factor of 10. They conclude that within the limits of the experimental results

TABLE 11

Effects of Cisplatin[a] Combined with Radiation Therapy Against the P388 Tumor in Mice

γ Radiation	mg kg^{-1}				
rad	0	0.5	1	2	4
	(% ILS)				
0	0	18	27	36	64
100	9	18	27	46	73
200	18	27	46	55	73
400	36	55	64	68	91
600	36	64	46	82	114

[a]Administered 4 hr prior to γ radiation.

Note: P388 10^6 cells ip.

Source: Adapted from Ref. 62.

there is little difference due to the schedule of treatments. This is of significance in limiting the possible mechanisms of action. It immediately allows us to discount the possibility that the platinum drug acts in concert with the short-lived chemical species created in the cell by irradiation.

Zak and Drobnik [63] had earlier published results on the lethality of combined cisplatin and x-rays. In mice the lethality was least for cisplatin administered at 24 hr (27% deaths) or 1 hr (30%) before irradiation, and was highest 1 hr (100%) and 24 hr (78%) after irradiation. This is consistent with the decrease in %ILS described above.

Douple et al. [64] tested the combination of cisplatin and local x-ray irradiation on two localized tumors in mice and rats, the transplantable mouse mammary adenocarcinoma (MTG-B) with a significant hypoxic cell fraction (see below); and a well-vascularized intracerebrally implanted, rat brain tumor. For both tumors, the mean survival times, the tumor cell kill, and the amount of tumor regression were increased by the combination therapy over each therapy alone. There was also a schedule dependence. The best results were obtained with at least a 24-hr delay between cisplatin administration and irradiation.

This work had been stimulated by the prior discovery of Richmond and Powers [65] of the radiosensitization of *B. megaterium* spores by cisplatin. Most intriguing, however, was the observation that the radiosensitization by cisplatin was twice as high in the absence of oxygen (anoxic condition) than in its presence (oxic condition). This is of some significance, since anoxic tumor cells are more difficult to kill with irradiation than oxic cells.

Analog structures to cisplatin are presently being studied for combination with radiation by Nias and Szumiel [66], using Chinese hamster ovary cells in tissue culture. This allows them to study the combination effects as a function of the cell cycle. The first analogs were cis-dichlorobis(cyclopentylamine)platinum(II) and cis-dichlorobis(isopropylamine)-trans-dihydroxyplatinum(IV). Both act synergistically with x-rays to increase lethality. This action is most pronounced for the first drug in the G_1 and late S phases of the cell cycle, and less so in the mid-S phase.

The use of radiation in combination with chemotherapy is in an early stage of development, but there is already a growing interest in applying it to human cancer patients.

6. DEVELOPMENT OF RESISTANCE TO PLATINUM DRUGS IN ANIMALS

6.1. Tissue Culture: Whole Animal Studies

Current chemotherapy protocols for a number of human cancers can put significant numbers of patients into complete remission with no detectable evidence of disease. Yet a sizable fraction of these patients will (months to years) later relapse. A major frustration of clinicians is that most of these relapsed patients no longer respond to the therapy that initially worked; the tumors have become resistant. If one assumes, as is likely, that drug resistance is mainly an intracellular phenomenon, then this result may be rationalized as due to some noncycling tumor cells (G_o state) that were initially exposed to the therapy and not killed, but were changed in their intracellular biochemistry. Cycling, beginning some time later, will lead to new tumors that are now resistant to this chemotherapy. It is of great significance, therefore, to understand the mechanism of resistance induced by a drug, or of cross-resistance developed to other drugs.

The first requirement is for a model system that develops such a resistance. Burchenal and coworkers [67,68] have done just that with an L1210 line in C57BLxDBA/$2F_1$ mice by repeatedly transferring cells through successive generations in the presence of continued treatment with cisplatin. This tumor line is designated L1210/PDD. A resistant subline of P388 was also developed. These lines in tissue culture also show resistance to the cytotoxicity of cisplatin. This indicates that resistance is an intracellular, rather than a host, characteristic.

TABLE 12

Tissue Culture Studies of Cross-Resistance of Platinum Coordination Compounds (PtA_2X_2)

A	X	LD_{50} dose ($\mu g\ ml^{-1}$)	
		L1210/0	L1210/DDP
Diammino	Dichloro	0.05	2.50
Diammino	Malonato	0.40	6.50
1,2-Diaminocyclohexane	Malonato	0.25	0.25
1,2-Diaminocyclohexane	Dichloro	0.14	0.14
1,2-Diaminocyclohexane	Sulfato	0.20	0.36
1,2-Diaminocyclohexane	Carboxyphthalato	0.30	0.23
1,2-Diaminocycloheptane	Malonato	0.50	0.09
1,2-Diaminocycloheptane	Dichloro	0.06	0.20
1,2-Diaminocycloheptane	Sulfato	0.30	0.50
1,2-Diaminocyclopentane	Malonato	0.60	5.0
1,2-Diaminocyclopentane	Dichloro	0.30	4.6
1,2-Diaminocyclopentane	Sulfato	0.60	4.9
Ethylenediamine	Dichloro	0.35	29.0
Orthophenylenediamine	Dichloro	0.69	2.4
Bis(Isopropylamino)	Dichloro	0.6	4.4
Bis(Isopropylamino)	Sulfato	0.6	5.3

Source: Adapted from Ref. 67.

This tissue culture technique provides a simple test procedure for determining the cross-resistance, or lack thereof, to other analogs of cisplatin. Table 12 shows one set of results of the dose necessary to kill 50% of the cells in the normal (L1210/0) and the resistant (L1210/PDD) lines. Unsurprisingly, cross-resistance is not a function of the leaving groups (chloride, malonato, or sulfato), since once they are inside the cell, they must all aquate in order to react. The determinant of cross-resistance is primarily a function of the nonexchanging amine ligands.

TABLE 13

In Vivo Studies of Cross Resistance in Leukemia L1210

Compound	Dose mg kg^{-1} (days 1, 5,9,13)	L1210		L1210/DDP		L1210/Pt155[a]	
		Survival (days)	ILS (%)	Survival (days)	ILS (%)	Survival (days)	ILS (%)
Control		8.8		12.0		12.2	
Dichlorodiamminoplatinum	6.7			12.3	3	18.9	55
Dichlorodiamminoplatinum	4.5	14.9	69	13.5	13	16.8	38
(1,2-diaminocyclohexane)dichloro Pt	5.0	20.4	132	22.9	90.8	12.3	1
(1,2-diaminocycloheptane)dichloro Pt	5.0	13.2	50	18.9	57.5	14.9	22
(1,2-diaminocyclopentane)dichloro Pt	15.0	12.7	44	18.6	55	12.8	5
Ethylenediaminedichloro Pt	20	14.3	63	13.2	10		
Ethylenediaminedichloro Pt	13	12.8	45	13.7	14		

[a] Pt 155 = (1,2-diaminocyclohexane)dichloro Pt.

Source: Adapted from Ref. 67.

The ethylenediamine, orthophenylenediamine, isopropylamine, and 1,2-diaminocyclopentane ligands are all cross-resistant in the L1210/PDD line. However, the 1,2-diaminocyclohexane and 1,2-diaminocycloheptane ligands are not cross-resistant--they kill the resistant line at the same dose levels that kill the normal line. These are striking results, first because the 1,2-diaminocyclohexane structures are quite good antitumor agents and may, therefore, play a significant role in cisplatin-sensitive cancers that relapse and are now resistant to cisplatin. The second striking result is the theoretical implications of the structure-activity relations. They suggest that large, nonplanar, floppy amine structures allow the molecules to avoid their disablement by the cell that has developed resistance to cisplatin. This suggests to the author that the intracellular disabling event may involve a large molecule reacting with the platinum drug--perhaps a protein. If further structure-activity studies confirm this relation, then evidence for such a "sequestering" protein, which can apparently be induced in such cells, should be sought.

To verify that the cross-resistance phenomenon occurs in vivo as well as in vitro, the data of Burchenal et al. [67] is presented in Table 13. Here they have also included an L1210/Pt155 line, which is resistant to molecules containing the 1,2-diaminocyclohexane ligand, and which is not cross-resistant to the diammine or 1,2-diaminocycloheptane ligands, but does show some cross-resistance with the 1,2-diaminocyclopentane ligand. This last result is somewhat uncertain and requires cleaner data or a confirmation. It also indicates the "muddiness of the waters" at this stage.

7. MECHANISMS OF ACTION

7.1. Host Activities for Responses

The central problem of cancer chemotherapy is selective action. When an animal or human body abounds with masses of tumor tissue, in total quantity larger than many of the major sensitive organs, and these tumors all disappear under chemotherapy without significant

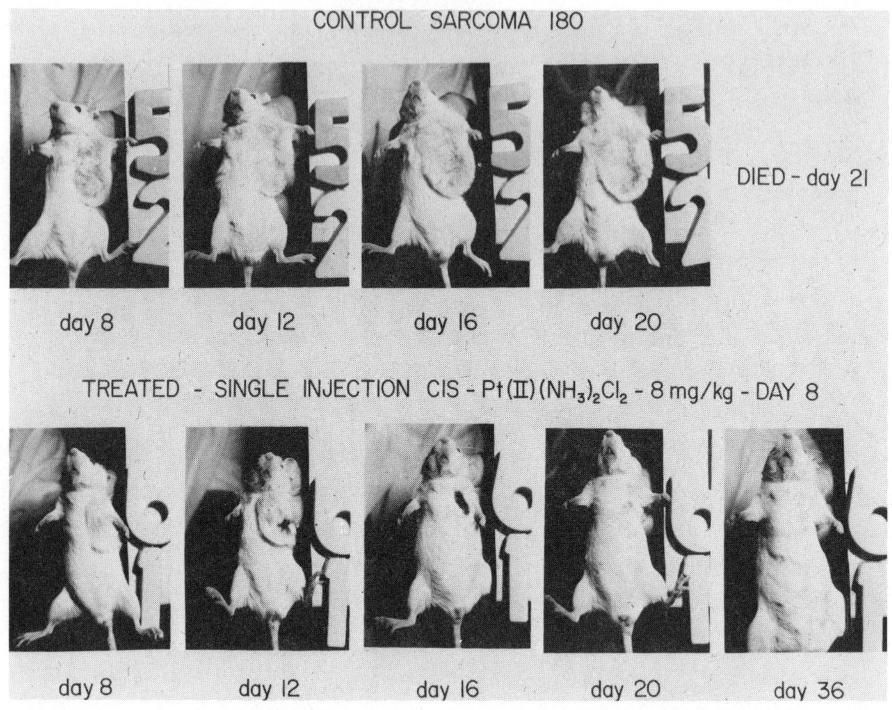

FIG. 6. Cisplatin chemotherapy of advanced solid sarcoma 180 in ICR mice. Top row, progression of tumor (~1 g on day 8) to death of animal in untreated control; bottom row, regression of tumor in treated animal.

damage to the normal tissue, one must be impressed with the selective destruction. Were this not the case, of course, chemotherapy in cancer would be a failure. An early, but clear, example of this phenomenon in the case of cisplatin is shown in Fig. 6. ICR mice with transplanted solid sarcoma 180 (10 mg initial inoculum) grow large tumors locally by day 8--approximately 1 g in weight in a 20-g mouse. If untreated, the tumors grow to about 3 g and kill the animals by days 20-35, as illustrated in the upper section of the figure. If, however, the mouse is treated with a single intraperitoneal injection of 8 mg kg^{-1} of cisplatin on day 8, the tumor becomes necrotic and either falls out or is groomed out by the animal. The wound heals and the animal survives to die of old age of natural causes. This occurs in almost all animals so treated [69]. Similar or even more striking results can be obtained with other tumors in mice (e.g., ADJ/PC6A) [70], rats (Dunning leukemia or Walker 256 carcinosarcoma) [71], and humans (see below).

It has been pointed out that there is no selective uptake of the cisplatin by the tumor tissue when compared to many other organs of the body, so this cannot be the cause of selective tumor destruction. Inevitably, when faced with a high degree of selectivity, one turns to a host immunologic reaction as the source of selectivity. This was suggested early in this field by Rosenberg [72]. Unfortunately, direct, hard evidence is difficult to obtain and he presented only an accretion of weak arguments to buttress his case. Basically Rosenberg's argument runs that the treatment with cisplatin enhances the antigenicity of the tumor cells and this triggers an increased immunologic reaction by the host against the tumor.

One molecular mechanism for enhanced antigenicity has emerged from studies on the staining of cells with electron-dense "platinum-pyrimidine blues" [19] for electron microscopy [73]. These complexes of unknown structures, but of an ionic, polymeric nature) react selectively with nucleic acids. The cells exhibited electron density in the nuclear chromatin (DNA), the nucleolus (RNA), and the cytoplasmic ribosomes (RNA). Additional sites of staining occurred in tumor cells on the cell surface. These stainable sites were removed by a gentle pretreatment of the cells with DNase, which infers that tumor cells, but not normal cells, have DNA associated with the cell surface. Rosenberg [74] speculated that the low immunogenicity of nucleic acids may, by the trivial mechanism of masking antigens, serve to decrease the antigenicity of tumor cells. Treatment with cisplatin may remove the nucleic acids and thereby enhance antigenicity. Recent evidence by Juckett and Rosenberg [75] has verified, by the cell electrophoresis technique, that tumor cells treated in vivo with cisplatin have lost their cell-surface-associated nucleic acids.

Other evidence more relevant to immunosuppression by cell-surface-associated nucleic acids (DNA) has been reported by Russell and Golub [76]. Spleen cells (lymphocytes) from strain AKR mice with spontaneous leukemia can suppress the immunologic reaction of normal AKR antisheep erythrocyte antibody response. The same cells from nonleukemic mice do not suppress this reaction. If the cells from the leukemic mice are gently treated with DNase, they no longer suppress. Treatment of the leukemic animal with cisplatin also abrogates the suppressor activity of these cells. Thus, a direct link in at least one tumor system is established between cell-surface DNA and immunosuppression that is consistent with the Rosenberg speculation.

A further piece of evidence along these lines was obtained by Hollinshead [77], working with human cancers of the lung (oat-cell type). She isolated "inhibitory antigens" from the membranes of these cells and showed, by DNase digestion and gel electrophoresis, that they contain DNA. These "inhibitory antigens" suppress a variety of immunologic reactions.

It is not yet known whether the cell-surface-associated nucleic acids are DNA, RNA, nucleoprotein, or mixtures of these. Their source, mode of production, and migration is still to be studied; their involvement with cancer cells is still to be proved; and their mechanism of immunosuppression is still to be determined.

The work of Golub and Hollinshead may provide the firm basis for a quite new development in immunology and cancer research.

7.2. Molecular Mechanism of Action

A second hypothesis to explain the selective destruction of cancer cells by cisplatin arises from the study, at the molecular level, of the reactions of cisplatin and its cellular target, DNA. It is known that cisplatin (and its aquated products) do not intercalate between the base pairs of DNA [78]. There is also good reason to believe that reactions with the phosphate and sugar moieties of the double-helical strands either do not occur, or if they do to any extent, are not significant for the anticancer activity. All other possible reactions do occur, and these must include the significant lesion. These are interstrand cross-links [79,80], DNA-protein cross-links [81], intrastrand cross-links [82], and finally, reactions with the individual bases.

It is necessary here to make a clearcut distinction between two not necessarily related phenomena--cytotoxicity and anticancer activity. The former is nonselective, while the latter is very selective. Not all cellular poisons are anticancer agents. For example, both cis- and trans-dichlorodiammineplatinum(II) are cytotoxic at appropriate dose levels, but the former is an anticancer agent and the latter is not. In fact, all anticancer-active platinum(II) complexes are invariably in the cis configuration; and also, invariably, the corresponding trans configurations are not active. Table 14, from the work of Connors et al. [70], illustrates this stereoselectivity very nicely. This well-established structure-activity relationship provides a powerful tool for attacking the problem of which particular DNA lesion(s) lead to anticancer activity. Those lesions that are common to both isomers may be involved in cytotoxicity; but only a lesion that the cis isomer can form but the trans cannot may be considered relevant to anticancer activity. Thus, we must look for the differences between the two isomers' reactions with DNA. Pascoe and Roberts [83] found one significant difference; for equitoxic doses in tissue culture, more trans molecules are bound to DNA than are cis. A further study by Zwelling and coworkers [81] showed that this was due to an excess of DNA-protein cross-links caused by the trans isomer, and that when these are removed by proteinase treatment, both the cis and trans isomers at equitoxic doses cause about the same number of interstrand cross-links on DNA. This suggests, as Roberts had earlier reported [83], that DNA cross-links are probably responsible for cellular toxicity, but not for anticancer activity.

TABLE 14

Anticancer Activity of Cis and Trans Isomers of $Pt(II)(Amine)_2Cl_2$ Against the ADJ/PC6A Tumor in C^- Mice

	Isomer	LD_{50} (mg kg^{-1})	ID_{90} (mg kg^{-1})	Therapeutic index (LD_{50}/ID_{90})
NH_3	cis	13	1.6	8.1
	trans	27	>27	
NH	cis	56.5	2.6	21.7
	trans	18.0	>18.0	
NH	cis	240	17.5	13.7
	trans	72	>72	
NH_2	cis	57	2.3	25
	trans	27	>27	
NH_2	cis	90	2.9	31
	trans	250	>110	
NH_2	cis	480	2.4	200
	trans	180	72	2.5
NH_2	cis	<3200	12	>267
	trans	680	260	2.6
H_3C — NH_2	cis	990	1180	
	trans	360	>800	

Source: Adapted from Ref. 70.

The stereoselectivity argues for a rigid identifying structure to adequately distinguish the two isomers. DNA is a relatively plastic structure and can make accommodations to small reacting molecules readily. The bases themselves are quite rigid, however. This urges that the cisplatin-base reaction be searched for the stereoselectivity. The search was narrowed by the work of Stone and coworkers [84], who showed that cisplatin preferentially reacted with the G-C rich regions of DNA. Shortly afterwards, three groups of researchers independently proposed a first solution to the problem [85-87]; a closed-ring chelate of the aquated cisplatin with both N-7 and O-6 of guanine. Only the cis isomer can form this, the trans cannot; this is shown in Fig. 7. However, the

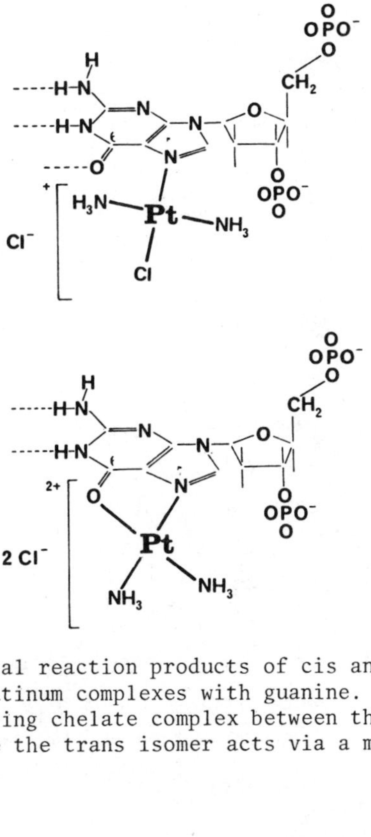

FIG. 7. Hypothetical reaction products of cis and trans isomers of aquated diammineplatinum complexes with guanine. The cis isomer may form a closed-ring chelate complex between the N7-O6 nucleophilic sites, while the trans isomer acts via a monodentate bond to N-7 [85].

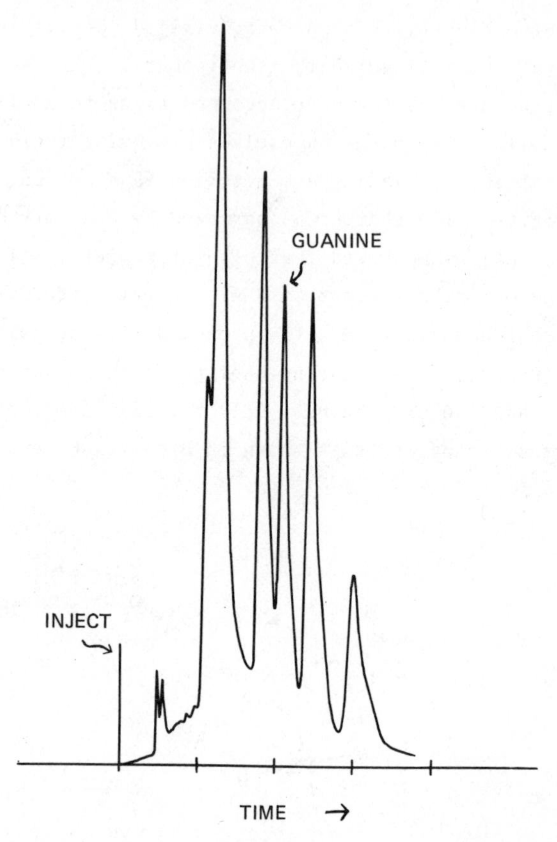

FIG. 8. High-pressure liquid chromatogram of the reaction mixture of aquated cisplatin, with guanine showing the multiplicity of products occurring [88].

chemical evidence for the existence of this chelate is not yet sufficiently hard. The N-7 position of guanine is the most nucleophilic, and it is reasonable to expect that an electrophilic agent such as the aquated cisplatin would attack this position preferentially. There are, of course, other nucleophilic sites of guanine that may also react. These include the N-3, O-6, 2-NH_2, and C-8 positions. Indeed, a reaction mixture of 1:1 aquated cisplatin and guanine maintained at a pH of 6.2 and a temperature of 37°C exhibits a multiplicity of peaks in an HPLC chromatogram. One such set of peaks is illustrated in Fig. 8 [88]. The amplitude of the peaks changes with time and eventually (two months) the number of peaks sharply decreases [89]. These results imply that initially a wide variety of reaction products occur (within the biologically significant time of days), but at equilibrium only a few of the products are stable. The identification of these peaks is now proceeding.

The problem now resolves itself down to the question, Why should any particular reaction site on guanine be any more significant for anticancer activity than any other? One possible answer was proposed by Rosenberg [88]. The O-6 position of guanine is involved in hydrogen bonding to cytosine. If its hydrogen-bonding capacity were impaired by the chelate formation (as it is in alkylation by certain carcinogens), then on base pairing during cell replication the guanine would mispair with thymine if the chelate were not removed or repaired before the replication occurred. On a second round of replication, at least one of each four such lesions would lead to the thymine correctly pairing with adenine. These two tandem events constitute formation of a base-substitution mutation, GC → AT. Thus, he predicted [90] that the cis isomer should be a base substitution mutagen, but the trans isomer should not. This prediction has been confirmed, both in bacteria by Beck et al. [91] and in mammalian cells by Zwelling et al. [92].

There is yet another possible bonus in this proposal. In addition to explaining the stereoselectivity of cis versus trans isomers, it has, with one additional assumption required, the possibility of explaining the selectivity of action of cisplatin against cancer cells. The assumption is that at least some cancers are caused by carcinogens creating O-6 lesions which cannot be repaired in time in those cells that become cancer cells. A large body of evidence exists from studies with alkylating agent carcinogens that supports this assumption [93]. Therefore, cisplatin lesions on O-6 guanine in normal cells are repaired before replication, while in cancer cells, which are so because of a deficiency in this repair process, the lesions are not removed and the burden of mutations increases beyond the limits of survivability. This part of the proposal has no direct evidence to support it at this time. Further, it may well be that base interactions, other than 0-6 of guanine with cisplatin, may also lead to a similar sequence of events.

8. MUTAGENESIS AND CARCINOGENESIS OF PLATINUM DRUGS

8.1. Mutagenesis in Bacteria and Mammalian Cells

Among the many effects of cisplatin on bacteria, the latest to be discovered was its mutagenicity. Beck and Brubaker [94], using *E. coli* auxotrophic mutants (requiring special growth substances in the media), demonstrated that cisplatin was very effective in reverting these mutants back to the prototrophic condition (not requiring special growth substances in the media). The spontaneous reversion frequency in these test conditions was 10^{-7}, whereas cisplatin increased it under maximal conditions to a value of 10^{-2}. The trans isomer produced only a minimal effect on the reversion rate, and was reported to be an ineffective mutagen.

Monti-Bragadin and coworkers [95] used the Ames bacterial screening test for mutagens--*Salmonella typhimurium* carrying the hisG46 missense mutation, which requires histidine in the growth medium. A series of such strains have been isolated. The TA1535 and TA100 strains are both sensitive only to mutagens which cause base substitution mutations; the latter strain also carries a plasmid R factor pKM101, which enhances the mutagenic activity. No liver microsomal fractions (required usually to metabolically activate drugs) were used. The cisplatin drug was shown to be an active mutagen in one strain (TA100). They reported that cisplatin was not an active mutagen in those Ames tester strains sensitive to frame shift mutagens. $RuCl_2(DMSO)_4$, which effects bacteria similarly to cisplatin, was also shown to be active as a mutagen. Activity in this system was confirmed by Benedict et al. [96].

Lecointe and coworkers [97] extensively studied the mutagenic activity in the TA100 strain of a variety of platinum complexes in order to look for structure-activity relations. The cisplatin was the most active among 13 complexes tested. The charged species (such as K_2PtCl_4, etc.) showed very little activity. The trans isomer could not be evaluated as it was too toxic to the bacterial strain. The author believes this result was due to contamination, since in early studies on cytotoxicity in *E. coli*, the trans isomer of cisplatin was much less toxic than cisplatin.

No definite correlations were found in this study. Later work by this group, however, has shown a reasonable correlation between mutagenicity and cytotoxicity [98].

Beck and Fisch [91] also studied the mutagenicity of a variety of anticancer-active platinum complexes (and included the trans isomer of cisplatin) against a broad range of Ames tester strains (TA100, TA98, TA1537, TA1538, and TA1535). Their conclusions are that the platinum drugs are direct mutagens (not requiring metabolic activation); that they induce base-pair substitution mutations rather than frame-shift mutations; that cisplatin is the most efficient mutagen in the group; and finally that the trans isomer is not an effective mutagen. The trans isomer was, in this study as in earlier work, not as toxic to the bacteria as cisplatin. Their data on the mutagenic potencies are depicted in Fig. 9. From these results, they have concluded that there is a rough correlation between mutagenic potency and antitumor activity.

Zwelling et al. [92] have reported mutagenicity assays performed with mammalian V79 cells. Cisplatin increased mutation frequencies 15 times over that of the control, whereas the trans isomer (which was 40 times less cytotoxic to these cells) produced no detectable change above control. They suggested a correlation for the two isomers between the cytotoxicity and mutagenicity on one hand, and DNA interstrand cross-links on the other.

In summary then, these studies prove that cisplatin is a moderate mutagen of the base-substitution type, and that the trans

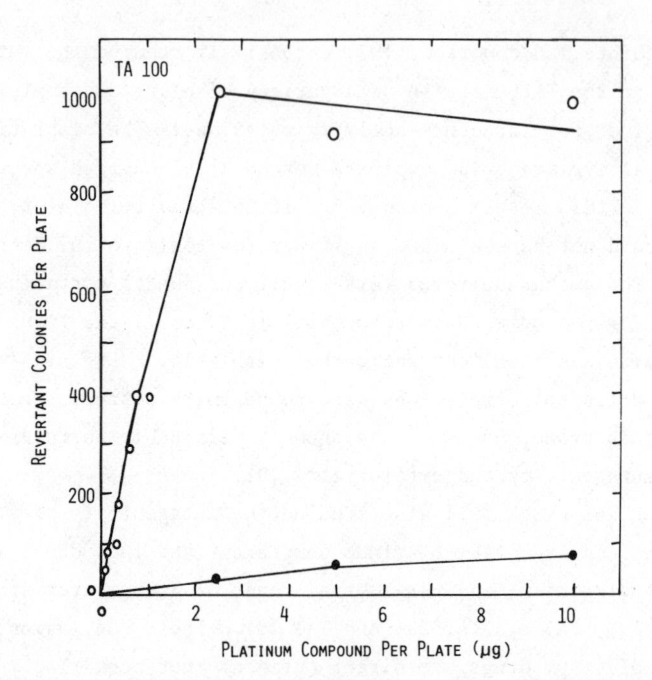

FIG. 9. Reversion of *Salmonella typhimurium* strain TA100 to histidine prototrophy by exposure of bacteria to cisplatin (open circles) and trans-dichlorodiammineplatinum(II) (closed circles) in Ames plate incorporation mutagenesis assays. Data are averages of nine plates. Spontaneous revertants have been subtracted from dose response curves [91].

isomer is not detectably mutagenic. For small sample sets of platinum complexes, a rough correlation exists between their mutagenicity and both cytotoxicity and anticancer activity.

8.2. Carcinogenesis of Platinum Drugs

Most anticancer drugs are mutagenic and carcinogenic [96]. Cisplatin has been shown in repeated tests to be moderately mutagenic. Until recently, however, no study has been published regarding its carcinogenicity. Leopold et al. [99] have now rectified this omission. Cisplatin and three analogs, cis-dichlorobis(cyclopentylamine)platinum(II), cis-dichlorobis(pyrrolidine)platinum(II), and cis-dichlorobis(dimethylsulfoxide)platinum(II), were administered at multiple, low-level doses, intraperitoneally, over periods up to 41 weeks in A/Jax mice, CD-1 mice, and Fischer rats. The animals were sacrificed and autopsied. Suspected tumors were diagnosed by histopathologic examination.

Cisplatin was very active in inducing lung adenomas in the A/Jax mice (100%). This is a benign tumor, but its appearance is taken as indicative, in that over longer times some of these would become malignant. The cyclopentylamine analog produced similar results in this system (95%), while the isotonic saline negative controls produced a small number of adenomas in 65% of the animals.

In the CD-1 mice, cisplatin induced skin papillomas (again, a benign tumor, but also indicative of carcinogenicity), but only when given in conjunction with croton oil as a promoter substance. It did produce eight malignant tumors in a group of 38 mice after 52 weeks.

Sarcomas developed in the Fischer rats after subcutaneous injections of cis-dichlorobis(cyclopentylamine)platinum(II) and cis-dichlorobis(pyrrolidine)platinum(II). Cisplatin was not tested in this system.

The conclusion drawn from these studies is that cisplatin is a moderately active carcinogen.

9. ANTICANCER ACTIVITY OF CISPLATIN IN HUMANS

9.1. Early Trials of Cisplatin as a Single Agent

Phase I clinical trials of the drug cisplatin began in 1971 at the Wadley Institutes of Molecular Medicine, and somewhat later at a variety of institutions under the sponsorship of the National Cancer Institute. A detailed description of the phases of clinical trials, with particular relevance for cisplatin has been provided by Carter and Goldsmith [100]. The purpose of Phase I trials is to determine the safe dose levels to be administered, and the types of toxicities that will occur. The patients are generally terminally ill and no longer responsive to existing therapies. An informed consent for entering experimental protocols is required. The patients are debilitated both by the advanced disease and by the prior treatments. Observations of anticancer activity ("responses"), while neither expected or necessary for progress to Phase II, are interesting and encouraging in these poor-prognosis patients.

The first clinical observations were reported by Hill et al. [5]; of 12 patients treated, 6 survived to the evaluation day. The cisplatin was given iv rapid push (5-15 min) at what is now considered very high dose levels (2-4 mg kg^{-1}). The toxicities found included deafness in one patient; tinnitus in two patients; loss of appetite (anorexia) and nausea and vomiting in all patients; bone marrow depression at higher dose levels (greater than 4 mg kg^{-1}) which was dose-limiting; and renal failure due to tubular degeneration at higher dose levels. No hair loss (alopecia), pancreatitis, or liver toxicity was found. Low doses (less than 2.5 mg kg^{-1}) or divided doses (up to 8 mg kg^{-1}) produced no serious side effects.

Clinical responses were noted in two patients with lymphosarcoma and in one patient with advanced Hodgkin disease. Measurable responses (less than 50% regression of tumors) were seen in two other patients. They also noted that a large portion of the injected dose was eliminated in the urine, and that recovery of some kidney damage could occur. Thus in this first report, many of the clinical problems in using cisplatin were depicted, and some favorable responses were noted.

A follow-up report was given by Hill et al. in 1973 [101]. With 67 patients evaluated, partial responses (greater than 50% reduction in tumor mass) were noted in endometrial carcinomas, hepatomas, and squamous cell carcinomas, using cisplatin as a single agent. Again, kidney toxicity and bone marrow depression were the dose-limiting side effects. Reports from other groups were by then appearing with a general concensus emerging that, as a single agent, cisplatin was of limited value because of the kidney toxicity at doses and schedules necessary for the manifestation of responses. Many had already begun combination chemotherapy in the hope of an improvement in results. But at this time, the picture for cisplatin was depressing.

Three bright spots emerged at the Second International Symposium on the cisplatin drug at Oxford, England in 1973. The first was the report by Wallace and Higby [6] of the single-agent activity of cisplatin against testicular cancers. Their results are summarized in Table 15. A second encouraging note was struck by Wiltshaw and Carr [7], who obtained good therapeutic responses against ovarian cancers in 7 of 19 assessable patients, all of whom were resistant to prior therapy with alkylating agents or multiple-drug regimens.

A third portentous paper was that of Krakoff and Lippman [102], which indicated a significant degree of activity against epidermoid carcinomas of the head and neck. But more importantly, the paper foreshadowed Krakoff's interest in the amelioration of the kidney toxicity, which was later to lead to the technique primarily responsible for turning the picture around.

TABLE 15

Responses of Advanced Testicular Cancer by Histologic Type to Cisplatin (Single-drug, Low-dose) Therapy

Response	Embryonal	Choriocarcinoma	Seminoma	Mixed
Complete	1/5 (5)[a]	1/2 (2+)	1/1 (10)	
Partial	1/5 (3+)	1/2 (2+)		1/3 (3)
Improved	2/5 (2,2)			1/3 (3)
No change	1/5			1/3

[a]Duration of response in months.

Source: Adapted from Ref. 6.

9.2. Nephrotoxicity in Humans and Its Amelioration

By 1973, it was quite clear that the extensive damage to the kidneys was limiting the possible utility of cisplatin. Several attempts were made to minimize this toxicity either by administering drugs which would chelate the cisplatin or block the receptor sites in the kidney (unknown), or by altering the administration schedule of the cisplatin itself. These have been reviewed recently by Krakoff [103].

Cvitkovic, working with Krakoff, was struck by the idea that hydration, followed by mannitol diuresis, mitigates the kidney damage due to mercury; and since mercury and platinum are both kidney-toxic metals, that this technique could be tried for cisplatin. The initial tests were done on dogs [104] and clearly demonstrated a significant protection of the renal function by vigorous hydration before, during, and after administration of cisplatin with mannitol-forced diuresis to flush the large volume of fluid through the kidney. This technique did not affect the rate of excretion of the cisplatin (although it did dilute its concentration in the urine), the serum levels, or the toxicity to blood elements as measured by leukocyte and platelet counts. Thus, it was clear that hydration would not impair the anticancer activity by prematurely removing the cisplatin from the body. The animals tolerated 3-5 times the amount of cisplatin that could be given without hydration. Human clinical trials were then undertaken by this group [8]. The dose level that could now be administered to patients before unacceptable toxicity was increased from about 1 mg kg^{-1} (40 mg m^{-2}) to 3 mg kg^{-1} (120 mg m^{-2}). At this same time, Hill et al. [101] tried a slow infusion of cisplatin at a rate of 1 mg kg^{-1} hr^{-1}, which apparently ameliorated the kidney toxicity. Merrin [9] gave cisplatin as a slow infusion over 6-8 hr with improved tolerance in the patients. This still remains an accepted technique for administration of the drug [105].

There are still some unresolved questions concerning the kidney toxicity. Does adding mannitol provide extra protection over hydration alone? Would furosemide, a diuretic which acts principally in the distal tubules be better than mannitol, an osmotic diuretic acting primarily in the proximal tubules? Is the site of damage localized in either the proximal or distal tubules? Would still slower infusion rates decrease the kidney toxicity further?

Partial answers to some of these questions do exist. In most animals, including man, the evidence supports damage localized in the proximal tubules [41]. Studies with 24-hr slow infusions do suggest a further decrease in kidney damage with no loss of anticancer activity [106,107]. In rats given cisplatin and either furosemide or mannitol, the histopathologically evident damage to the kidney was persistent in the furosemide group, but repaired in the mannitol group, while measures of renal function such as blood urea nitrogen (BUN) levels indicated that both diuretics exerted protective action [26]. Pera and Harder [108] showed that mannitol but not furosemide also provided a sparing effect on other rapidly proliferating tissue in the rat such as bone marrow and intestinal mucosa.

One unexpected advantage of the 24-hr slow infusion protocol was a decrease in nausea and vomiting [106,107]. This is an invariant accompaniment of cisplatin therapy, and is usually so intensive that compazine-type antiemetics are of limited value.

10. ACTIVE COMBINATION CHEMOTHERAPIES OF HUMAN CANCERS WITH CISPLATIN

10.1. Testicular Cancers

There are three major therapy regimens for the treatment of advanced metastatic testicular cancers. Each has its origin in the findings of Samuels [11] that bleomycin and vinblastine produce a significant number of complete remissions, even if of short duration only. With a report by Wallace and Higby [6] of the high activity of cisplatin as a single agent, it was quite natural to add cisplatin to these other drugs in a three-drug combination. This was done in slightly different fashions at the Memorial Sloan Kettering Cancer Center, the University of Indiana Medical School, and the Roswell Park Institute. While differing in details, the success rates are quite comparable.

A review of the results achieved at Sloan Kettering was given recently by Burchenal [109]. The VAB III therapy consists of an induction regimen as follows: day 1--cyclophosphamide (600 mg m^{-2}, iv), vinblastine (4 mg m^{-2}, iv), and actinomycin D (1 mg m^{-2}, iv); days 1-7--bleomycin (20 mg m^{-2}/day, by continuous infusion); day 8--cisplatin (120 mg m^{-2}, iv, with mannitol-induced diuresis). This induction phase is followed by a maintenance regimen given every 3 weeks, followed by a repeat of the induction regimen and a final long-term maintenance regimen. A variation on this theme to include adriamycin in the maintenance regimen is the VAB IV therapy. The tabulated results of these two therapies are given in Table 16.

Einhorn [12] used a lower dose of cisplatin (20 mg m^{-2}) every day for 5 days, and then repeated every 3 weeks, in combination with vinblastine and bleomycin. He too had recently added adriamycin as

TABLE 16

Combination Chemotherapy with Cisplatin Against Advanced Testicular Cancers (Memorial Sloan-Kettering Cancer Center Protocols)

Response status	VAB III	VAB IV
Number of patients adequate to evaluate	90	50
Complete	54 (60%)	30 (60%)
Partial	23 (26%)	14 (28%)
Minor regression and progression	13 (14%)	6 (12%)
Complete and partial	77 (86%)	44 (88%)
Relapse rate of complete responders	15/54 (28%)	4/30 (13%)

Source: Adapted from Ref. 109.

TABLE 17

Combination Chemotherapy with Cisplatin (P) + Vinblastine (V) + Bleomycin (B) + Adriamycin (A) Against Advanced Testicular Cancers

Response status	PVB	PVB + A
Number of patients adequate to evaluate	26	26
Complete	18 (69%)	19 (73%)
Partial	8 (31%)	6 (23%)
Disease-free after surgery	5 (19%)	2 (8%)
Relapse rate of complete responders	4 (17%)	1 (5%)
Number alive	24 (92%)	21 (81%)
Number with no evidence of disease	18 (69%)	20 (77%)

Source: Adapted from Ref. 48.

a variation, but gave it with the other drugs during the induction phase. His later results are given in Table 17 [48].

Merrin [110] uses still another variation which has an induction phase of 6 weeks of bleomycin, vincristine, cisplatin, and prednisone; a consolidation phase of 9 weeks' duration of actinomycin D, vincristine, and cisplatin; and finally, a maintenance phase of 2 years' duration of actinomycin D and cisplatin. His results are given in Table 18.

It is quite clear from these studies that 100% of the patients respond with either complete responses (no evidence of disease) (60-80%), or partial responses (20-40%) to these combination therapies. The remissions are of long duration--many greater than 2 years. Since relapses usually occur within 9 months of initiation of therapy if they are going to occur at all, we now can safely specify cure rates for advanced metastatic testicular tumors of 50% or better.

One quite interesting result which has emerged in all three studies is that about half of the partial responders have large tumor masses which neither increase nor decrease in size with treatment. Surgical excision of these masses, and histopathologic

TABLE 18

Combination Chemotherapy with Cisplatin, Bleomycin, Vincristine, Prednisone, and Actinomycin D Against Advanced Testicular Cancers

Response status	Number of evaluable patients	%
Complete		
Chemotherapy alone	16	24
Chemotherapy and surgery	37	56
Chemotherapy and surgery + radiation	1	1.5
Partial		
Chemotherapy alone		
Chemotherapy and surgery	12	18.2
No response	0	0
Total	66	100
Currently alive with no evidence of disease	41	62

Source: Adapted from Ref. 110.

analyses of these tissues indicate that they consist of benign tumors. The implication is that therapy has caused a malignant-to-benign transformation of these tumors. While this has been known to occur on occasion with other chemotherapies, it seems to be a fairly common occurrence in these newer therapies. After surgical removal, these patients are now in a disease-free status.

Recently, Williams and coworkers have reported [111] that VP-16, an epipodophyllotoxin derivative, was very active in combination with cisplatin, adriamycin, and bleomycin in the remaining number of patients refractory to the Einhorn therapy described above. For example, in 16 such patients so treated there were 6 complete responses and 10 partial responses, with 6 of these latter still in the induction phase.

The author has gone into some detail here to illustrate the complexity of the evolution of combination chemotherapy protocols.

One can readily imagine the astronomical number of variations possible with six drugs, with different dose levels and schedules, different induction and maintenance combinations and dose levels, and with at least four major different histologic types of testicular cancers, each having idiosyncratic responses. With results as good as have now been found, it would require very large numbers of patients to provide adequate statistical evidence to substantiate improved therapies. The ever smaller number of failures still, however, permit some new variations to be tested, and such tests are currently under way. It is not overly optimistic to suggest that, in the near future, 100% of testicular cancers may be cured. For a disease that was almost invariably fatal only five years ago, and incidentally, was the major cause of cancer deaths in men between the ages of 15 and 35 years, the present progress is very encouraging and attests to the power of chemotherapy in the treatment of metastatic cancers.

10.2. Ovarian Cancer

There are now two major protocols (and a number of less-studied ones) for the treatment of advanced ovarian cancer with cisplatin. The first set of protocols, formulated by Wiltshaw et al. [112], involves two regimens. One regimen is cisplatin (20 mg m^{-2}, day 1) with chlorambucil (0.15 mg kg^{-1} day^{-1}, days 1-7), repeated every three weeks--regimen B; while the second regimen has the addition of adriamycin (50 mg m^{-2}, day 1) to the previous combination--regimen C. The resulting responses, including "second look" surgery to determine the patient's status more exactly, are compiled in Table 19. The overall response rates for the two regimens are 59% and 66%, respectively. The complete response rates are 32% and 41%. The median duration of the complete remissions is now greater than 15 months, with the longest complete remission now out to over 34 months and continuing.

TABLE 19

Combination Chemotherapy with Cisplatin Against Advanced Ovarian Cancer

Response status	Chlorambucil + cisplatin	Chlorambucil + cisplatin + adriamycin
Number of evaluable patients	34	24
Complete	11 (32%)	10 (41%)
Partial	9 (27%)	6 (25%)
Complete and partial	20 (59%)	16 (66%)
Second-look operations	4	8
No pathologic evidence of tumor	3	3

Source: Adapted from Ref. 112.

The second major protocol was developed by Bruckner et al. [15] and consists of cisplatin (50 mg m^{-2}, iv day 1) and adriamycin (50 mg m^{-2}, iv. day 1) given every 3 weeks. Their response rates exceed 65%, with a complete response rate of 40%. The median duration of the complete responses is greater than 22 months and continuing to increase.

The major toxicities occurring in these patients are hematologic (47% of the patients), renal (29%), neurologic (peripheral neuropathy, 11%), and cardiac (17%).

These protocols now represent the optimal therapy for advanced ovarian cancers. Newer modifications presently being studied include higher dose levels of cisplatin (up to 120 mg m^{-2}); the addition of cytoxan to the Bruckner protocols; and combinations of cisplatin with hexamethylmelamine and 5-fluorouracil added to adriamycin and cytoxan. An example that may reflect an improved response rate, although it is still in an early stage; Ehrlich et al. [113] added cytoxan to cisplatin (20 mg m^{-2}, days 1-5) and adriamycin with the results: objective responses, 92%; complete responses, 46%.

10.3. Head and Neck Cancers

Advanced epidermoid carcinomas of the head and neck have in the past responded poorly to chemotherapy with single agents or combinations. Active clinical research programs in the last few years have altered the picture slightly for the better. Three protocols look promising: a high-dose methotrexate with leucovorin rescue regimen [114]; a complex, kinetically based, multiple-drug therapy [115]; and cisplatin alone or in combinations with other drugs [116,117]. This subject has been reviewed recently by Wittes et al. [116].

Wittes' group at Memorial Sloan-Kettering Cancer Center has studied four regimens in the preoperative treatment of Stage-III and -IV head and neck cancers. The first combined high-dose cisplatin (120 mg m^{-2}, iv, day 1) with bleomycin (10 mg m^{-2}, iv, fast push, day 3, and continuous slow infusion, days 3-10). The responses obtained are tabulated in Table 20. This was followed by either surgery or radiation or both. Two of 21 entering, generally inoperable, patients are alive and without evidence of disease at 24 and 25 months after inception of therapy. A third was alive with disease at 23 months.

TABLE 20

Combination Chemotherapy with Cisplatin, Bleomycin, and Methotrexate Against Advanced Cancers of the Head and Neck

Response status	Cisplatin + bleomycin	Cisplatin, bleomycin, + methotrexate
Number of evaluable patients	21	8
Complete	4 (19%)	0
Partial	11 (52%)	4 (50%)
Minor	5 (24%)	0
No response	1 (5%)	4 (50%)
Complete and partial	15 (71%)	4 (50%)

Source: Adapted from Ref. 116.

Attempts to improve these results included adding high-dose methotrexate with leucovorin rescue, and the addition of vinblastine to the cisplatin therapy. Neither change increased the response rate, and both were increasingly toxic and, therefore, discarded. Cisplatin alone produced only a 40% response rate compared to a 71% response rate in combination with bleomycin. A major cooperative study supported by the NCI is presently under way which will compare the standard therapy of surgery and radiation alone with this preceded by chemotherapy with cisplatin and bleomycin, or these two followed by maintenance chemotherapy with 24-hr infusions of cisplatin for 6 months.

The 24-hr infusion of cisplatin was developed by Jacobs et al. [107]. With 18 patients (6 previously untreated) given cisplatin at 80 mg m^{-2}, 24 hr, the response rates were complete + partial responses = 39%; and complete + partial + minor responses = 72%--not much different from that of cisplatin given by rapid push. However, the kidney toxicity was minimal and nausea and vomiting were absent in 8 of 32 courses.

10.4. Squamous Cell Carcinoma of the Cervix

The gynecologic oncology group sponsored by the NCI has undertaken a large group study of the effectiveness of cisplatin as a single drug against gynecologic cancers. A preliminary report by Thigpen et al. [118] showed cisplatin to be the most effective single agent against squamous cell carcinoma of the cervix. This is a fairly common malignancy with few prior effective chemotherapies. Twenty-five patients with advanced or recurrent measurable disease had been entered by the time of the report. All had received therapy previous to this; 24 had radiotherapy, 16 surgery, and 4 chemotherapy. The physiologic performance status of each was good.

The treatment consisted of cisplatin (50 mg m^{-2}, iv) administered at a rate of 1 mg min^{-1}, with adequate hydration, but no mannitol. This was repeated every 3 weeks for up to at least six courses. Eight percent of the patients demonstrated continued increase of disease; 44% responded with either complete responses (12%) or partial responses (32%). The other 48% had stable disease. The median survival is 7 months and continuing. These results are preliminary to beginning combination chemotherapy with cisplatin against cervical cancer.

10.5. Bladder Cancers

The early report of Wallace and Higby [6] of one bladder patient put into complete remission by cisplatin, whose duration of response is now over four years, stimulated a strong interest in applying cisplatin alone or in various combinations to this cancer. A number of reports have appeared and a review of the literature is available by Yagoda [119]. He also describes his experience with 95 patients treated with any of four protocols: cisplatin alone; with adriamycin; with cytoxan; or with both.

Because of the difficulty of evaluating complete responses, his results are given as complete (CR) + partial (PR) responses. With cisplatin alone the results are: 28 evaluable patients, 36% CR + PR; cisplatin + cytoxan, 35 evaluable patients, 43% CR + PR; cisplatin + adriamycin, 26 evaluable patients, 54% CR + PR; cisplatin + cytoxan + adriamycin, 6 evaluable patients, 50% CR + PR. These are consistent with the results of other studies: Merrin [120], 1 CR and 8 PR in 19 patients; Soloway [121], 1 CR and 5 PR in 10 patients; Sternberg et al. [122], 83% CR + PR in 12 patients with their CISCA regimen (cisplatin, cytoxan, adriamycin); Williams et al. [123], 65% PR in 17 patients treated with cisplatin, adriamycin, and 5-fluorouracil.

The duration of remissions in all these cases must be considered disappointingly short, although in the last case, two patients are alive at 33+ and 63+ weeks.

10.6. Oat Cell Lung Cancers

Oat cell (or small cell) carcinoma of the lung, one of the four major histologic types of lung cancer, is proving to be a fairly drug- and radiation-sensitive tumor. One strikingly successful induction therapy is described here. Sierocki and associates [124] at Memorial Sloan-Kettering Cancer Center have studied a combination therapy consisting of cisplatin (60 mg m^{-2}, iv, with hydration, days 1 and 22) and VP-16 (120 mg m^{-2}, iv, days 4, 6, 8, 24, 26, 28). A maintenance phase included cytoxan, adriamycin, and vincristine given iv on days 42, 63, 84, and 105. The induction phase with cisplatin and VP-16 was repeated on day 126 and the whole schedule recycled. 3000 rad of whole brain irradiation was given in ten fractions between days 42 and 63.

A total of 38 patients with clearly diagnosed oat cell carcinoma of the lung and no prior chemotherapy were entered into the study. The responses obtained in this study show a clearly impressive 95% response rate as a whole. Of the 21 patients with limited disease, 52% had a complete response which, with one exception, are continuing from 5.75+ months to 15.25+ months. The partial remission rate was 48%, but the duration of these was 2.25-10.5 months. The 17 patients with extensive disease had a complete remission rate of 41% (4.75-11+ months duration) and a partial remission rate of 47% (3.75-11.5 months duration).

While longer-term follow-ups are necessary to evaluate the survival rates, the cisplatin-VP-16 combination is a very effective induction regimen. Further variations of this theme are called for.

10.7. Other Cancers

Cisplatin alone or in combination has also shown an interesting degree of activity against prostate carcinomas, non-oat-cell lung carcinomas, osteogenic sarcomas, and neuroblastomas. All of this work is in too early and too rapidly changing phases to comment on.

The major failures are cancers of the stomach, colon, and kidney. Since cisplatin does not penetrate the blood-brain barrier, it should not be expected to be active in brain cancers, yet a few reports of such activity are known and additional work is continuing.

11. GENERAL CONCLUSIONS

Cisplatin is the first of a new class of anticancer drugs based on platinum coordination complexes. It has already become of significant utility in the treatment of a number of human cancers. Like most other anticancer drugs, it is best used in combination chemotherapy, and the development of best combinations for each type of cancer is proceeding vigorously. Its use in the treatment of advanced metastatic testicular cancers may shortly provide a cure for this disease.

The major toxic side effects are damage to the kidney and nausea and vomiting. The simple pharmacologic trick of hydrating the patient greatly ameliorates the kidney toxicity. Administration of the drug as a slow infusion over 24 hr ameliorates both side effects.

The molecular mechanism of action is not yet clear, but it is generally accepted that it involves an interaction of the aquated cisplatin with cellular DNA. How such a reaction leads to selective destruction of cancer cells is under intense study.

Newer analogs of cisplatin are now in human clinical trials. We are particularly interested in the 1,2-diaminocyclohexane ligand as a substitute for the diammine ligands, since such compounds are active in cancers that develop resistance to cisplatin. Malonate, sulphate, and carboxyphthalate leaving groups instead of chloride represent the most promising modification for higher efficacy and less toxicity. Clinical trials are of necessity notoriously slow, and it may be many years before an accurate assessment of these analogs will be complete. Meanwhile, a heavy responsibility devolves upon coordination chemists to advance this new field, for who else can do it?

REFERENCES

1. B. Rosenberg, L. Van Camp, and T. Krigas, *Nature*, *205*, 4972 698 (1965).

2. B. Rosenberg, L. Van Camp, E. B. Grimley, and A. J. Thomson, *J. Biol. Chem.*, *242*, 6, 1347 (1967).

3. S. Reslova, *Chem.-Biol. Interactions*, *4*, 66 (1971-72).

4. B. Rosenberg, L. Van Camp, J. E. Trosko, and V. H. Mansour, *Nature*, *222*, 385 (1969).

5. J. Hill, R. J. Speer, E. Loeb, A. MacLellan, N. O. Hill, and A. Khan, in Advances in Antimicrobial and Antineoplastic Chemotherapy, Vol. II, University Park Press, Baltimore, 1972, p. 255.

6. H. J. Wallace and D. J. Higby, in Recent Results in Cancer Research: Platinum Coordination Complexes in Cancer Chemotherapy (T. A. Connors and J. J. Roberts, eds.), Springer-Verlag, New York, 1974, p. 167.

7. E. Wiltshaw and B. Carr, in Recent Results in Cancer Research: Platinum Coordination Complexes in Cancer Chemotherapy (T. A. Connors and J. J. Roberts, eds.), Springer-Verlag, New York, 1974, p. 178.

8. D. M. Hayes, E. Cvitkovic, R. B. Golbey, E. Scheiner, L. Helson, and I. H. Krakoff, *Cancer*, *39*, 1372 (1977).

9. C. Merrin, Abstr. C-26, AACR Ann. Mtg., May, 1976.

10. R. J. Woodman, A. E. Sirica, M. Gang, I. Kline, and J. M. Venditti, *Chemother.*, *18*, 169 (1973).

11. M. L. Samuels, E. E. Johnson, and P. Y. Holoye, *Cancer Chemother. Rep.*, *59*, 563 (1975).

12. L. H. Einhorn and J. Donohue, *Ann. Intern. Med.*, *87*, 293 (1977).

13. E. Cvitkovic, D. Hayes, and R. Golbey, *Proc. Amer. Soc. Clin. Oncol.*, *17*, Abstr. C-237, p. 296 (1976).

14. E. Wiltshaw, S. Subramarian, C. Alexopoulos, and G. Barker, *Cancer Treat. Rep.*, Proc. NCI Conf. on Cis-platinum and Testicular Cancer, *63*, 1545 (1979).

15. H. W. Bruckner, C. C. Cohen, G. Deppe, B. Kabakow, R. C. Wallach, E. Greenspan, S. B. Gusberg, and J. F. Holland, *J. Clin. Hematol. Oncol.*, *7*, 619 (1977).

16. C. G. Zubrod, *Cancer Res.*, *38*, 4377 (1978).

17. "Instruction 14," Screening Data Summary Interpretation and Outline of Current Screen, Drug Evaluation Branch, National Cancer Institute, Bethesda, 1979.

18. M. K. Wolpert-DeFilippes, *Cancer Treat. Rep.*, Proc. NCI Conf. on Cis-platinum and Testicular Cancer, *63*, 1453 (1979).

19. J. P. Davidson, P. J. Faber, R. G. Fischer, Jr., S. Mansy, H. J. Peresie, B. Rosenberg, and L. Van Camp, *Cancer Chemother. Rep.*, Part 1, 59, 287 (1975).

20. S. Yolles, J. F. Morton, and B. Rosenberg, *Acta Pharm. Svec.*, *15*, 382 (1978).

21. J. Toth-Allen, Doctoral dissertation, Michigan State University, E. Lansing, Michigan, 1970.

22. R. C. Kociba and S. D. Sleight, *Cancer Chemother. Rep.*, Part 1, *55*, 1 (1971).

23. U. Schaeppi, I. A. Heyman, R. W. Fleischman, H. Rosenkrantz, V. Ilievski, R. Phelan, D. A. Coonan, and R. D. David, *Toxicol. Appl. Pharmacol.*, *25*, 230 (1973).

24. A. M. Guarino, D. S. Miller, S. T. Arnold, J. B. Pritchard, R. D. Davis, M. A. Urbanek, T. J. Miller, and C. L. Litterst, *Cancer Treat. Rep.*, Proc. NCI Conf. on Cis-platinum and Testicular Cancer, *63*, 1475 (1979).

25. J. M. Ward, M. E. Grabin, E. Berlin, and D. M. Young, *Cancer Res.*, *37*, 1238 (1977).

26. M. F. Pera, Jr., B. C. Zook, and H. C. Harder, *Cancer Res.*, *37*, 1269 (1977).

27. M. Zak, J. Drobnik, and Z. Rezny, *Cancer Res.*, *32*, 595 (1972).

28. R. W. Fleischman, S. W. Stadnicki, M. F. Ethier, and U. Schaeppi, *Toxicol. Appl. Pharmacol.*, *33*, 320 (1975).

29. J. D. Hoeschele and L. Van Camp, in Advances in Antimicrobial and Antineoplastic Chemotherapy, Vol. II, University Park Press, Baltimore, 1972, p. 241.

30. W. Wolf and R. C. Manaka, *J. Clin. Hematol. Oncol.*, 7, 79 (1977).

31. C. L. Litterst, I. J. Torres, and A. M. Guarino, *J. Clin. Hematol. Oncol.*, 7, 169 (1976).

32. R. C. Lange, R. P. Spencer, and H. C. Harder, *J. Nucl. Med.*, *13*, 328 (1972).

33. C. L. Litterst, A. F. LeRoy, and A. M. Guarino, *Cancer Treat. Rep.*, Proc. NCI Conf. on Cis-platinum and Testicular Cancer, *63*, 1485 (1979).

34. R. J. Belt, K. J. Himmelstein, T. F. Patton, S. J. Bannister, L. A. Sternson, and A. J. Repta, *Cancer Treat. Rep.*, Proc. NCI Conf. on Cis-platinum and Testicular Cancer, *63*, 1515 (1979).

35. J. W. Reishus and D. S. Martin, Jr., *JACS*, *83*, 2457 (1961).

36. M. J. Cleare, P. C. Hydes, B. W. Malerbi, and D. M. Watkins, *Biochim.*, *60*, 835 (1978).

37. P. Horacek and J. Drobnik, *Biochim. Biophys. Acta*, *254*, 341 (1971).

38. J. Drobnik, A. Krekulova, and A. Kubelkova, *Folia Microbiol.*, *16*, Abstr. 9th Ann. Mtg. (1971).

39. B. W. Wilkinson and J. Toth-Allen, *Nucl. Technol.*, *13*, 103 (1972).

40. F. Cavalli, L. Tschopp, R. W. Sonntag, and A. Zimmerman, *Cancer Treat. Rep.*, *62*, 2125 Let. (1978).

41. N. E. Madias and J. T. Harrington, *Amer. J. Med.*, *65*, 313 (1978).

42. G. J. Goldenberg, C. L. Vanstone, L. G. Israels, D. Ilse, and I. Bihler, *Cancer Res.*, *30*, 2285 (1970).

43. G. R. Gale, C. R. Morris, L. M. Atkins, and A. B. Smith, *Cancer Res.*, *33*, 813 (1973).

44. H. C. Harder, and B. Rosenberg, *Int. J. Cancer*, *6*, 207 (1970).

45. J. A. Howle and G. R. Gale, *Biochem. Pharmacol.*, *19*, 757 (1970).

46. H. W. van den Berg, H. N. A. Fraval, and J. J. Roberts, *J. Clin. Hematol. Oncol.*, *7*, 349 (1977).

47. J. A. Howle, H. S. Thompson, A. E. Stone, and G. R. Gale, *Proc. Soc. Exp. Biol. Med.*, *137*, 820 (1971).

48. L. H. Einhorn, Chemotherapy Foundation Symposium III, p. 52, abstract (1978).

49. J. J. Roberts and A. J. Thomson, *Prog. Nucl. Acid Res. Mol. Biol.*, *22*, 71 (1979).

50. M. E. Friedman, B. Musgrove, K. Lee, and J. E. Teggins, *Biochim. Biophys. Acta*, *250*, 286 (1971).

51. M. E. Friedman and J. E. Teggins, *Biochim. Biophys. Acta*, *350*, 263 (1974).

52. R. J. Woodman, J. M. Venditti, S. A. Schepartz, and I. Kline, *Proc. Amer. Assoc. Cancer Res.*, *12*, 24 (1971).

53. R. J. Speer, S. Lapis, H. Ridgway, T. D. Meyers, and J. M. Hill, in Advances in Antimicrobial and Antineoplastic Chemotherapy, Vol. II, University Park Press, Baltimore, 1972, p. 253.

54. L. Van Camp and B. Rosenberg, in Advances in Antimicrobial and Antineoplastic Chemotherapy, Vol. II, University Park Press, Baltimore, 1972, p. 239.

55. G. R. Gale, L. M. Atkins, S. J. Meischen, A. B. Smith, and E. M. Walker, Jr., *Cancer Treat. Rep.*, *61*, 445 (1977).

56. G. R. Gale, L. M. Atkins, S. J. Meischen, and P. Schwartz, *Cancer*, *41*, 1230 (1978).

57. G. R. Gale, L. M. Atkins, P. Schwartz, and S. J. Meischen, *Bioinorg. Chem.*, *8*, 445 (1978).

58. M. S. Soloway, *Cancer*, *36*, 333 (1975).

59. V. K. Ghanta, M. T. Jones, D. A. Woodard, J. R. Durant, and R. N. Hiramoto, *Cancer Res.*, *37*, 771 (1977).

60. F. M. Schabel, Jr., M. W. Trader, W. R. Laster, Jr., T. H. Corbett, and D. P. Griswold, Jr., *Cancer Treat. Rep.*, Proc. NCI Conf. on Cis-platinum and Testicular Cancer, *63*, 1459 (1979).

61. I. Wodinsky, J. Swiniarsky, C. J. Kensler, J. M. Venditti, in Recent Results in Cancer Research: Platinum Coordination Complexes in Cancer Chemotherapy (T. A. Connors and J. J. Roberts, eds.), Springer-Verlag, New York, 1974, pp. 134.

62. P. C. Merker, I. Wodinsky, J. Mabel, A. Branfman, and J. M. Venditti, *J. Clin. Hematol. Oncol.*, *7*, 301 (1977).

63. M. Zak and J. Drobnik, *Strahlenther.*, *142*, 112 (1971).

64. E. B. Douple, R. C. Richmond, and M. E. Logan, *J. Clin. Hematol. Oncol.*, *7*, 2, 585 (1977).

65. R. C. Richmond and E. L. Powers, *J. Clin. Hematol. Oncol.*, *7*, 580 (1977).

66. A. H. W. Nias and I. I. Szumiel, *J. Clin. Hematol. Oncol.*, *7*, 562 (1977).

67. J. H. Burchenal, K. Kalaher, T. O'Toole, and J. Chisholm, *Cancer Res.*, *37*, 3455 (1977).

68. J. H. Burchenal, K. Kalaher, K. Dew, L. Lokys, and G. Gale, *Biochim.*, *60*, 961 (1978).

69. B. Rosenberg and L. Van Camp, *Cancer Res.*, *30*, 1799 (1970).

70. T. A. Connors, M. Jones, W. C. J. Ross, P. D. Braddock, A. R. Khokhar, and M. L. Tobe, *Chem.-Biol. Interact.*, *5*, 415 (1972).

71. R. J. Kociba, S. D. Sleight, and B. Rosenberg, *Cancer Chemother. Rep.*, Part 1, *54*, 325 (1970).

72. B. Rosenberg, in Advances in Antimicrobial and Antineoplastic Chemotherapy, Vol. II, University Park Press, Baltimore, 1972, p. 101.

73. S. K. Aggarwal, R. W. Wagner, P. K. McAllister, and B. Rosenberg, *Proc. Nat. Acad. Sci. U.S.*, *72*, 928 (1975).

74. B. Rosenberg, *Cancer Chemother. Rep.*, Part 1, *59*, 589 (1975).

75. D. A. Juckett and B. Rosenberg, *Biophys. J.*, *25*, 29(a) (1979).

76. J. L. Russell and E. S. Golub, *Proc. Nat. Acad. Sci. U.S.*, *75*, 6211 (1978).

77. A. C. Hollinshead, *Proc. Amer. Assoc. Cancer Res.*, *20*, Abstr. 590 (1979).

78. M. Howe-Grant, K. C. Wu, W. R. Bauer, and S. J. Lippard, *Biochem.*, *15*, 4339 (1976).

79. J. J. Roberts and J. M. Pascoe, *Nature*, *235*, 282 (1972).

80. H. C. Harder, in Recent Results in Cancer Research: Platinum Coordination Complexes in Cancer Chemotherapy (T. A. Connors and J. J. Roberts, eds.), Springer-Verlag, New York, 1974, p. 69 ff.

81. L. A. Zwelling and K. W. Kohn, *Cancer Treat. Rep.*, Proc. NCI Conf. on Cis-platinum and Testicular Cancer, *63*, 1439 (1979).

82. S. Mansy, B. Rosenberg, and A. J. Thomson, *JACS*, *95*, 1633 (1973).

83. J. N. Pascoe and J. J. Roberts, in Recent Results in Cancer Research: Platinum Coordination Complexes in Cancer Chemotherapy (T. A. Connors and J. J. Roberts, eds.), Springer-Verlag, New York, 1974, p. 108.

84. P. J. Stone, A. D. Kelman, and F. M. Sinex, *Nature*, *251*, 736 (1974).

85. J.-P. Macquet and T. Theophanides, *Inorg. Chim. Acta*, *18*, 189 (1976).

86. D. M. L. Goodgame, I. Jeeves, F. L. Philips, and A. C. Skapski, *Biochim. Biophys. Acta*, *378*, 153 (1975).

87. J. Dehand and J. Jordanov, *J. C. S. Chem. Comm.*, 598 (1976).

88. R. J. Pollock and B. Rosenberg, *Bioch.*, *60*, 859 (1978), reference to unpublished data.

89. A. R. Kausar and B. Rosenberg, unpublished results.

90. B. Rosenberg, *J. Clin. Hematol. Oncol.*, *7*, 817 (1977).

91. D. J. Beck and J. E. Fisch, *Mutation Research*, *77*, 45 (1980).

92. L. A. Zwelling, K. W. Kohn, and T. A. Anderson, *Proc. Amer. Assoc. Cancer Res.*, *19*, 233 (1978).

93. P. Kleihues, P. L. Lantos, and P. N. Magee, in International Review of Experimental Pathology, Vol. 15 (G. W. Richter and M. A. Epstein, eds.), Academic Press, New York, 1976, p. 153.

94. D. J. Beck and R. R. Brubaker, *Mut. Res.*, *27*, 181 (1975).

95. C. Monti-Bragadin, M. Tamaro, and E. Banfi, *Chem.-Biol. Interac.*, *11*, 469 (1975).

96. W. F. Benedict, M. S. Baker, L. Haroun, E. Choi, and B. N. Ames, *Cancer Res.*, *37*, 2209 (1977).

97. P. Lecointe, J.-P. Macquet, J.-L. Butour, and C. Paoletti, *Mut. Res.*, *48*, 139 (1977).

98. P. Lecointe, J.-P. Macquet, and J.-L. Butour, CNRS Conf. on Coordination Chemistry and Cancer Chemotherapy, July, 1978, Toulouse.

99. W. R. Leopold, E. C. Miller, and J. A. Miller, *Cancer Res.*, *39*, 913 (1979).

100. S. K. Carter and M. Goldsmith, in Recent Results in Cancer Research: Platinum Coordination Complexes in Cancer Chemotherapy (T. A. Connors and J. J. Roberts, eds.), Springer-Verlag, New York, 1974, p. 137.

101. J. M. Hill, E. Loeb, A. S. MacLellan, N. O. Hill, A. Khan, and J. Kogler, in Recent Results in Cancer Research: Platinum Coordination Complexes in Cancer Chemotherapy (T. A. Connors and J. J. Roberts, eds.), Springer-Verlag, New York, 1974, p. 145.

102. I. H. Krakoff and A. J. Lippman, in Recent Results in Cancer Research: Platinum Coordination Complexes in Cancer Chemotherapy (T. A. Connors and J. J. Roberts, eds.), Springer-Verlag, New York, 1974, p. 183.

103. I. H. Krakoff, *Cancer Treat. Rep.*, Proc. NCI Conf. on Cis-platinum and Testicular Cancer, *63*, 1523 (1979).

104. E. Cvitkovic, J. Spaulding, V. Bethune, J. Martin, W. F. Whitmore, *Cancer*, *39*, 1357 (1977).

105. Platinol™ (Cisplatin) product monograph, Bristol Laboratories, 1978.

106. T. L. Loo, S. W. Hall, P. Salem, R. S. Benjamin, and K. Lu, *Biochim.*, *60*, 957 (1978).

107. C. Jacobs, J. R. Bertino, D. R. Goffinet, W. E. Fee, and R. L. Goode, *Cancer*, *42*, 2135 (1978).

108. M. F. Pera and H. C. Harder, *Cancer Res.*, *39*, 1279 (1979).

109. J. H. Burchenal, *Biochim.*, *60*, 915 (1978).

110. C. Merrin, *Cancer Treat. Rep.*, Proc. NCI Conf. on Cis-platinum and Testicular Cancer, *63*, 1579 (1979).

111. S. D. Williams, L. H. Einhorn, A. Greco, R. Oldham, R. Fletcher, and W. H. Bond, *Proc. Amer. Assoc. Cancer Res.*, *20*, Abstr. 291, p. 72 (1979).

112. E. Wiltshaw, S. Subramarian, C. Alexopoulos, and G. Barker, *Cancer Treat. Rep.* Proc. NCI Conf. on Cis-platinum and Testicular Cancer, *63*, 1545 (1979).

113. C. E. Ehrlich, L. H. Einhorn, and J. L. Morgan, *Proc. Amer. Soc. Clin. Oncol.*, *19*, Abstr. C-292, p. 379 (1978).

114. N. H. Goldberg, P. B. Chretien, E. G. Elias, et al., *Proc. Amer. Soc. Clin. Oncol.*, *18*, C-103, p. 292 (1977).

115. L. A. Price, B. T. Hill, A. H. Calvert, et al., *Brit. Med. J.*, *3*, 10 (1975).

116. R. Wittes, K. Heller, V. Randolph, J. Howard, A. Vallejo, H. Farr, C. Harrold, F. Gerold, J. Shah, R. Spiro, and E. Strong, *Cancer Treat. Rep.*, Proc. NCI Conf. on Cis-platinum and Testicular Cancer, *63*, 1533 (1979).

117. W. K. Hong, R. Bhutani, S. Shapshay, M. L. Craft, A. Ucmakli, M. N. Snow, C. Vaughan, S. Strong, *Proc. Amer. Soc. Clin. Oncol.*, *19*, Abstr. C-59, p. 321 (1978).

118. T. Thigpen, H. Shingleton, H. Homesley, L. LaGasse, and J. Blessing, *Cancer Treat. Rep.*, Proc. NCI Conf. on Cis-platinum and Testicular Cancer, *63*, 1549 (1979).

119. A. Yagoda, *Cancer Treat. Rep.*, Proc. NCI Conf. on Cis-platinum and Testicular Cancer, *63*, 1565 (1979).

120. C. Merrin, *J. Urol.*, *119*, 493 (1978).

121. M. S. Soloway, *Proc. Amer. Assoc. Cancer Res. and ASCO*, *19*, Abstr. C-239, p. 366 (1978).

122. J. J. Sternberg, R. B. Bracken, P. B. Handel, and D. E. Johnson, *JAMA*, *238*, 2282 (1977).

123. S. D. Williams, J. P. Donohue, and L. H. Einhorn, *Cancer Treat. Rep.*, Proc. NCI Conf. on Cis-platinum and Testicular Cancer, *63*, 1573 (1979).

124. J. S. Sierocki, B. S. Hilaris, S. Hopfan, N. Martini, D. Barton, R. B. Golbey, and R. E. Wittes, *Cancer Treat. Rep.*, Proc. NCI Conf. on Cis-platinum and Testicular Cancer, *63*, 1593 (1979).

Chapter 4

CARCINOSTATIC COPPER COMPLEXES

David H. Petering
Department of Chemistry
University of Wiconsin-Milwaukee
Milwaukee, Wisconsin

1. INTRODUCTION 198
2. BIS(THIOSEMICARBAZONATO) COPPER COMPLEXES 199
   2.1. Antitumor Studies 199
   2.2. Activation of Bis(thiosemicarbazones) by Cu^{2+} . . . 202
   2.3. Physical and Chemical Properties 205
   2.4. Cellular Biochemistry 207
3. α-N-HETEROCYCLIC CARBOXALDEHYDE THIOSEMICARBAZONATO COPPER COMPLEXES 212
   3.1. Cytotoxic Properties 212
   3.2. Chemical Studies 214
   3.3. Cellular and Biochemical Studies 216
4. COPPER BLEOMYCIN 218
   4.1. Antitumor Studies 218
   4.2. Physical and Chemical Properties 220
   4.3. Biochemical and Cellular Reactions 223
ACKNOWLEDGMENTS 225
ABBREVIATIONS 225
REFERENCES 226

1. INTRODUCTION

For years in drug design, the focus of the medicinal chemist has been on organic compounds and natural products, with little attention being given to inorganic compounds such as metal complexes. Yet, as a class, metal complexes offer a remarkably rich chemistry which could be used to affect biological processes. That this area has not been explored in any breadth may be a reflection of the general lack of basic research into the pervasive role of essential metals in the biochemistry of living systems.

During the past decade, for the first time, transition metal complexes have been studied as cancer chemotherapeutic agents. The success of the cis-dichlorodiammine platinum complex in animals and humans has generated a major effort to discover other platinum complexes with significant pharmacological properties and to understand these compounds' mechanisms of reaction with biological systems. Secondarily, a number of investigations have begun into the use of other biologically unessential metals. A principal example has been that of the studies indicating marked antineoplastic activity of rhodium carboxylates [1]. The choice of rather exotic metals for this work may be due in part to an assumption that toxic, heavy metals are necessary if complexes are to be active. However, a variety of these metal complexes may share similar problems, such as acute renal toxicity and possible long-term accumulation in the organism, leading to dose-limiting side effects.

A complementary alternative is to consider the properties of complexes involving biologically essential metals such as copper, zinc, and iron. Since any essential metal which escapes its normal metabolic pathways can be very toxic to the organism (as for example, copper in Wilson's disease and iron in thalassemia patients), complexes of such metals may serve as effective cytotoxic agents [2,3]. Nevertheless, the fact that normal homeostatic mechanisms do exist for such metals suggests that long-term side effects may be avoided as the metal complexes break down and the metal ions interact with the organism.

Just as there are complexes of many metal ions which have demonstrated cytotoxic properties, so there may be a variety of general mechanisms by which they manifest their activities. The studies with the platinum complexes emphasize a binding reaction between cis-diammine-Pt and nitrogen of DNA pyrimidine and purine bases [4]. In fact, the ligands of these complexes have been rather simple in nature and generally serve the function of providing the metal complex with characteristics necessary for it to reach the site, and to react and bind. Other mechanisms can be envisioned. Kirschner synthesized a silver 6-mercapto purine complex which might deliver both a toxic metal and toxic antimetabolic ligand to the tumor cell [5]. Indeed, such a metal complex may alter the properties of the ligand to enhance its cytotoxicity, just as the ligand may provide the necessary chemical environment to activate the metal. Thus in the limiting case where interesting ligands are used in conjunction with metal, the metal complex may be a vehicle for activation of the ligand as the principal cytotoxic entity. These and other mechanistic possibilities appear as a result of examination of the mode of reaction of carcinostatic copper complexes, the subject of this chapter.

2. BIS(THIOSEMICARBAZONATO) COPPER COMPLEXES

2.1. Antitumor Studies

In their seminal work in this area, French and Freelander examined the antitumor activity of a large number of bis(thiosemicarbazones), whose generalized structure is shown in Fig. 1 [6,7]. They suggested that the structures possessed good ligands for the binding of transition metals and that a plausible mechanism of action might be the removal of essential metal ions from sites (MX) critical to tumor growth by such ligands as portrayed in Eq. (1).

$$H_2KTS + MX_{cell} \longrightarrow MKTS + H_2X \qquad (1)$$

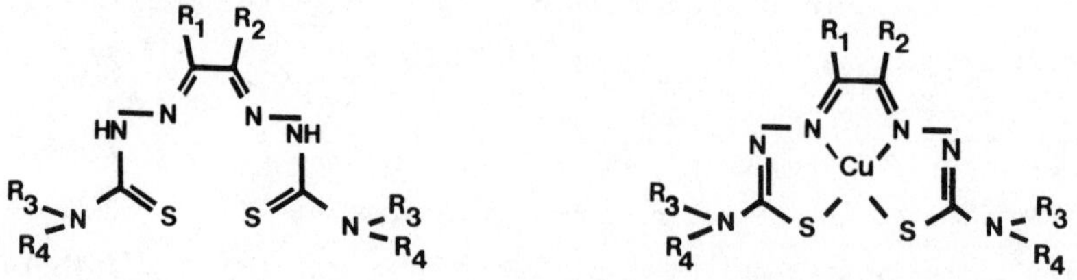

FIG. 1. General structures of bis(thiosemicarbazones) and their copper complexes. The principal compound, 3-ethoxy-2-oxobutyraldehyde bis(thiosemicarbazone) [left] is abbreviated as H_2KTS (R_1 = ethoxyethyl, R_2-R_4 = H) and its copper complex [right] as CuKTS.

Later H. G. Petering and coworkers [8-11] investigated in detail the antitumor activity of 3-ethoxy-2-oxobutyraldehydebis(thiosemicarbazone) and its related compounds. Their findings as well as those of other groups are listed in Table 1. Summarizing its salient features, H_2KTS causes complete regression of established Walker 256 and Walker 256 nitrogen mustard resistant carcinosarcomas, complete regression of the solid form of sarcoma 180, and has activity against a host of transplantable solid tumors, as well as the spontaneous mammary tumor of DBA_2 mice. It does not seem to be effective against leukemias, however. Based on its low host toxicity, Mihich and Nichol concluded that H_2KTS is a remarkably specific agent for tumors [12]. This compound significantly increases the number of animal tumor cures following surgical removal [13]. Of particular interest is the study showing that H_2KTS is not an immunosuppressive agent [14]. In this context, Ferrer and Mihich used thymectomized rats bearing sarcoma 180 tumors to demonstrate that the antitumor effect observed with this compound is due in part to host immune response to the neoplasm [15]. A recent review expands on this discussion [16].

TABLE 1

Antitumor Effects of H_2KTS and CuKTS

Rat tumors	Dose[a]	Schedule[b] (days)	Effect	Ref.
Walker 256 nitrogen-mustard-resistant	50 mg kg^{-1} intraperitoneal	9-22	85% inhibition of tumor growth (I)	8
	100 oral	9-22	85% I	8
	50 oral	5-16	100% complete regression of tumors (R)	8
	16 CuKTS oral	4-17	60% R	9
	12 oral 4 CuKTS	4-17	60% R	9
	25 oral (Cu-deficient diet)	5-17	0 I	10
	25 oral (Cu-deficient diet + 320 ppm Cu in water)	5-17	100% R	10
Walker 256	25 oral	4-15	50% R	11
Murphy-Sturm lymphosarcoma	50 oral	5-15	44% I	8
Jensen sarcoma	50 oral	9-22	54% I	8
Mouse tumors				
C3H mammary carcinoma	0.1% diet	4-20	44% I	11
Adenoma sarcoma 755	0.2% diet	2-8	62% I	12
Ridgeway osteosarcoma	0.2%	2-8	60% I	12
Mecca lymphosarcoma	100 mg kg^{-1}	2-8	30% I	12
Carcinoma 1025	100 mg kg^{-1}	2-8	58% I	12

TABLE 1 (continued)

Mouse tumors	Dose[a]	Schedule[b] (days)	Effect	Ref.
Sarcoma 180	0.1% diet	2-8	50% R (after 6 weeks)	12
	0.5% (pyridoxine-deficient diet)		83% R (after 6 weeks)	
Spontaneous mammary carcinoma (DMB_2 mice)	50 mg kg^{-1}	Treatment begun on day 3 of consecutive increment of growth	Survival time T/C = 46/20	12

[a] H_2KTS, unless indicated otherwise.
[b] Experiments begun on day 1 with injection of tumor cells. Drug given over the period indicated.

2.2. Activation of Bis(thiosemicarbazones) by Cu^{2+}

The question then arises as to whether or not H_2KTS acts according to the mechanism suggested by French. Again, H. G. Petering and colleagues carried out several definitive experiments strongly indicating that it is the copper complex CuKTS which is the active, cytotoxic species in vivo [10]. Figure 2 shows that although W256 tumors grow well in animals maintained on a copper-deficient diet, H_2KTS has no activity and only becomes active when Cu^{2+} is titrated back into the drinking water of the animals. In a comparative study, zinc deficiency has been shown to have no comparable effect upon the properties of H_2KTS [10]. This suggests that H_2KTS chelates copper in the organism, probably from the organism's diet, to become activated as described in Eqs. (2) and (3).

$$H_2KTS + Cu^{2+} \rightarrow CuKTS + 2H^+ \quad (2)$$

$$CuKTS + cell \rightarrow cytotoxic\ Rxn \quad (3)$$

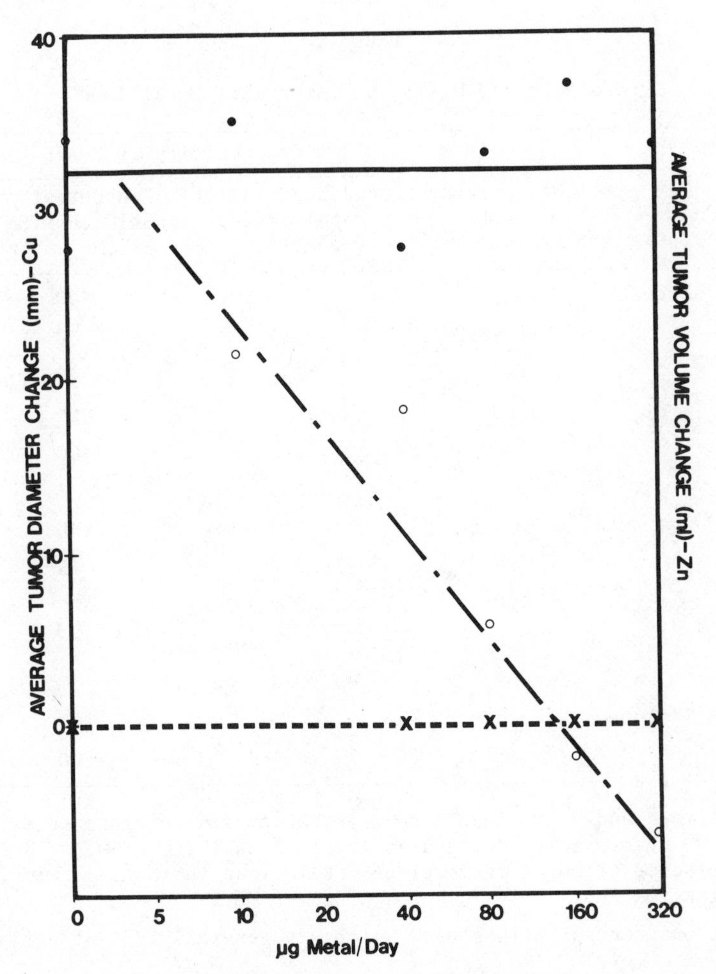

FIG. 2. Effect of nutrient copper and zinc upon antitumor activity of H_2KTS against established Walker 256 nitrogen-mustard-resistant tumors [10]. H_2KTS dosage was 25 mg kg^{-1} day^{-1}. Treatment periods were days 5-19 for (•) and (○) and 4-17 for (X). Tumors were measured on last day of treatment. (•) = controls fed copper-deficient diet plus H_2KTS; (○) = groups given copper-deficient diet, level of Cu^{2+} in water indicated and H_2KTS; (X) = animals provided with zinc-deficient diet, level of Zn^{2+} in the water indicated and H_2KTS.

TABLE 2

Cytotoxicity of H_2KTS in Presence of Metal Ions[a]

Metal ion	Cytotoxicity of metal ($\mu g\ ml^{-1}$)	Cytotoxicity of H_2KTS: Concentration of metal tested ($\mu g\ ml^{-1}$)[b]	Cytotoxicity of H_2KTS: Concentration of H_2KTS, H_2KTSM, or H_2KTSM_2 ($\mu g\ ml^{-1}$)[c]
H_2KTS alone			800
Cu^{2+}	1.6-3.2	0.4	0.8
Zn^{2+}	100	40	25
Co^{2+}	>200	40	>400
Fe^{2+}	200	40	>400
Cd^{2+}	>200	50	3.2
Hg^{2+}	1.6	0.4	>400
Au^{3+}	25	6.3	25
Ni^{2+}	100	20	200
H_2KTSM[d] alone			800
Cu^{2+}		0.4	1.6
H_2KTSM_2[d] alone			800
Cu^{2+}		0.4	>100

[a]Metal ions and H_2KTS incubated with Walker 256 carcinosarcoma cells. Cytotoxicity defined as level of H_2KTS, H_2KTSM, or H_2KTSM_2 in presence of metal or level of metal alone inhibiting tumor cell respiration 20%.

[b]Metal ion concentrations used which are generally 25% of defined cytotoxic level.

[c]Concentration H_2KTS, H_2KTSM, or H_2KTSM_2 in presence of metal giving defined cytotoxic response.

[d]H_2KTSM, R_1 = ethoxyethyl, R_2 and R_4 = H, R_3 = methyl; H_2KTSM_2, R_1 = ethoxyethyl, R_2 = H, R_3 and R_4 = methyl.

Source: Adapted from Ref. 17.

Table 1 indicates that CuKTS is also a very effective antitumor agent, although more toxic to the host. Finally, these workers developed a short-term in vitro assay for drug cytotoxicity which minimized the exposure of cells and drugs to extraneous metals in the growth medium [17]. Table 2 summarizes the ability of a number of metal ions to activate H_2KTS. Among essential metals, only Cu^{2+} is effective. However, as methyl groups are added to the thiosemicarbazone moieties, the effect is lost. It will also be noted that among the metals tested, copper was most effective and besides Cu^{2+}, only Cd^{2+} had a marked influence on H_2KTS activity. Another study by Booth and Sartorelli also shows that the ascites mouse sarcoma 180 was inhibited by $H_2KTS + Cu^{2+}$, but not by H_2KTS alone [18].

2.3. Physical and Chemical Properties

The crystal structure of CuKTS has been determined [19]. The in-plane ligands and copper form an approximately planar structure, as illustrated in Fig. 1. In the solid state, axial contacts occur between copper and sulfurs of other molecules lying above and below the plane of the thiosemicarbazonato-copper complex. Considering the R_1 substituent, the structural analysis shows that the ethoxyethyl group is large enough to occupy some space near the axial positions of the complex.

The electron paramegnetic spectrum of CuKTS is typical of copper complexes with $g_{\parallel}$ of 2.14, $A_{\parallel}$ of 560 mc sec^{-1} [20]. The $g_{\perp}$ resonance is split into nine lines by two approximately equivalent nitrogens bound to copper. Although an initial analysis of superhyperfine splitting in the $g_{\parallel}$ segment of the spectrum argued that the unpaired electron also interacts with the N-2 nitrogens of the thiosemicarbazone side chain, thus indicating a large delocalization of the electron onto the ligand, subsequent studies with ^{63}CuKTS showed that there were no further electron-nuclear interactions than observed in the $g_{\perp}$ region of the spectrum [21].

To survey the reactivity of CuKTS, the basic chemistry of the molecule was assumed to be described by the typical reactions of copper complexes, axial adduct formation with Lewis bases, ligand substitution reaction and redox reactions, as portrayed in Eqs. (4)-(6). The reactivity of the coordinated bis(thiosemicarbazone) ligand was not considered.

$$CuKTS + B \rightleftharpoons CuKTS \cdot B \qquad (4)$$

$$CuKTS + H_2X \rightleftharpoons CuX + H_2KTS \qquad (5)$$

$$Cu(II)KTS + X_{reduced} \rightleftharpoons Cu(I)KTS^{1-} + X^{1+}_{oxidized} \qquad (6)$$

There is no spectral evidence for significant adduct formation between CuKTS and Lewis bases [22]. Substitution reactions also seem unlikely, for CuKTS is a very stable complex at pH 7.4 with a conditional log stability constant of 18.40 [23]. Furthermore, reaction kinetics of CuKTS with ligands such as ethylenediamine are slow, so that Eq. (5), involving typical cellular ligands, is favorable neither thermodynamically nor kinetically [24].

Polarographic and potentiometric studies indicate that the complex undergoes a one-electron reversible reduction to a pale blue species with an E_o' of -120 mV at pH 6.6 and -170 mV at pH 9.45 [22]. This is certainly accessible to biological reducing agents. Because of the presence of thiols in all cells which have reduction potentials for the disulfide form at or below -200 mV, the reaction of CuKTS with sulfhydryl groups was investigated as a thermodynamically possible reaction [25].

The following general reaction between the copper complex and thiols has been defined, using dithiothreitol and glutathione as reducing agents:

$$Cu(II)KTS + 2RSH \rightarrow Cu(I)SR + \tfrac{1}{2}RSSR + H_2KTS \qquad (7)$$

The overall reaction has been described in terms of the mechanism

$$RS^{1-} + CuKTS \underset{k_2}{\overset{k_1}{\rightleftharpoons}} CuKTS \cdot RS^{1-} \qquad (8)$$

$$CuKTS \cdot RS^{1-} + RSH \xrightarrow{k_3} Cu(I)SR + HKTS^{1-} + RS^{\bullet} \qquad (9)$$

$$2RS^{\bullet} \rightarrow RSSR \qquad (10)$$

in which the second mole of thiol is the other sulfhydryl group in dithiothreitol or another molecule of glutathione [25]. The mechanism accounts for the second-order dependence of the reaction involving RSH, the first-order dependence dithiothreitol, the pH dependences of these reactions, and the rapid reaction of Cu(I)KTS with thiols.

The observed rate constant for the reaction of CuKTS with dithiothreitol is independent of the concentration of O_2 in solution. Although the mechanism need not be altered when O_2 is present, it is expected that the copper ion which is released in the reaction can serve as a redox catalyst for the oxidation of sulfhydryl groups by oxygen [26]. However, this aspect of the problem has not been examined.

The relative rate constants for reaction of a series of bis-(thiosemicarbazonato)copper(II) complexes differing in peripheral substitution with dithiothreitol are listed in Table 3, along with the half-wave reduction potentials for the complexes measured at pH 9.1 [24]. The difference in rate constants between complexes with similar R_2 and R_3 substitution is attributed to the steric hindrance which the R_1 ethoxyethyl contributes to the axial interaction of thiol and copper. Two linear free-energy relationships between log k/k_0 and $E_{1/2}$ are defined by the complexes containing the ethoxyethyl group or the groups methyl or hydrogen in the R_1 position: $\log k/k_0 = 20.90\ E_{1/2} + 3.76$ ($-CH(OC_2H_5)CH_3$) and $\log k/k_0 = 22.27\ E_{1/2} + 4.64$ ($-CH_3$ or $-H$). These reactions are described as inner-sphere electron transfers [24]. Furthermore, with the exception of glyoxal bis(thiosemicarbazonato)copper(II), the relative rates of thiol oxidation can be correlated with the in vitro cytotoxicity of the complexes set forth in Table 3. Those complexes which react much slower than CuKTS are considered inactive.

2.4. Cellular Biochemistry

Because of the intense charge-transfer absorption band at 469 nm as well as its characteristic epr spectrum, the reaction of CuKTS with cells can be studied in detail [27]. The complex is accumulated rapidly by Ehrlich cells and then reacts with cellular

TABLE 3

Comparative Properties of Copper Complexes

R_1	R_2	R_3	R_4	Relative in vitro cytotoxicity (Ref. 17)	$E_{1/2}$[a] (V)	log K (9.1)	k/k_0[b] (dtt)	k/k_0(en)[c]	k/k_0 cells (Ref. 30)
$CH(OEt)CH_3$	H	H	H	+	-0.178	22.15[d]	1.0	1.0	1.0
$CH(OEt)CH_3$	H	CH_3	H	+	-0.188	21.93[d]	0.84	0.17	0.43
$CH(OEt)CH_3$	H	C_2H_5	H		-0.195	22.30	0.43	0.091	
$CH(OEt)CH_3$	H	CH_3	CH_3	-	-0.283		0.007		0.0036
CH_3	H	H	H	+	-0.188	21.60	3.7	1.3	0.48
CH_3	H	CH_3	H	+	-0.208	21.79	3.7	0.35	0.52
CH_3	H	CH_3	CH_3	-	-0.278		0.015		0.0012
H	H	H	H	-	-0.098	20.23	165	>570	0.0132
CH_3	CH_3	H	H	-	-0.243	23.48	0.12	0.0022	

[a]Range (several runs) = ±0.010V.

[b]Ratio of pseudofirst-order rate constants when k_0 represents constant for CuKTS; ddt is dithiothreitol.

[c]En is ethylenediamine.

[d]Taken from Ref. 23.

constituents as shown by the first-order decay of the 469-nm band and the loss, over time, in intensity of the electron paramagnetic resonance spectrum. The ligand H_2KTS is released from the complex and equilibrates into the extracellular medium. All of the copper remains in the cell. Booth and Sartorelli also observed a net dissociation of metal and ligand upon reaction of CuKTS with sarcoma 180 cells in vivo. Similarly, ligand was lost from the cells and metal retained over a 24-hr period [28]. During the reaction of CuKTS with Ehrlich cells, and extending on over the course of about 60 min, the cells show a stimulated rate of oxygen consumption [29]. Thereafter the rate falls off slowly to less than 20% of the control level. According to analyses of homogenates of the cells after the reaction is complete, most of the copper is recovered as Cu(I) by chelation with bathocuproine disulfonate and the internal concentration of cellular thiols has been substantially depleted [27]. Furthermore, other agents which react with thiols, such as N-ethylmaleimide, inhibit the reaction of CuKTS with cells. Mild heating of Ehrlich cells to 50°C, which markedly reduces thiol content, also has the same effect. Therefore, Eq. 7 also represents a summary reaction which is consistent with the known information about the initial reaction of CuKTS with cells.

A kinetic analysis of this reaction shows it to be first-order in CuKTS and in cells [30]. When the series of bis(thiosemicarbazonato)copper complexes listed in Table 3 is reacted with Ehrlich cells and the logarithms of the relative observed rate constants, log k/k_0 cells are plotted versus $E_{1/2}$ of the complexes, a linear relationship is obtained. The slope is virtually the same as that generated in the plot of the log-relative rate constants for the reaction of these complexes with dithiothreitol as a function of $E_{1/2}$. This is secondary evidence that these complexes react initially with thiols. More importantly, it shows that the cellular chemistry is closely related to the inorganic properties of these complexes and again divides complexes into active and inactive species, according to previous studies with Walker 256 carcinosarcoma

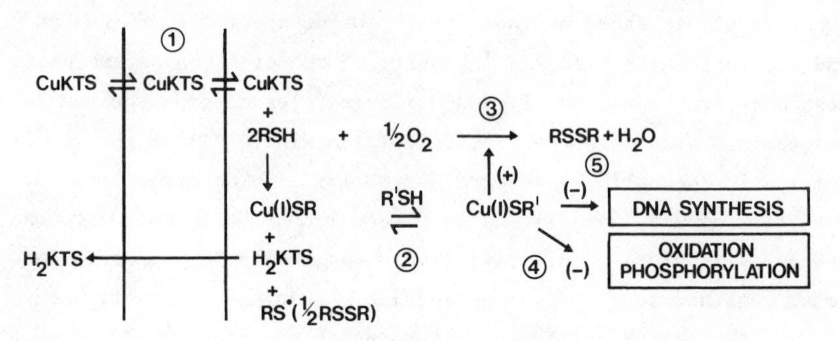

FIG. 3. Reaction of CuKTS with Ehrlich cells; (+) = reaction stimulated by Cu(I)SR'; (-) = processes inhibited by Cu(I)SR'.

cells (Table 2). It is noted that glyoxal bis(thiosemicarbazonato)-copper(II) does not react rapidly in the Ehrlich cell model, although it is rapidly reduced by dithiothreitol. The reason is that the complex is hardly taken up into their cells. Thus the lack of cytotoxicity of this material, according to Table 3, may be due to its inability to penetrate the cell. To be sure that these results could be transferred to Ehrlich cells, the cytotoxicity of CuKTS and $CuKTSM_2$ toward Ehrlich cells (incubated with these complexes and then injected into host mice) was compared [30]. At one-tenth the concentration of $CuKTSM_2$, CuKTS completely destroys the viability of the cells, while the methylated complex has no effect on cell growth.

Shown in Fig. 3 is a scheme which elaborates on the cellular nature of the reaction. CuKTS is rapidly taken up into cells in a nonrate-limiting reaction [Reaction (1)]. This is supported by temperature-dependence studies of the reaction, which reveal a break in the Arrhenius plot at 15°, indicative of a transition temperature below which the reaction rate may be limited by the rate of membrane transport [30]. Above this temperature, the energy of activation is virtually identical to that for the reaction of CuKTS with dithiothreitol. When reaction does not occur, as exemplified by the interaction of CuKTS with mildly heat-treated Ehrlich cells, the complex remains undissociated, locating in lipophilic regions of the cell (perhaps the outer membrane).

Secondly, once Cu(I)SR forms, it is likely that the copper ion is distributed about to a variety of cellular thiols as portrayed in Reaction (2). Certainly, copper seems evenly distributed in all cell fractions when an analysis is done on homogenates of Ehrlich cells treated with CuKTS [27]. Naturally, redistribution may be occurring during the manipulations of this work.

In agreement with this view is the finding that diverse biochemical processes are affected by these reactions [Reactions (3)-(5) of Fig. 3]. The cells show a stimulated oxygen uptake which extends well beyond the first period of reaction. This is not due to some uncoupling phenomenon involving CuKTS and Ehrlich mitochondria [29], but probably is an indication of an extended oxidation of thiols caused by oxygen catalyzed by copper that has been released from CuKTS. In addition, DNA synthesis is inhibited rapidly within 5 min of mixing of the complex with cells. There is also a much slower inhibition of mitochondrial respiration in the period following the reaction of CuKTS with cellular thiols [27,29].

What is the critical site or sites of reaction for the cytotoxic process? In experiments measuring the concentration dependence of cytotoxicity relative to the inhibition of DNA synthesis and mitochondrial oxidation phosphoxylation, it is clear that cell viability is completely destroyed at concentrations of CuKTS comparable to those necessary to affect DNA synthesis, but at an order of magnitude less than needed to depress cellular respiration significantly [30].

Whether DNA synthesis is the ultimate site of cytotoxicity has not been determined. Nor can one define the precise nature of the cytotoxic entity; it may consist of copper bound to the sensitive site or critical thiols oxidized to disulfides. Alternatively, reduced-oxygen radicals may be generated in abundance in the cells as copper ion serves as a redox catalyst for the oxidation of sulfhydryl groups of oxygen, and this is a reasonable possibility for the monothiosemicarbazonato-copper complexes examined in the next section.

3. α-N-HETEROCYCLIC CARBOXALDEHYDE THIOSEMICARBAZONATOCOPPER COMPLEXES

3.1. Cytotoxic Properties

There is now an extensive literature on the antitumor properties of many α-N-heterocyclic carboxaldehyde thiosemicarbazones having the generalized structures illustrated in Fig. 4 [31]. French and his colleagues showed that certain of these structures such as 1-formylisoquinoline thiosemicarbazone and 5-hydroxy-2-formylpyridine thiosemicarbazone had excellent antitumor activity in animal models [32,33]. Following the view developed with bis(thiosemicarbazones), French considered that these structures act as tridentate, NNS, ligands in vivo [34]. It was postulated that this ligand structure binds to iron at the active site of ribonucleotide reductase, a key enzyme in the synthesis of precursors of DNA, to cause cytotoxicity [35,36]. Experiments in animals and clinical trials in humans have documented that representative thiosemicarbazones mobilize large amounts of iron in the body which is excreted in urine, presumably as the iron complex [37-39]. Hence at least two species, ligand and iron complex, must be considered as possible cytotoxic species in vivo. In fact, one brief report has indicated that the iron complex of 1-formylisoquinoline thiosemicarbazone is two orders of

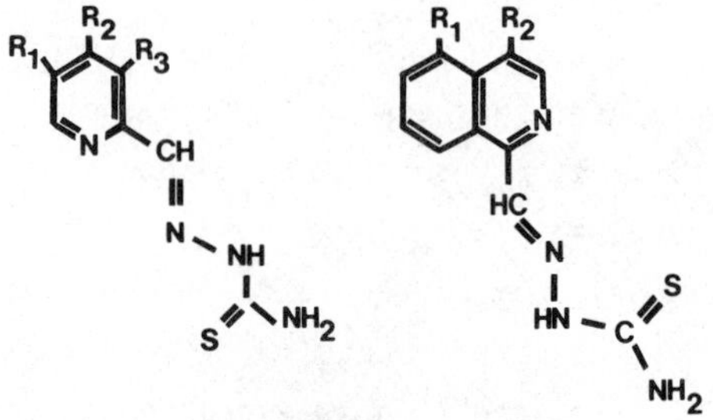

FIG. 4. Generalized thiosemicarbazone structure which contains variably substituted pyridine or isoquinoline rings.

TABLE 4

Cytotoxicity of Ligands and Metal Complexes

Drug	Cytotoxicity: Concentration (nmol mg^{-1})	Cytotoxicity: T/C[a]	50% inhibition of DNA synthesis (nmol mg^{-1})
HL[b]	1800	1.1	1.4
IQ-1	500	1.1	0.35
HL(5-OH)	200	1.1	
FeL_2	530	1.5	0.05
$Fe(IQ\text{-}1)_2$	360	1.6	0.09
$Fe[L(5\text{-}OH)]_2$	540	0.9	
CuL	50	1.6	0.7
Cu(IQ-1)	48	2.8	>0.6
Cu[L(5-OH)]	138	2.8	

[a] T/C = average survival of animal with treated cells/average survival of controls; C = 18 days. Experiments terminated at 50 days. T/C maximum = 2.8.

[b] Abbreviation: HL = 2-formylpyridine thiosemicarbazone; IQ-1 = 1-formylisoquinoline thiosemicarbazone; HL(5-OH) = 5-hydroxy-2-formylpyridine thiosemicarbazone.

Source: Adapted from Ref. 42.

magnitude more effective against the ascites form of sarcoma 180 than is the free ligand [40].

In order to inquire about the relationship of ligand and metal complexes to cytotoxicity, a number of thiosemicarbazones and their copper and iron complexes have been incubated for 1 hr with Ehrlich cells in a medium which minimizes the exposure of the ligands to metals and then injected into mice in order to assay the effects of these materials on cell viability [41,42]. According to Table 4, no ligands were active in this assay system, despite their accumulation by cells and complete ability to inhibit DNA synthesis at much lower drug concentrations. In contrast, some iron complexes and all copper complexes examined were found to be active in cell destruction, as well as in the inhibition of DNA synthesis [42]. Although innocuous in this 1-hr incubation procedure, it is possible that ligands might be effective during long, continuous exposures of cells. Nevertheless, this experiment demonstrates that copper and iron complexes are different than the ligands alone and may represent independent cytotoxic entities. The remainder of this discussion will consider only the copper complexes. The reader may refer to Ref. 43 for additional information about the iron complexes of these thiosemicarbazones.

3.2. Chemical Studies

Using Eqs. (1)-(3) as a general guide to the chemistry of copper complexes, some chemical properties have been defined for 2-formylpyridine thiosemicarbazonato Cu(II), designated CuL, and for a number of related chelates substituted on the five positions of the pyridine ring. The log stability constant of CuL is 16.90 [44]. In competition with H^+ at pH 7.4, the ligand binds Cu^{2+} with a log conditional stability constant of 13.37. In order to make these determinations, ethylenediamine (en) was used as a competitive ligand for copper to dissociate CuL. The details of the calculation of the stability constant and an epr analysis of the reaction of CuL with en both imply that CuL forms an adduct with en (Fig. 5).

The shift in the hyperfine lines in the $g_{||}$ part of the spectrum is a clear indication of the formation of a new structure, which is believed to have the en bound through an inplane nitrogen and perhaps an axial nitrogen. Thus, in contrast to CuKTS, adduct formation is a significant reaction with these complexes. The log stability constant for CuL•en is 5.53.

The half-wave reduction potential of CuL at pH 7 is +2 mV [44]. Accordingly, the redox reaction between CuL and thiols is thermodynamically favorable. The reaction stoichiometry follows Eq. (7) under anaerobic conditions. However, there are significant

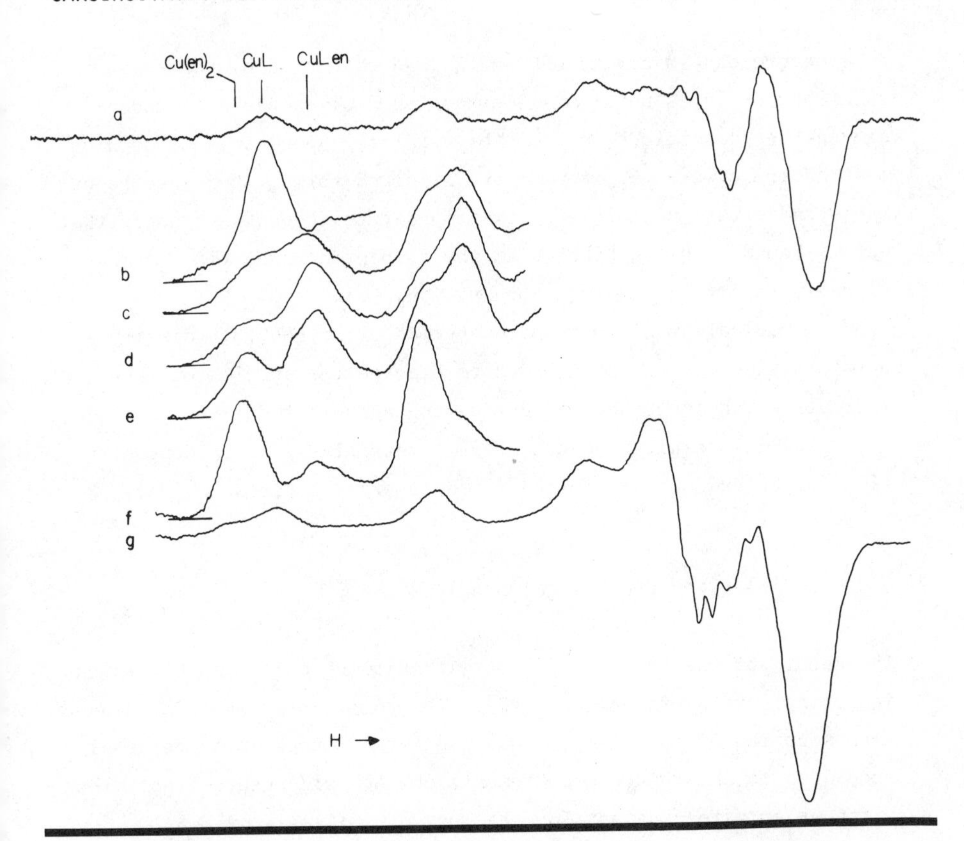

FIG. 5. Epr spectra of CuL^+ at 77K. Titration of 6.13×10^{-4} M CuL^+ at (a)-(f) pH 8.2 ± 0.1 and (g) pH 1.85 in 0.10 M KCl and 30% Me_2SO. Microwave power = 2 mW; microwave frequency = 9.147 GHz; modulation amplitude = 5 G; (b)-(f), x10 gain: (b) CuL^+; (c) CuL^+ + 1 en; (d) CuL^+ + 2 en; (e) CuL^+ + 10 en; (f) CuL^+ + 100 en. [Reproduced with permission of W. C. Antholine et al., Inorganic Chemistry *16*, 569 (1977). Copyright by American Chemical Society.]

differences in detail between the reaction of thiols with CuL and CuKTS. Utilizing glutathione as a model cellular thiol, the reduction of CuL under oxygen-free conditions is first-order in both complex and thiol with a second-order rate constant at pH 7 of 6.25 M^{-1} sec^{-1} [45]. When oxygen is present, initial rate data lead to a smaller rate constant, 1.95 M^{-1} sec^{-1}. However, the reaction does not proceed to completion, but reaches a steady state in which a redox process involving GSH and O_2 and catalyzed by copper in the solution continues until O_2 is exhausted. Given the data under oxygen-free conditions, one of the catalytic species is CuL, which must cycle between the reduced and oxidized forms. Epr spectra of the steady-state mixture at room temperature suggests strongly that CuL is bound to glutathione with the tripeptide supplying an in-plane sulfhydryl group for coordination.

A radical trap for reduced species of oxygen, 5,5-dimethyl-1-pyrolline-N-oxide, is being used to examine the species of free radicals which are generated during the aerobic redox reaction [46, 47]. Initial results indicate that comparatively large concentrations of O_2^- and $OH^{\bullet}$ are formed in the overall process.

3.3. Cellular and Biochemical Studies

The nature of the reaction of CuL with Ehrlich cells has been studied using several techniques [45]. The complex is readily taken up from solution by the cells. Subsequently CuL appears quite stable, according to absorbance measurements on the cell suspensions. The first-order rate constant for the loss of Cu(II)L is 4.5×10^{-5} sec^{-1} when 10^{-4} M CuL is incubated with 1-15 mg ml^{-1} cells. Electron paramagnetic resonance spectra of such suspensions at 77K also indicate that the complex is intact (Fig. 6). However, a new species dominates the spectrum; its epr parameters are distinctly different from CuL in solution, indicating that adduct formation has occurred. Not only has $g_{\|}$ shifted from 2.206 to 2.13 and $A_{\|}$ from 185 to 178 gauss, but $g_{\perp}$ is now clearly split by two approximately equivalent nitrogens. The magnitude of $g_{\|}$ and $A_{\|}$ are consistent with a N_2S_2 coordination site for the copper [49]. The nitrogens of the thiosemicarbazone are inequivalent in the planar structure in the presence of Lewis bases, which supply oxygen or nitrogen atoms for the vacant in-plane site [Fig. 7(a)]. However, when sulfur occupies the fourth site, the symmetry is raised and the splitting of $g_{\perp}$ seen [Fig. 7(b)]. The mixture of CuL and GSH yields a very similar epr spectral

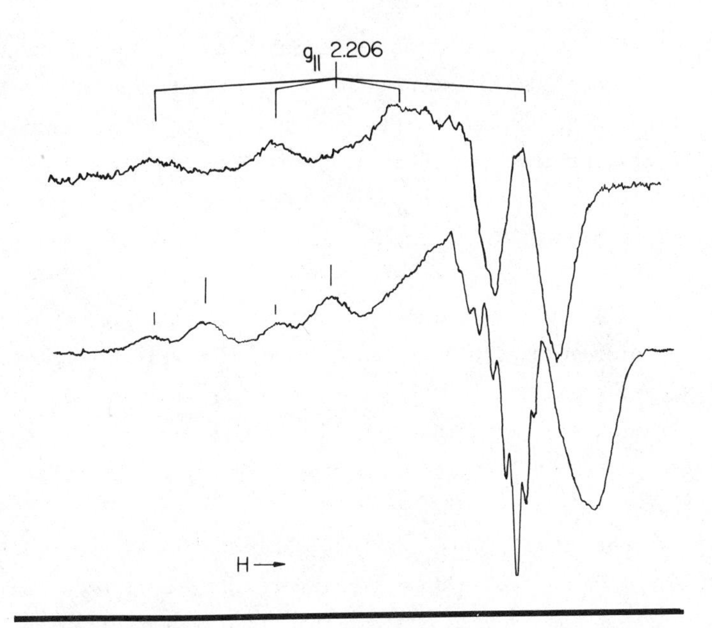

FIG. 6. Epr spectra of CuL. Top spectrum: CuL in aqueous solution. Lower spectrum: CuL in Ehrlich cells. Modulation frequency = 100 kHz; modulation amplitude = 8 G; microwave frequency = 9.108 GHz; temperature = 77K. (Reproduced from Ref. 48 with permission of Plenum Press.)

envelope at 77K, albeit without the distinct splitting in the $g_{\perp}$ region. Nevertheless, the room temperature spectrum does have this superhyperfine structure. It may well be, therefore, that CuL is binding glutathione or other thiols within the Ehrlich cell.

There is a marked stimulation of oxygen utilization of Ehrlich cells by CuL [45]. At the same time, the thiol content of the cells is decreasing, as was found with CuKTS. Except at very large

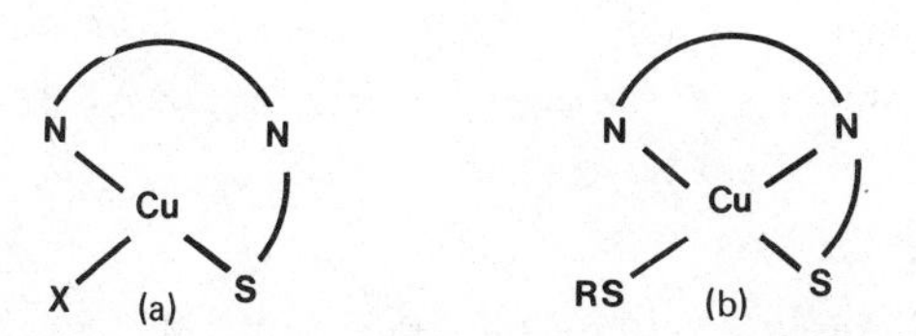

FIG. 7. In-plane coordination of copper.

concentrations of complex, there is no inhibition of oxygen consumption relative to a control suspension of cells not exposed to the agent. CuL does not stimulate electron transport in Ehrlich mitochondria, implying that the enhancement of O_2 consumption observed with whole cells is not due to a direct effect on mitochondria. Thus the properties of the reaction of CuL with Ehrlich cells are consistent with the oxidation of thiols by oxygen catalyzed by the complex. Similarly, the stimulation of oxygen consumption, depletion of cellular sulfhydryl groups, the apparent stability of CuL, and the adduct formation thought to involve CuL and a thiol donor are closely modeled by the reaction of CuL with GSH.

In many respects the initial interaction of CuL with Ehrlich cells mirrors the reaction involving CuKTS. The interesting difference is that very little CuL undergoes net reduction and dissociation, while CuKTS is broken down completely and the ligand released from the cells. Considering present knowledge, cytotoxicity from CuL results from thiol oxidation reactions involving superoxide ion and hydroxyl radical with cell constituents, or perhaps the binding of CuL to some critical site within the cell. What the molecular sites may be have not been determined. CuL does inhibit DNA synthesis at low concentration. However, inhibition is complete at concentrations less than one-tenth those necessary to cause a significant decrease in Ehrlich cell viability (Table 4).

4. COPPER BLEOMYCIN

4.1. Antitumor Studies

Bleomycin comprises a family of glycopeptides isolated from *Streptomyces verticillis* as a set of copper complexes (Fig. 8). The representatives shown are the principal components of the clinically used mixture. As currently employed, the copper has been removed from the antibiotic. As shown in the figure, the structure was revised very recently in a minor but important way [50]; the original

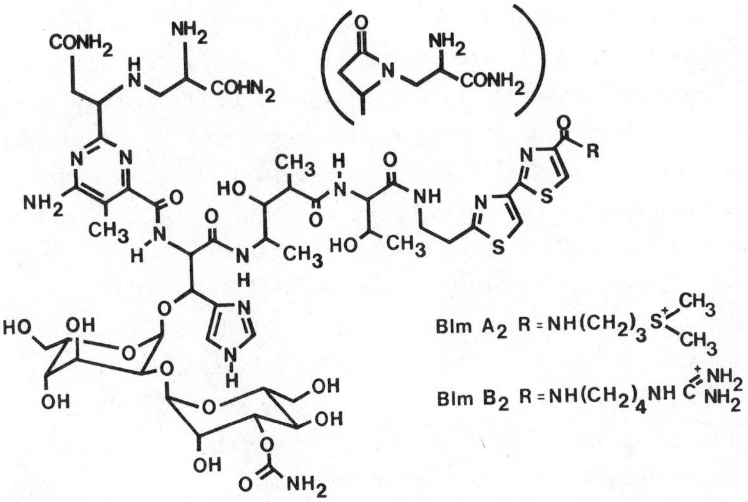

FIG. 8. Structure of bleomycin (H_4Blm).

version of the fragment is portrayed along side the revision [51]. Initial animal studies of the antitumor properties of bleomycin utilized the natural copper complexes of the mixture as well as individual, isolated components [52,53]. This work demonstrated that the copper complexes were effective against ascites and solid forms of Ehrlich carcinoma and sarcoma 180.

Later, the copper-free materials were also found to be active in animal models and to be less toxic to the host than the metal complex [53]. Following this discovery, virtually all of the antitumor studies on bleomycin have focused on the free ligands. Currently, bleomycin is used in a number of drug combination regimens for the treatment of human cancers [54].

This introduction leads to the same questions which were posed with the two classes of thiosemicarbazone ligands. Is bleomycin, itself, the proximate cytotoxic species in vivo, or must the natural product react with a metal (such as copper) to generate the active entity? Recent reports suggest that iron bleomycin, not the species containing copper, may be the activated form [55-58].

Using an incubation of Ehrlich cells with bleomycin or copper bleomycin under metal-free conditions as a way to avoid possible metal exchange or metal-binding processes, it has been found in initial studies that both the ligand and metal complex inhibit subsequent growth of the cells in animals or in cell culture [48,59,60]. The animal model shows CuBlm to be somewhat more active, while in cell culture it has been noted that the cytotoxicity of bleomycin, but not the copper complex, is inhibited if fetal calf serum is present during the incubation of cells with drugs [60]. Therefore, in contrast to the clear discrimination between ligands and complexes which this method of assay of cytotoxicity provides with the two classes of thiosemicarbazones discussed above, both bleomycin and copper bleomycin have activity according to this analysis. It is not yet resolved by these studies whether bleomycin chelates a metal within the Ehrlich cell to become activated or copper bleomycin is broken down intracellularly to yield a free, active bleomycin. Suzuki et al. found copper and metal-free bleomycin to have equivalent effects against HeLa cells in culture containing 10% fetal calf serum [61]. Bleomycin and copper bleomycin have also been incubated with Chinese hamster V79 cells containing 15% fetal calf serum and found that both agents similarly inhibited cell survival [62].

4.2. Physical and Chemical Properties

According to newly published conclusions of an x-ray diffraction study on crystals of the copper complex of a fragment of bleomycin, the copper-binding site has the structure drawn in Fig. 9 [50]. It appears as a distorted square pyramid with in-plane ligands comprised of a secondary amino group, the N-1 nitrogen of the substituded pyrimidine, N-1 of the imidazole, and a peptide bond nitrogen. Occupying an axial position is the primary amine group of the bleomycin molecule. This structure of a fragment of copper bleomycin is consistent with results accumulated on the whole molecule.

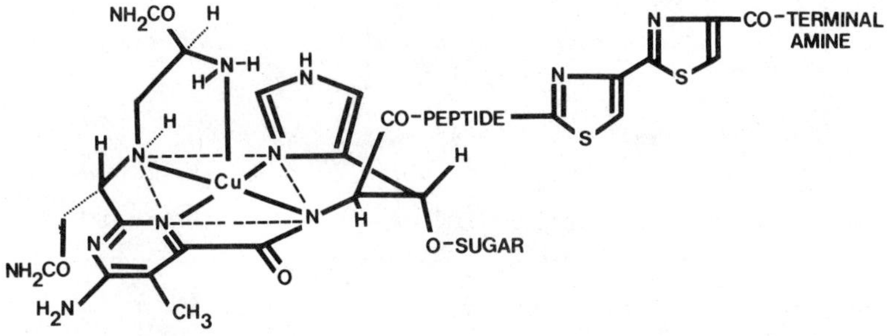

FIG. 9. Proposed structure of copper bleomycin (CuBlm).

Thus Dabrowiak and coworkers have reported the electron paramagnetic resonance spectral parameters for the copper complex as $g_{\parallel} = 2.203$, $A_{\parallel} = 0.0186\ cm^{-1}$, and $g_{\perp} = 2.058$ and have concluded that they are consistent with a structure of axial symmetry containing four in-plane nitrogen ligands bound to copper [63]. On the basis of changes in the ultraviolet spectrum of the ligand upon complexation, they also suggested that the pyrimidine, but not the bithiazole, moiety was part of the coordination site. This assignment was confirmed by 13-C nmr spectroscopy on CuBlm, in which all the ligand resonances eliminated by the paramagnetic line broadening by Cu^{2+} come from the pyrimidine-imidazole region of the structure [64].

Work in the author's laboratory indicates that neither the imidazole nor one of the amine groups is observed in potentiometric titration of the complex between pH 3 and 10, whereas pKa values of 4.93 and 7.50 are defined for the free ligand [65]. These have been previously assigned to the imidazole and primary amine groups, the imidazole on the basis of the pD dependence of proton nmr spectra, and the primary amine because, in the original structure, it was the strongest Lewis base [66]. The binding of a pyrimidine nitrogen is inferred from the quenching of the pyrimidine fluorescence emission by Cu^{2+} as complex formation occurs [65]. That a fourth group with a very large protonation constant is bound is consistent with the finding that one proton is released from bleomycin as the copper complex forms in the pH range 8.5-10, in which the pyrimidine, imidazole, and amine groups are essentially unprotonated [65]. Other data on the stability constant of copper bleomycin (to be discussed below) also support this assignment.

In order to dissociate CuBlm, the pH value must be decreased well below 1. The properties of the pH-dependent formation of CuBlm have been described [65]. Changes in the ultraviolet and d-d band regions of the electronic spectrum of CuBlm, and corresponding changes in its epr and fluorescence spectra support a two-step formation of the complex:

$$Cu^{2+} + H_4Blm \rightleftharpoons CuH_2Blm + 2H^+ \qquad \log K^{Cu}_{CuH_2Blm} = 1.32 \qquad (11)$$

$$CuH_2Blm \rightleftharpoons CuBlm + 2H^+ \qquad \log K^{CuH_2Blm}_{CuBlm} = -4.31 \qquad (12)$$

The intermediate complex, formulated as CuH_2Blm, has epr $g_{\|}$ and $A_{\|}$ values of 2.25 and 170 gauss, respectively, which is indicative of the formation of an N_2O_2 in-plane coordination for copper [49]. The binding of pyrimidine to copper occurs in the second step as seen by the pH dependence of the quenching of pyrimidine fluorescence, as well as by the finding that the shoulder in the ultraviolet spectrum assigned previously to a copper-pyrimidine interaction only develops in the second step [63].

Conditional stability constants for CuBlm over the pH range 2.94-5.90 have been determined by establishing measurable equilibrium between the complex and ligands such as EDTA, for which conditional formation constants can be calculated from literature data [65,67]. At pH 7, the extrapolated conditional formation constant is $10^{18.1}$ M^{-1} [60]. At present none of these results suggests the presence of an amine group bound axially to the copper center. It can be shown that these data are consistent with the binding site containing amine, imidazole, and pyrimidine groups as well as a site with an undefined, very large pKa from which protons dissociate upon complexation in the appropriate pH ranges. Clearly CuBlm is a complex with a very large thermodynamic stability.

A complement to the studies on the thermodynamic stability of Cu(II)Blm has been the measurement of the kinetics of several reactions of CuBlm [60,68]. The formation of the complex at pH 7.0 from Cu^{2+} and bleomycin occurs rapidly with a second-order rate constant of $3{\cdot}8 \times 10^5$ sec^{-1} M^{-1}. The site is relatively inert to substitution. At pH 7, the second-order rate constant for the reaction of EDTA with CuBlm is 6×10^{-3} M^{-1} sec^{-1}. The reaction of CuBlm with H_2KTS is considerably slower. From these studies it is apparent that biological reactions of the type

$$\mathrm{CuBlm} + \mathrm{H_2X_{cell}} \underset{k-1}{\overset{k_1}{\rightleftharpoons}} \mathrm{CuX_{cell}} + \mathrm{H_2Blm} \tag{13}$$

are not expected to occur either on a thermodynamic or kinetic basis.

Cu(II)Blm will react slowly with thiols in redox reactions, which have the same overall stoichiometry as seen with thiosemicarbazonato copper complexes [60,68]. The rate constants at pH 7 for the reaction of CuBlm with glutathione, cysteine, and 2-mercaptoethanol are 1.2×10^{-2} M^{-1} sec^{-1}, 1.2×10^{-1} M^{-1} sec^{-1}, and 9.5×10^{-3} M^{-1} sec^{-1}, respectively. The reactions are slow and there is no evidence either from oxygen consumption studies or the search for reduced, radical species of oxygen that the complex is acting as a catalyst for the oxidation of thiols by oxygen [60]. Hence, as in the analysis of the thiosemicarbazonato copper complexes, the chemical studies point to redox processes as reasonable reactions for CuBlm with biological systems.

4.3. Biochemical and Cellular Reactions

The study of the reaction of CuBlm with cells is still in its initial stages. It is known that bleomycin participates in the strand cleavage of DNA, showing specificity for reaction at thymine residues, and that reducing agents such as thiols greatly stimulate the reaction and that Cu^{2+} inhibits this reaction [69,70]. However, the chemistry of these interactions is still in question. Under the assumption that strand cleavage is the basic cytotoxic reaction in vivo, the formation of CuBlm in the host organism has been viewed as an adverse process [62]. Nevertheless, the tests of CuBlm as a cytotoxic agent have been uniformly positive [52,53,59-62]. Having shown that CuBlm reacts with homogenates of AH66 rat hepatoma ascites cells in a redox reaction involving thiols, Takahashi et al. argued that the transformations of bleomycin in vivo are as follows [70]:

$$\text{Bleomycin} + \text{Cu}^{2+}_{\text{host}} \rightarrow \text{CuBlm} \qquad (14)$$

$$\text{CuBlm}_{\text{out}} \rightarrow \text{CuBlm}_{\text{in cells}} \qquad (15)$$

$$\text{Cu(II)Blm} + 2\text{RSH} \rightarrow \text{Cu(I)SR} + \tfrac{1}{2}\text{RSSR} + \underset{\text{(cytotoxic species)}}{\text{H}_2\text{Blm}} \qquad (16)$$

Upon entering the organism, bleomycin binds available Cu^{2+} and enters tumor cells in this form. There it is reduced and dissociated, releasing bleomycin for the cytotoxic-strand cleavage reaction. In support of this interpretation, CuBlm is shown to cause DNA strand scission in whole cells, implying that free bleomycin is generated [70]. However, in another recent study the concentration dependence of CuBlm cytotoxicity is not correlated with strand cleavage [62]. That is, at concentrations reducing surviving cells by 99%, there is within error no increase in strand cleavage. Even with free bleomycin, the correlation between cytotoxicity and strand cleavage is poor.

If Reactions 14-16 do occur in organisms, then one must also consider that copper released from bleomycin in tumor cells can act as a major cytotoxic species, just as it does in the mechanism of action of CuKTS (Sec. 2). However, to provide strength for this hypothesis it must be shown that CuBlm is reductively dissociated in whole cells as well as in homogenates.

In summary, the role of copper in the pharmacology of bleomycin has not been resolved. Nevertheless, as the effort continues to enhance the activity of this drug, a reasonable, simple modification of bleomycin will be the formation of new transition metal complexes which can bring cancer cells into contact with a variety of metal centers that may exhibit cytotoxic properties.

ACKNOWLEDGMENTS

Current research included in this review is supported by U. S. National Institutes of Health Grants CA-16156 and CA-22184.

ABBREVIATIONS

CuBlm	copper bleomycin
Cu(IQ-1)	copper complex of IQ-1
CuKTS	copper complex of H_2KTS
CuL	copper complex of HL
CuL(5-x)	copper complex of HL(5-x)
EDTA	ethylenediamine tetraacetic acid
$Fe(IQ\text{-}1)_2$	iron complex of IQ-1
FeL_2	iron complex of HL
$Fe[L(5\text{-}x)]_2$	iron complex of HL(5-x)
GSH	reduced glutathione
H_xBlm	bleomycin with x-dissociable proteins
H_2KTS	3-ethoxy-2-oxobutyraldehyde bis(thiosemicarbazone)
H_2KTSM	3-ethoxy-2-oxobutyraldehyde bis(N^4-methylthiosemicarbazone)
H_2KTSM_2	3-ethoxy-2-oxobutyraldehyde bis(N^4-dimethylthiosemicarbazone)
HL	2-formylpyridine thiosemicarbazone
HL(5-x)	5-substituted-2-formylpyridine thiosemicarbazone
IQ-1	1-formylisoquinoline thiosemicarbazone
RSH	generalized thiol
RSSR	generalized disulfide

REFERENCES

1. A. Erck, L. Rainen, J. Whileyman, I. M. Chang, A. P. Kimball, and J. Bear, *Proc. Soc. Exp. Biol. Med.*, *145*, 1278 (1974).
2. H. Scheinberg, *Med. Clin. N. Amer.*, *60*, 705 (1976).
3. R. W. Grady, in Iron Metabolism and Thalassemia (D. Bergsma, A. Cerami, C. M. Peterson, and J. H. Graziano, eds.), Alan R. Liss, New York, 1976, pp. 161-176.
4. B. Rosenberg, in Inorganic and Nutritional Aspects of Cancer (G. N. Schrauzer, eg.), Plenum Press, New York, 1978, pp. 129-150.
5. S. Kirschner, Y. Wei, D. Francis, and J. G. Bergman, *J. Med. Chem.*, *9*, 369 (1966).
6. F. A. French and B. L. Freelander, *Cancer Res.*, *18*, 1298 (1958).
7. F. A. French and B. L. Freelander, *Cancer Res.*, *21*, 505 (1960).
8. H. G. Petering, H. H. Buskirk, and G. E. Underwood, *Cancer Res.*, *24*, 267 (1964).
9. J. A. Crim and H. G. Petering, *Cancer Res.*, *27*, 1278 (1967).
10. H. G. Petering, H. H. Buskirk, and J. A. Crim, *Cancer Res.*, *27*, 1115 (1967).
11. H. G. Petering, H. H. Buskirk, and J. A. Crim, *Cancer Chem. Ther. Rep.*, *50*, 557 (1966).
12. E. Mihich and C. A. Nichol, *Cancer Res.*, *25*, 1410 (1965).
13. D. S. Martin, *Proc. Amer. Assoc. Cancer Res.*, *4*, 42 (1962).
14. H. H. Buskirk, J. A. Crim, H. G. Petering, K. Merritt, and A. G. Johnson, *J. Nat. Cancer Inst.*, *34*, 747 (1965).
15. J. F. Ferrer and E. Mihich, *Proc. Soc. Exp. Biol. Med.*, *124*, 939 (1967).
16. D. H. Petering and H. G. Petering, in Handbook of Experimental Pharmacology, Vol. 38/2 (A. C. Sartorelli and D. G. Johns, eds.), Springer Verlag, New York, 197, pp. 841-849.
17. G. J. Van Giessen, J. A. Crim, D. H. Petering, and H. G. Petering, *J. Natl. Cancer Inst.*, *51*, 139 (1973).
18. B. A. Booth and A. C. Sartorelli, *Nature*, *210*, 104 (1966).
19. M. R. Taylor, J. P. Glusker, E. J. Gabe, and J. A. Minkin, *Bioinorg. Chem.*, *3*, 189 (1974).
20. W. E. Blumberg and J. Peisach, *J. Chem. Phys.*, *49*, 1793 (1968).
21. L. E. Warren, S. M. Horner, and W. E. Hatfield, *J. Amer. Chem. Soc.*, *94*, 6392 (1972).
22. D. H. Petering, *Bioinorg. Chem.*, *1*, 255 (1972).

23. D. H. Petering, *Biochem. Pharmacol.*, *23*, 567 (1974).

24. D. A. Winkelmann, Y. Bermke, and D. H. Petering, *Bioinorg. Chem.*, *3*, 261 (1974).

25. D. H. Petering, *Bioinorg. Chem.*, *1*, 273 (1972).

26. P. C. Jocelyn, The Biochemistry of the SH Group, Academic Press, New York, 1972, pp. 95-100.

27. D. T. Minkel and D. H. Petering, *Cancer Res.*, *38*, 117 (1978).

28. B. A. Booth and A. C. Sartorelli, *Mol. Pharmacol.*, *3*, 290 (1967).

29. C. H. Chan-Stier, D. Minkel, and D. H. Petering, *Bioinorg. Chem.*, *6*, 206 (1976).

30. D. T. Minkel, L. A. Saryan, and D. H. Petering, *Cancer Res.*, *38*, 124 (1978).

31. K. C. Agrawal and A. C. Sartorelli, in Handbook of Experimental Pharmacology, Vol. 38/2 (A. C. Sartorelli and D. G. Johns, eds.), Springer Verlag, New York, 1975, pp. 793-807.

32. F. A. French and E. J. Blanz, Jr., *Cancer Res.*, *25*, 1454 (1965).

33. E. J. Blanz, Jr. and F. A. French, *Cancer Res.*, *28*, 2419 (1965).

34. F. A. French and E. J. Blanz, Jr., *J. Med. Chem.*, *9*, 585 (1966).

35. F. A. French, E. J. Blanz, Jr., J. R. DoAmaral and D. A. French, *J. Med. Chem.*, *13*, 1117 (1970).

36. A. C. Sartorelli, K. C. Agrawal, and E. C. Moore, *Biochem. Pharmacol.*, *20*, 3119 (1971).

37. F. A. French, A. E. Lewis, A. H. Sheena, and E. J. Blanz, Jr., *Fed. Proc.*, *24*, 402 (1965).

38. R. C. DeConte, B. R. Toftness, K. C. Agrawal, R. Tomchick, J. A. R. Mead, J. R. Bertino, A. C. Sartorelli, and W. A. Creasey, *Cancer Res.*, *32*, 1455 (1972).

39. I. H. Krakoff, E. Etcubanas, C. Tan, K. Mayer, V. Bethune, and J. H. Burchenal, *Cancer Chemother. Rep.*, *Part 1*, *58*, 207 (1974).

40. K. C. Agrawal, B. A. Booth, E. C. Moore, and A. C. Sartorelli, *Proc. Amer. Assoc. Cancer Res.*, *15*, 289 (1974).

41. W. E. Antholine, J. M. Knight, and D. H. Petering, *J. Med. Chem.*, *19*, 339 (1976).

42. L. A. Saryan, E. Ankel, C. Krishnamurti, and D. H. Petering, *J. Med. Chem.*, *22*, 1218 (1979).

43. W. Antholine, J. Knight, H. Whelan, and D. H. Petering, *Mol. Pharmacol.*, *13*, 89 (1977).

44. W. E. Antholine, J. M. Knight, and D. H. Petering, *Inorg. Chem.*, *16*, 569 (1977).

45. L. A. Saryan, K. Mailer, C. Krishnamurti, W. E. Antholine, and D. H. Petering, submitted to *Biochem. Pharmacol.*

46. J. R. Harbour, V. Chow, and J. R. Bolton, *Can. J. Chem.*, *52*, 3549 (1974).

47. W. E. Antholine and D. H. Petering, unpublished data.

48. D. H. Petering, in Inorganic and Nutritional Aspects of Cancer (G. N. Schrauzer, ed.), Plenum Press, New York, 1978, pp. 179-189.

49. J. Peisach and W. E. Blumberg, *Mol. Pharmacol.*, *5*, 200 (1969).

50. T. Takita, Y. Muraoka, T. Nakatani, A. Fujii, Y. Umezawa, H. Naganawa, and H. Umezawa, *J. Antibiot.*, *31*, 801 (1978).

51. H. Umezawa, *Fed. Proc.*, *33*, 2296 (1974).

52. M. Ishizuka, H. Takayama, T. Takeuchi, and H. Umezawa, *J. Antibiot., Ser. A.*, *20*, 15 (1967).

53. H. Umezawa, M. Ishizuka, K. Kimura, J. Iwanaga, and T. Takeuchi, *J. Antibiot.*, *21*, 592 (1968).

54. P. Pietsch, in Handbook of Experimental Pharmacology, Vol. 38/2 (A. C. Sartorelli and D. B. Johns, eds.), Springer-Verlag, New York, 1975, pp. 850-876.

55. R. Ishida and T. Takahashi, *Biochem. Biophys. Res. Commun.*, *66*, 1432 (1975).

56. E. A. Sausville, J. Peisach, and S. B. Horwitz, *Biochem. Biophys. Res. Commun.*, *73*, 814 (1976).

57. E. A. Sausville, J. Peisach, and S. B. Horwitz, *Biochem.*, *17*, 2740 (1978).

58. E. A. Sausville, R. W. Stein, J. Peisach, and S. B. Horwitz, *Biochem.*, *17*, 2746 (1978).

59. D. Solaiman, W. E. Antholine, L. A. Saryan, and D. H. Petering, *Lloydia*, *39*, 470 (1976).

60. D. Solaiman, E. A. Rao, D. H. Petering, R. C. Sealy, and W. E. Antholine, *Int. J. Rad. Oncol. Biol. Phys.*, *5*, 1519 (1979).

61. G. Suzuki, K. Nagai, H. Yamaki, N. Tanaka, and H. Umezawa, *J. Antibiot.*, *21*, 379 (1968).

62. A. D. Nunn and J. Lunec, *Eur. J. Cancer*, *14*, 857 (1978).

63. J. C. Dabrowiak, F. T. Greenaway, W. E. Longo, M. V. Husen, and S. T. Crooke, *Biochim. Biophys. Acta*, *517*, 517 (1978).

64. J. C. Dabrowiak, F. T. Greenaway, and R. Grulich, *Biochem.*, *17*, 4090 (1978).

65. D. Solaiman, E. A. Rao, W. E. Antholine, and D. H. Petering, *Bioinorg. Chem.*, in press.

66. D. M. Chen, B. I. Hawkins, and J. D. Glickman, *Biochem.*, *16*, 2731 (1977).

67. L. G. Sillén and A. E. Martell, in Stability Constants of Metal-Ion Complexes, The Chemical Society, Burlington House, London, 1964.

68. D. Solaiman, W. E. Antholine, and D. H. Petering, submitted to *Bioinorg. Chem.*

69. W. E. G. Müller and R. K. Zahn, *Prog. Nucl. Acid Res. Mol. Biol.*, *20*, 21 (1977).

70. K. Takahashi, O. Yoshioka, A. Matsuda, and H. Umezawa, *J. Antibiot.*, *30*, 861 (1977).

Chapter 5

ONCOLOGICAL IMPLICATIONS OF THE CHEMISTRY OF RUTHENIUM

Michael J. Clarke
Department of Chemistry
Boston College
Chestnut Hill, Massachusetts

1. INTRODUCTION 232

2. CHEMICAL PROPERTIES OF RUTHENIUM COMPLEXES PERTINENT TO THE DEVELOPMENT OF ANTICANCER AGENTS 233

2.1. General 233

2.2. Differences in the Affinities of Ru(II) and Ru(III) for Various Ligands 234

2.3. Ligand Substitution Reactions 236

2.4. Electrochemistry 239

2.5. Electron Transfer Reactions 240

3. RUTHENIUM(III) COMPLEXES AS ONCOSTATIC PRODRUGS 241

4. COORDINATION OF RU(II) AND RU(III) AMMINE COMPLEXES TO BIOLOGICALLY IMPORTANT MOLECULES AND FUNCTIONAL GROUPS 244

4.1. Synthetic Methods 244

4.2. Amino Acid Complexes 244

4.3. Pyridine Complexes 246

4.4. Nucleoside and Nucleotide Complexes 246

4.5. Effects of Metal Complexation of Nucleosides and Nucleotides 250

4.6. Migration of the Metal Ion on Purines 254

4.7. Coenzyme Complexes 255

5. COMPLEXES WITH NUCLEIC ACIDS 256
5.1. Binding Sites 256
5.2. Coordination Under Equilibrium Conditions 257
5.3. Possible Effects of Ruthenium Binding to Nucleic Acids 257
6. DISPOSITION OF RUTHENIUM COMPLEXES IN LIVING ORGANISMS 259
6.1. Modes of Ru(III) Reduction by Cellular Components 259
6.2. Tissue Distribution of Ruthenium Trichloride and Nitrosyl Complexes 261
6.3. Biological Studies with Ruthenium Red 263
6.4. Oncological Properties of Ruthenium Coordination Compounds 265
6.5. Oncological Properties of Organoruthenium Complexes 271
7. CONCLUSION 273
ACKNOWLEDGMENTS 274
ABBREVIATIONS 275
REFERENCES 276

1. INTRODUCTION

While the use of metal salts in medicine had been indicated as early as Hippocrates [1], in 1965 Dwyer and coworkers were still substantially ahead of their time in suggesting that chelates of ruthenium might function as oncostatic and viruscidal agents [2,3]. Later work in this area has been pursued by several research efforts scattered over the globe [4-7]; however, it was not until recently that a concerted, rational approach to the design of anticancer pharmaceuticals had been made with a firm basis in both tumor metabolism and the unique chemistry of ruthenium. Moreover, little advantage has been taken from the numerous studies of the fate of radioruthenium (which is prevalent in the wastes from nuclear reactors) in animals [8-23]. Recent interest on the distribution of ruthenium complexes in the body has centered on their potential use as tumor-specific radioscintigraphic agents, since some ruthenium isotopes possess good-to-excellent radiophysical properties for organ imaging [24-29].

It is the purpose of this chapter to outline those aspects of ruthenium chemistry which provide the basis for the design of anticancer chemotherapeutic and radiodiagnostic agents and to critically review the biomedical work already done in these areas. Finally, it should also be noted that many metal ions, including some of those used as oncostatic drugs, are mutagenic [30-36] and the chemical interactions of ruthenium complexes with nucleic acids indicate possible mechanisms for metal-induced mutagenesis.

2. CHEMICAL PROPERTIES OF RUTHENIUM COMPLEXES PERTINENT TO THE DEVELOPMENT OF ANTICANCER AGENTS

2.1. General

The chemistry of ruthenium has been under investigation since its discovery in 1844 by Klaus, who named the element after a district in Russia [37]. Although rare, ruthenium is the least expensive of the platinum group metals and is commercially available in several forms. Most syntheses begin with $[(NH_3)_6Ru]Cl_3$ or $RuCl_3 \cdot 3H_2O$, which is a soluble mixture of Ru(III) and Ru(IV) complexes that can be readily converted entirely to the Ru(III) form by refluxing in ethanol [37]. In aqueous solution the Ru(II) and Ru(III) oxidation states predominate, although Ru(IV) is also accessible [37-38]. Ruthenium (II) and Ru(III) complexes are almost invariably six-coordinate, low-spin complexes of octahedral geometry. As such, the three (t_{2g}) d orbitals oriented toward the faces of the octahedron are fully populated in Ru(II) complexes and have a single electronic vacancy in Ru(III) species [39,40]. The difference of a single electron between the two oxidation states significantly affects their chemistry, so that Ru(II) usually behaves as a relatively "soft" [41], π-donor metal ion [42-44] and Ru(III) as a borderline "hard" metal ion, which can function as a π-acceptor of ligand electron density [45].

The stable complexes of ruthenium in aqueous solution are dominated by those of nitrogen-containing ligands, particularly those involving nitrosyls [46], ammines, organic amines [39,42,47], imines [48], and chelating agents [49,50]. A number of ligand-bridged, dinuclear species of ruthenium are known [51,52] and several organometallic complexes also persist in aqueous solution [53-59].

2.2. Differences in the Affinities of Ru(II) and Ru(III) for Various Ligands

The fully populated (t_{2g}^6) d orbitals of Ru(II) extend relatively far out into space and so allow this ion to enter into strong back-bonding interactions with common π-acceptor ligands such as CO [60], NO^+ [46,61], CN^-, and N_2 [39,42,44]. Retrodative bonding also occurs with heterocyclic aromatic ligands such as pyridines [43,62], pyrimidines [43,63], purines [63-67], and isoalloxazines [68] and appears to add about 5 kcal to the strength of the Ru-N bond in the case of a pyridine nitrogen [69,70]. Interactions of this type also account for the short Ru-N bond distances found in ruthenium(II) ammine complexes with pyrazine (1.980 Å) [71], isonicotinamide (2.058 Å) [72], and isoalloxazine (1.980 Å) [68] relative to that for an ammonia nitrogen (2.12 Å) [71-73]. Coordination of a π-acceptor ligand usually decreases the affinity of Ru(II) toward other such ligands [69,45,74,75], but can enhance its affinity for acido ligands.

Ruthenium(II) complexes with π-acid ligands invariably exhibit intense metal-to-ligand charge transfer transitions in their optical spectra, which provide convenient spectroscopic probes for quantitating these species in reactivity studies. The energies of these transitions correlate with the reduction potential of the ligand for a closely related series of ligands [43,69].

The higher charge and contracted size of Ru(III) relative to Ru(II) cause this ion to have a much higher affinity for anionic, less polarizable ligands such as chloride and anionic oxygen [76]. However, since Ru(III) possesses a partially empty d_π orbital, it may function more readily as a π-acceptor of electron density from fairly polarizable, π-donor ligands such as $R\text{-}S^-$ and imidazole [45, 74]. Table 1 summarizes the relative affinities of the two ions for a variety of ligands, when the metal is also complexed with five ammonia groups. Ruthenium(III) complexes of this type usually show strong ligand-to-metal charge transfer (LMCT) transitions corresponding to the excitation of an electron localized in a nonbonding or π-molecular orbital on the ligand to the partially empty d_π orbital on the metal [63,67,77].

TABLE 1

Comparison of Ligand Affinities for $[Ru(NH_3)_5H_2O]^{2+}$ and $[Ru(NH_3)_5H_2O]^{3+}$

Ligands	$[Ru(NH_3)_5H_2O]^{2+}$	$[Ru(NH_3)_5H_2O]^{3+}$	Ratio
N_2	3.3×10^4	4×10^{-13}	8×10^{16}
Thiophene	~ 10	$\sim 2.5 \times 10^{-8}$	4×10^8
H_2S	1.5×10^3	2.4×10^{-4}	6×10^6
$(CH_3)_2S$	>10	$>1.6 \times 10^{-2}$	$\sim 6 \times 10^6$
Pyridine	2.4×10^7	6×10^3	4×10^3
Imidazole	2.8×10^6	1.9×10^6	1.5
H_2O	1	1	1
NH_3	3.5×10^4	1.6×10^5	2×10^{-1}
Cl^-	0.4	1.1×10^2	4×10^{-1}
HS^-	1.5×10^6	2.4×10^{13}	6×10^{-3}
HO^-	6×10^2	6×10^{11}	1×10^{-9}

Source: Reprinted with permission from Ref. 45. Copyright by the American Chemical Society.

Ion pairs are also possible in solution between the usually cationic ruthenium complexes and anionic species. Equilibrium constants for ion-pair complexes of $(NH_3)_6Ru(III)$ and various halides have been determined to be on the order of 10 [78]. Ion pairing allows metal complexes to be solubilized in nonaqueous solvents of low dielectric constant [79], and so may affect membrane permeability. Association of the $(phen)_3Ru(II)$ complex with a variety of counter-ions is high both in water and a range of nonaqueous solvents, so that this species is readily extracted into organic phases [80].

2.3. Ligand Substitution Reactions

Ligand substitution reactions of Ru(II) ammine complexes generally occur by a dissociative (SN_1) mechanism proceeding through a five-coordinate intermediate [70,81-84]. The rate of aquation of $(NH_3)_5RuCl^{2+}$ to form $(NH_3)_5Ru(OH_2)^{2+}$ has variously been measured to be between 4.7 and 10 sec^{-1} and that for the corresponding bromo complex is also in this range [85-88]. The water exchange rate for $(NH_3)_5Ru(OH_2)^{2+}$ has been estimated at 3-10 sec^{-1} and most substitution rates of nitrogen ligands onto $(NH_3)_5Ru(II)$ are limited by this rate [70,81]. The loss of ligands from Ru(II) ions can also be acid-catalyzed via an associative (SN_2) attack of a proton onto a d_π orbital extruding through the octahedral face of the metal ion to form a seven-coordinate intermediate [89]. The loss of ammonia from $(NH_3)_6Ru(II)$ proceeds through acid-dependent and acid-independent pathways, so that the half-life for aquation in 1 M acid is approximately 1.5 hr while at neutral pH it is about 1 day [70,89].

The rates of nitrogen ligand substitution onto a variety of ruthenium(II) ammine complexes have been measured by several workers [70,81-84], leading to the conclusion that such processes usually proceed with stereospecific retention of the ligand configuration. Interestingly, the substitution of pyridine for ammonia in (pyr)$(NH_3)_5$Ru(II) proceeds stereospecificially to yield cis-$(pyr)_2$ $(NH_3)_4$Ru(II) [90].

The presence of a π-acceptor ligand usually decreases the substitution rate on Ru(II) ammine complexes and so allows control of ligand replacement. Marchant, Matsubara, and Ford [84] have recently shown that the rate of chloride loss in $L(NH_3)_4RuCl^+$ changes from 5.1 sec^{-1} when L = NH_3, to 0.1 sec^{-1} when L = acetonitrile. This effect appears to be due to delocalization of electron density from the metal ion through d_π-π^* overlap, so that the effective charge on the metal is enhanced, allowing it to function as a stronger Lewis acid toward σ-bond formation. Electron-withdrawing effects also affect the reduction potential of the metal center and Isied and Taube [91] have shown a free energy relationship between the Ru(III-II) reduction potential and the rate of substitution of isonicotinamide onto a series of cis- and trans-$L(H_2O)(NH_3)_4$Ru(II) complexes, where L is a π-acceptor ligand. For example, when L = dinitrogen (E^o = 1.1 V) the substitution rate constant is 1.2×10^{-5} M^{-1} sec^{-1}, but with L = imidazole (E^o = 0.1 V) the substitution rate is more rapid by a factor of 2.5×10^4.

Use of trans-labilizing ligands such as sulfito [74,92] and carbon-bound imidazole [66,93] can drastically enhance ligand substitution rates on Ru(II) ammine complexes. Although the factors affecting trans-labilization in this type of complex are not entirely clear, there is a correlation with the σ-donar ability of the labilizing ligand. Table 2 summarizes the effects of some labilizing and delabilizing ligands on Ru(II). Note that by changing a single ligand, the substitution rate can be varied over a range of 10^8.

The kinetic stability of ruthenium complexes for particular ligands is a function of the metal oxidation state, as well as the nature of the ligands. For example, $N_2(NH_3)_5$Ru(II) and $Cl(NH_3)_5$-Ru(III) aquate at rates on the order of 10^{-6} sec^{-1} at neutral pH [45,94-96]; however, when the oxidation states for these two species are switched, aquation proceeds approximately a million times more

TABLE 2

Substitution Rates of Isonicotinamide onto tr-$(H_2O)L(NH_3)_4Ru(II)$

Ligand	$k(M^{-1} sec^{-1})$
CO	3×10^{-6}
Isonicotinamide	5×10^{-3}
Pyridine	1×10^{-2}
OH^-	0.5
NH_3	1.0
CN^-	7
SO_3^{2-}	25
Imidazolylidene	60

Source: Reprinted with permission from Ref. 91. Copyright by the American Chemical Society.

rapidly [39,85]. In general, pentaammineruthenium(II) exhibits considerable lability for relatively "hard," nonpolarizable ligands such as halides and carboxylates, but forms substitution-inert bonds with most nitrogen, and some sulfur, ligands [39,45]. On the other hand, the pentaammineruthenium(III) ion usually forms kinetically stable complexes with both acido and nitrogen ligands [76].

In general, substitution reactions of Ru(III) ammine complexes are much slower than for the corresponding Ru(II) species, as indicated by the 1.6 year half-life for the uncatalyzed loss of ammonia from $(NH_3)_6Ru(III)$ [97], while that for the corresponding Ru(II) species is about 1 day [89]. The former reaction probably proceeds via base-dependent (SN1CB) and base-independent pathways, as does the loss of chloride from $Cl(NH_3)_5Ru(III)$ [94,95]. The half-life for the loss of chloride ion from $Cl(NH_3)_5Ru(III)$ under physiological conditions can be estimated at 2.5 days, while the loss of bromide from the corresponding bromo complex proceeds somewhat more rapidly. The aquation of cis-$Br_2(NH_3)_4Ru(III)$ appears to take place via an associative (SN_2) mechanism involving front-side attack so that the stereochemistry of the complex is maintained. This step has been shown to be rate limiting in the exchange of radiolabeled bromide ion onto the complex [98]. Computer stimulation has indicated that both associative and dissociative mechanisms are involved in the aquation of $X(NH_3)_5Ru(III)$, where X = monohalogenoacetate ions [99]. Trans-labilizing ligands such as imidazolylidene derivatives [66,93] are effective in increasing the rate of ligand substitution on Ru(III) ammine complexes.

Recent work with Ru(III)-EDTA complexes indicates that a deprotonated, pendant ligand may facilitate substitution at an open coordination position by intramolecular attack to form a seven-coordinate intermediate. This effect is not present with the analogous Ru(II) complex, since this ion has a lower affinity for the pendant EDTA carboxylate [50].

2.4. Electrochemistry

Reduction potentials for a number of ruthenium complexes have been measured, and these studies indicate that the value for the Ru(III-II) couple is a strong function of the ligand environment [45,64-68, 83,84,100]. In general, the rules governing the redox potentials of metal ions in solution hold for ruthenium complexes [101]; however, the effect of back donation of electron density is somewhat more prominent in stabilizing the lower valence state of ruthenium than for most other metals. The coordination of π-acceptor ligands such as carbonyl, pyridine, or dinitrogen increases the reduction potential, while the addition of anionic or good electron-donor ligands causes it to decrease. Table 3 serves to illustrate the wide range of reduction potentials that are possible by varying only a single ligand on the Ru(III) ion.

TABLE 3

Reduction Potentials for Selected $L(NH_3)_5Ru(III)$ Complexes

Ligand	E^o (V)
OH^-	-0.42
Cl^-	-0.042
NH_3	0.051
Pyridine	0.305
Thiophene	0.61
N_2	1.10

Source: Reprinted with permission from Refs. 45 and 88. Copyright by the American Chemical Society.

2.5. Electron Transfer Reactions

Electron transfer reactions involving ruthenium ammine complexes have been extensively studied and have been shown to proceed by both inner- and outer-sphere mechanisms [76,102-110]. Redox reactions involving the $(NH_3)_6Ru(II-III)$ couple usually proceed in an outer-sphere fashion, since the ammonia ligands are firmly bound in both oxidation states and do not serve as facile conduits for electron transfer. Since the Franck-Condon barrier to electron transfer is small in this couple, redox reactions involving either of these ions usually proceed rapidly. In general, the intrinsic reactivity of Ru(II) and Ru(III) complexes toward outer-sphere electron transfer reactions (as measured by self-exchange rates) is rapid, but varies over a range of 10^6, depending on the ligands. Lower reactivity ($k_{ex} = 4 \times 10^3\ M^{-1}\ sec^{-1}$) is shown for the Ru(II-III) couple in an insulating, hexaammine ligand environment, but increases with the presence of aromatic ligands such as pyridine, which can present extended π-electronic orbitals for overlap with the redox partner, so that the self-exchange rate for the trisbipyridyl complex is essentially diffusion-controlled ($k_{ex} = 2 \times 10^9\ M^{-1}\ sec^{-1}$) [108].

The oxidation of $(NH_3)_6Ru(II)$ by molecular oxygen has been studied with the probable mechanism involving the outer-sphere formation of superoxide ion followed by its reaction with a second Ru(II) ion to yield hydrogen peroxide and two molecules of $(NH_3)_6$ Ru(III). The rate of reduction of oxygen is first-order in both Ru(II) and O_2 and exhibits an effective rate constant of 1.26×10^2 $M^{-1}\ sec^{-1}$ over a wide pH range. The oxidation of $(NH_3)_6Ru(II)$ by H_2O_2 is slow, but is strongly catalyzed by the presence of $Fe(H_2O)_6^{2+}$. The rate constants for the reduction of O_2 by a variety of amine-ruthenium(II) ions can be correlated with the metal reduction potentials via a linear free energy relationship derived from the Marcus equation [110].

Since coordination of an anionic ligand usually lowers the reduction potential of a Ru(II) species and the electron transfer rates of these complexes are rapid, oxidation with subsequent retention of the ligand on the corresponding Ru(III) complex may take place. This method of redox catalysis allows many ligands such as Cl^-, carboxylates, and probably, phosphates to be easily substituted for water or acido ligands on Ru(III) ammine complexes, provided that small quantities of Ru(II) are present [76].

3. RUTHENIUM(III) COMPLEXES AS ONCOSTATIC PRODRUGS

Since ruthenium is a platinum group metal and in its lower oxidation states exhibits a high affinity for nitrogen ligands, several workers had suggested that some of its complexes may exhibit antitumor activity [6,7,111]. In fact, Rosenberg's early studies on the induction of filamentous growth in *E. coli* by transition metal complexes had indicated several ruthenium complexes to be active in this regard [4]. Later studies showed that fac-$(NH_3)_3RuCl_3$ and $(DMSO)_4RuCl_2$ are especially effective in causing this type of abnormal growth [5,112]. Evidence also accumulated to imply that the coordination of ruthenium ammine complexes to nucleic acids could be expected to occur in a fashion similar to that of diammineplatinum(II) complexes [64,65,113]. An early study of the effect of $(1,3Me_2Xan)(NH_3)_5RuCl_3$ on tumor-bearing animals further indicated that while this complex exhibited no antitumor activity, neither was it particularly toxic [114]. In general, however, little was done to capitalize on the wealth of ruthenium chemistry available to provide a rational approach to the design of antitumor agents which might also be relatively nontoxic to normal tissues.

The quantity of excellent kinetic, thermodynamic, and electrochemical data regarding ruthenium ammine complexes, much of it derived from the laboratory of Henry Taube at Stanford University, and recent work on tumor metabolism coupled with the precedent of the Pt(II) antitumor drugs and our own work on metal-nucleotide interactions, allows predictions as to the fate of ruthenium in living organisms and so provides the basis for the hypothesis that ruthenium ammine complexes may selectively attack a broad class of tumors [115-116]. This approach relies on the previously presented chemistry of ruthenium and the suggestion by Kelman et al. that the low oxygen content of many tumors may have an effect on the biological disposition of ruthenium [117].

Complexes such as cis-$(H_2O)_2(NH_3)_4Ru(II)$ readily substitute their water molecules to firmly bind nitrogen ligands in solution [39,70,81]. On the other hand, analogous Ru(III) complexes are much more inert to substitution. Thus a complex such as cis-Cl_2-$(NH_3)_4Ru(III)$, when introduced into an organism, might remain largely intact and so fairly innocuous until reduced yielding the "active" Ru(II) form, cis-$(H_2O)_2(NH_3)_4Ru(II)$ which could then bind to important cellular components to induce toxicity. This leads to the concept that some Ru(III) complexes could be administered as prodrugs which should be relatively nontoxic until activated by reduction. Since the reoxidation of Ru(II) by O_2 and biological oxidants [104, 110] proceeds fairly rapidly, the ratio of Ru(II) to Ru(III) should be highest in anoxic, reducing environments.

Tumor cells generally grow more rapidly than those of normal tissue with a concommitant greater uptake of glucose and oxygen [118-120]. In fact, the process of neovascularization usually cannot keep pace with tumor tissue development to adequately supply the growing tumor. This results in areas of hypoxia even only micrometers away from blood capillaries. Moreover, since the bulk of the oxygen content is depleted from the blood in tumors at a fraction of the distance along the capillary and capillaries tend to be longer in larger tumors, oxygen content decreases with increasing tumor size and the central areas of larger tumors tend to be anoxic [121-123].

Aerobic glycolysis has been noted as the hallmark of rapidly growing malignant cells [124]; however, in vivo, the assumption by glycolysis of the primary energy-producing role appears to be due to depressed respirating caused by tissue hypoxia [124]. This results in increased levels of lactic acid and concomitant lower pH [118]. Tissue hypoxia coupled with the glycolytic production of reductants such as NADH causes the interior of tumors to be a more reducing environment than the surrounding normal tissue [125,126]. Therefore, metal-containing prodrugs, which depend upon an activation by reduction mechanism to induce binding to nitrogen sites, should exist largely in the more active form inside of tumors. Because of this, these agents may also tend to concentrate in tumors.

The redox potentials of ruthenium ammine complexes make it thermodynamically possible to reduce the metal ion with a variety of biological reductants. Moreover, since complexes such as cis-$(H_2O)_2(NH_3)_4Ru(II)$ are expected to produce the same types of lesions on DNA as the cis-platinum(II) drugs, the mode of action of these agents should be similar once reduction has occurred.

4. COORDINATION OF RU(II) AND RU(III) AMMINE COMPLEXES TO BIOLOGICALLY IMPORTANT MOLECULES AND FUNCTIONAL GROUPS

4.1. Synthetic Methods

Ruthenium ammine complexes of biologically important molecules are normally formed under mild conditions in aqueous solution by direct combination with the Ru(II) starting material or by redox catalysis involving momentary reduction to induce Ru(III) to substitute the desired ligand. The physical and chemical properties of many of these complexes have been carefully studied with a view toward the effects of transition metals in biological systems. Since complexes with many nitrogen ligands tend to be inert to substitution on either Ru(II) or Ru(III), simply altering the oxidation state of the metal provides a convenient system to study the biochemical effects of both a fairly soft and a relatively hard transition metal ion firmly bound at a known position on a particular biomolecule.

4.2. Amino Acid Complexes

Diamond has shown that $(NH_3)_5Ru(II)$ coordinates the amine group of glycine; however, this complex undergoes linkage isomerization upon oxidation to Ru(III) in acid so that the carboxylate end becomes bound to the metal. Coordination of cis-$(H_2O)_2(NH_3)_4Ru(II)$ to the amine group of glycine or its methyl ester similarly occurs to anchor the amino acid to the metal ion. Subsequent oxidation of the metal ion causes the carboxylate end to bind as well, yielding a chelate complex. In the case of the glycine ester, coordination proceeds with hydrolysis of the ester bond. Hydrolysis of the ester in cis-$(H_2O)(L)(NH_3)_4Ru(II)$, where L = methyl glycinate, was also observed to proceed with a rate enhancement of several hundredfold over that of the free ester [48,127].

At neutral pH the $[(NH_3)_5RuNH_2CH_2CO_2]^+$ complex undergoes aerial oxidation of the amine to form a stable imine complex $[(NH_3)_5$-$RuNH{=}CHCO_2]^+$. In general, oxidation of Ru(III)-coordinated amines to yield imine or nitrile species proceeds readily at neutral or basic pH. Subsequent reduction of nitrilo complexes of the type $(NH_3)_5RuNC\text{-}R^{2+}$, followed by hydrolysis, yields $(NH_3)_5Ru(H_2O)^{2+}$ and an amide [128]. Carbonyls can also react with coordinated ammines to form imine complexes [129,130]. Air oxidation of coordinated ammonia in $(NH_3)_5Ru(III)$ at high pH results in $[NO(NH_3)_5Ru]^{3+}$ [131]. Ligands such as oxalate, 2-pyridinemethylamine and o-pyridinealdehyde bound in a monodentate fashion will displace an ammine ligand in a cis position to form chelate complexes [132,133].

Pentaammineruthenium in both the Ru(II) and Ru(III) oxidation states firmly coordinates the imidazole nitrogen of histidine. The Ru(II) complex undergoes an interesting linkage isomerization reaction in acid solution, yielding the C_2-bound isomer. Imidazolylidene complexes of this type are stable with both Ru(II) and Ru(III) [93,134].

The increase in acidity of the imidazole group on N-coordination of $(NH_3)_5Ru(III)$ ($pK_a = 8.5$) is significantly greater than that of the analogous Co(III) complex, even though the ionic radius of the latter ion is smaller. This has been taken as evidence of π-donation of electron density from an imidazole π-orbital to a partially filled d_π orbital on Ru(III). Binding of histidine and other imidazole ligands by tr-$(H_2O)(SO_3)(NH_3)_4Ru(II)$ has been shown to be much more labile than with $(H_2O)(NH_3)_5Ru(II)$ [74]. Selective labeling of histidine residues on ribonuclease A has recently been shown by direct combination of the latter ion with this protein [135]. The imidazolate anion can also serve as a bridge between ruthenium and other metal ions [136].

The pentaammineruthenium(II) complex of the methyl ester of methionine has been prepared. Backbonding to neutral coordinated sulfur(-II) ligands appears to be significant so that the Ru(II) oxidation state is stabilized. The affinity of $(H_2O)(NH_3)_5Ru(II)$ for this type of sulfur is approximately 10^7 greater than that of the corresponding Ru(III) species and is similar to that for amine ligands. However, when deprotonation of the sulfur is possible as with C_2H_5SH (and presumably cysteine), oxidation readily occurs yielding $[(NH_3)_5RuSC_2H_5]^{2+}$ [45], due to the high affinity of Ru(III) for RS^-.

4.3. Pyridine Complexes

The synthesis and properties of $(NH_3)_5Ru(II)$ complexes with pyridine derivatives were reported a decade ago in a now classic paper [43], which opened an experimental door to a range of ruthenium interactions with biomolecules. Metal backbonding interactions with pyridine, pyrazine, pyrimidine, and nicotinamide ligands are considerable so that the Ru(II) form is stabilized even in the presence of air for long periods. Since the Ru(II) increases the electron density in the π-system of the aromatic ligands, a decrease in the hydration of 4-pyridinealdehyde has been noted on coordination, whereas the effect of harder metal ions, such as Ru(III), is in the opposite direction [137]. Effects of metal binding on the reduction of biological redox agents, such as nicotinamide, have not been extensively studied; however, the two-electron reduction of $L(NH_3)_5Ru(II)$, L = pyrazine, has been investigated by cyclic voltammetry [88]. In this complex, metal coordination causes the two-electron reduction to occur at a more cathodic potential than that for free pyrazine at the same pH.

4.4. Nucleoside and Nucleotide Complexes

Pentaammineruthenium shows a high affinity for the heterocyclic bases of nucleosides and nucleotides, so that definitive studies of metal ions bound at various sites on these moieties can be undertaken. Since the synthesis of ruthenium-nucleotide complexes proceeds smoothly and quickly at room temperature in neutral solution, the sometimes delicate biomolecules remain intact. It has also been established that the monomeric complexes isolated as solids are the very ones present in solution and are indeed relevant to the complexes ruthenium and other metals form with nucleic acids.

Ruthenium(II-III) ammine complexes have been shown to bind to six separate nitrogen positions on purines. Guanosine and inosine coordinate $(NH_3)_5Ru(II)$ and the corresponding Ru(III) complex exclusively at the N-7 position in monomeric complexes generated at neutral or acidic pH [64,65,67]; however, if the sugar group is removed, coordination at N_3 and N_9 can also occur [67]. All three linkage isomers have been prepared and isolated for the hypoxanthine system (see Figs. 1 and 2), as well as the analogous N-7 and N-9 complexes of 2,2-dimethylguanine [64]. The use of charge transfer bands arising in the visible and near-ultraviolet regions of the spectrum has proven advantageous in assigning the linkage isomers of these complexes. In all cases to date [63-67] subsequently determined by x-ray diffraction methods [138-140], assignments based on spectral and pK_a considerations have been verified. Favorable hydrogen bonding interactions take place between coordinated ammonias and the

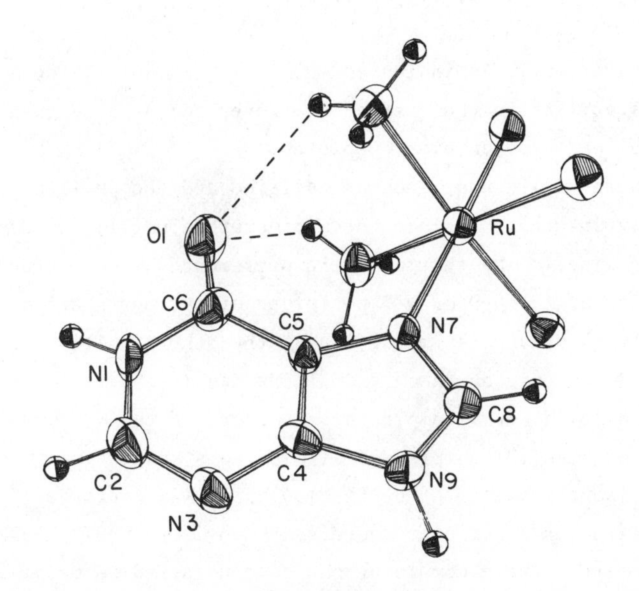

FIG. 1. Structure of the $N_7[(Hyp)(NH_3)_5Ru(III)]$ ion as determined by x-ray diffraction. --- indicates hydrogen bonds [67,139].

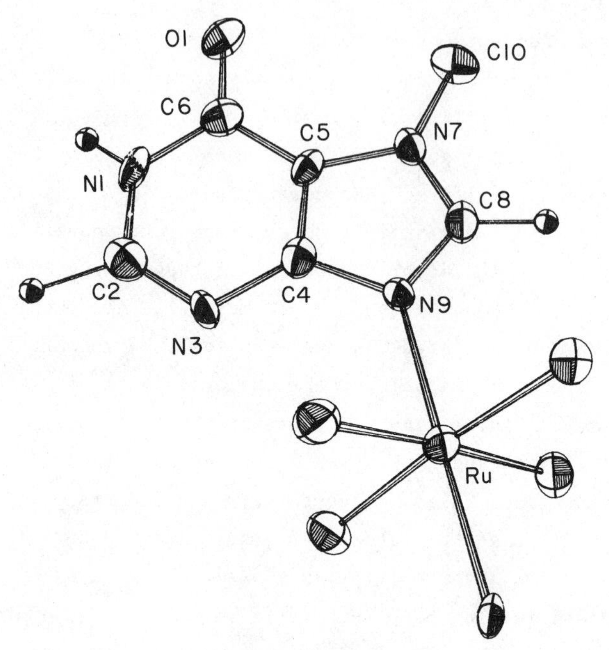

FIG. 2. Structure of the $N_9[(7MeHyp)(NH_3)_5Ru(III)]$ ion as determined by x-ray diffraction [67,139].

adjacent carbonyl, as indicated in Fig. 1, which provide some additional stability in the case of the seven-coordinated guanine, xanthine, and hypoxanthine complexes.

Xanthine ligands, such as caffeine and theophylline, are particularly interesting since the ammineruthenium(II) coordination sites of kinetic and thermodynamic preference are different [66,138]. Alkylation at the N-9 or N-3 positions of purines sterically prevents attack of a bulky metal ion at the other position, so that only the lone pair on the N-7 of $1,3Me_2Xan$ is readily attacked by most transition metal complexes. However, if the N-7 bound $(1,3\text{-}Me_2Xan)(NH_3)_5Ru(II)$ complex is allowed to stand in acidic solution linkage isomerization to the thermodynamically preferred C-8 position will occur. Caffeine coordinates ammineruthenium ions only at the C-8 site. These complexes can be considered as metal-stabilized carbenes with the ligand π orbital centered on the C-2 atom serving as a good acceptor of electron density from a d_π orbital on the metal.

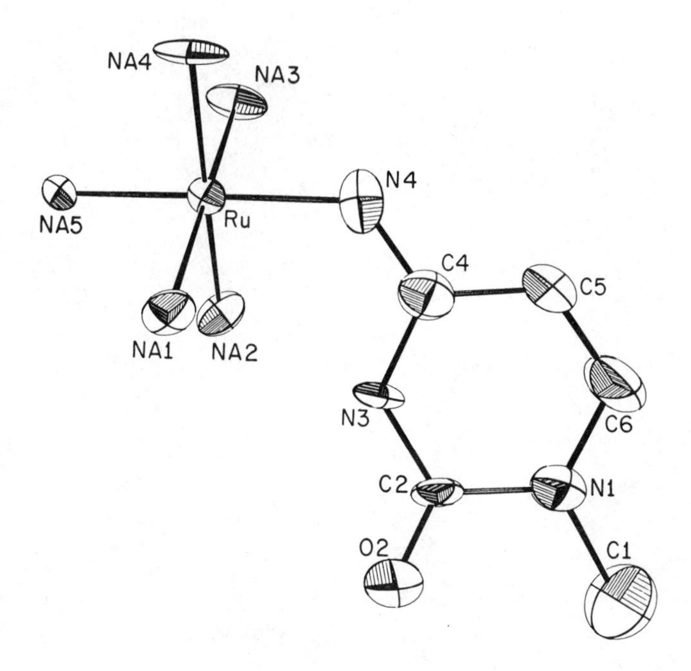

FIG. 3. Structure of the $N_4[(1MeCyt^-)(NH_3)_5Ru(III)]$ ion [140].

This is substantiated by the increased reduction potentials of C-bound over N-bound complexes with the same ligand and by the absence of a C-to-N linkage isomerization reaction. These complexes are also stable on oxidation to Ru(III), even though retrodative bonding does not take place in the higher valent state [66]. The carbon-bound ligands strongly labilize the trans position so that these species exist in solution largely as the trans-aquo species and have been isolated in solid form as the trans-chloro compounds. The association constant to form the trans-chloro complex has been measured as 15.

Complexes with adenosine and cytidine derivatives are formed by redox catalysis. In the case of adenosine, initial attack by Ru(II) probably occurs at the N-1 position, with subsequent formation of the exocyclic-nitrogen coordinated linkage isomer on oxidation to Ru(III). However, direct metal attack on the exo-N is also possible, since 3-methylcytidine reacts to yield an analogous complex.

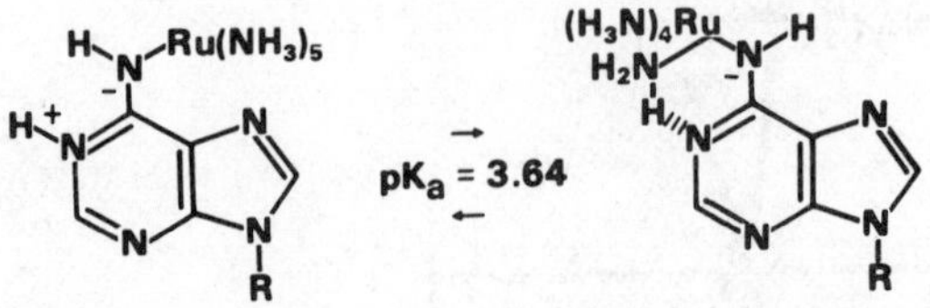

FIG. 4. Protonation equilibria and hydrogen bonding probably occurring with the $N_6[(Ado)(NH_3)_5Ru(III)]$ ion. (Reprinted with permission from Ref. 63. Copyright by the American Chemical Society.)

Coordination of the $(NH_3)_5Ru(III)$ group to the exo-N position was initially established by the close similarity in the charge-transfer spectra of complexes formed with adenosine, cytidine, 3-methylcytidine, and tubercidin and has since been verified by x-ray crystallography [66,140] (Fig. 3). This unusual series of compounds represents the only stable complexes reported with these nucleosides so bound. The driving force for complexation is provided by the binding of a tripositive metal ion to the exo-N in its deprotonated form, so that these compounds are also the first examples of stable adenosinate and cytidinate complexes. Moreover, the ligand can be brought back to its neutral form at low pH with full retention of stereochemistry in the Ru(III) form, except that the proton probably adds to the adjacent pyrimidine ring-nitrogen position. Hydrogen bonding between a coordinated ammonia and the adjacent ring nitrogen probably occurs in the deprotonated form, as depicted in Fig. 4. Similar modes of coordination are found for the entire series of mono-, di-, and triphosphate nucleotidyl complexes with Ru(III), although additional H-bonding interactions between coordinated ammines and phosphates may occur, as well [141].

4.5. Effects of Metal Complexation of Nucleosides and Nucleotides

4.5.1. *Spectral Changes*

The intense ligand-to-metal charge transfer (LMCT) absorption bands evident in all of the $(NH_3)_5Ru(III)$ purine and pyrimidine complexes provide a convenient probe into the energy levels of the highest occupied molecular orbitals on these heterocyclic bases. Since these energies are unique to each purine and pyrimidine ligand [142], the LMCT bands provide a useful indication of which nucleosides are coordinated in a complex mixture. Moreover, since the intensities of these absorptions are a function of the metal binding site [67], they also indicate the precise point of coordination on the individual bases. The LMCT transitions also depend on the protonation state of the ligand and the particular site of protonation or deprotonation, so that they yield information regarding these variables, as well [63-67].

4.5.2. Acid-Base Behavior of Coordinated Purines and Pyrimidines

Coordination of a metal ion by purines and pyrimidines can have a profound effect, attenuated by distance, on the acid-base properties of these ligands. In the most thorough study of this phenomenon to date, the effect of $(NH_3)_5$Ru(II and III) coordinated at the N-7 position of xanthine derivatives was investigated. The use of dimethylated xanthines allowed the individual assessment of the increase in acidity for each of the three-ring protons. These results, summarized in Table 4, indicate a strong inductive effect of Ru(III), which may be enhanced by a resonance contribution [63,74], to increase ring proton acidity. On the other hand, the effect exerted by Ru(II) is much lower. This difference results from not only the lower charge of Ru(II), but also from a small degree of back donation of electron density onto the purine ring. In the case of both ions, the effect on the approximately equidistant N-3 and N-1 positions is essentially the same but is much greater for the more proximal N-9 site. In an analogous study on hypoxanthine complexes, the deprotonation site was kept constant while the metal binding site was varied. Mapping out changes in acidity as a function of metal binding site in this manner allows the prediction of acid-base properties for newly synthesized complexes, and so facilitates in the assignment of linkage isomers on the basis of their pH behavior [65-67].

TABLE 4

Changes in Acidity of Purine and Pyrimidine Ligands on Coordination of $(NH_3)_5Ru(II)$ and $(NH_3)_5Ru(III)$

Coordination site	Ligand	Deprotonation site	Change in acidity relative to free ligand (ΔpK)	
			Ru(II)	Ru(III)
3	7MeHyp	1	0.9	4.18
7	Ino	1	0.4	2.11
9	7MeHyp	1	0.6	1.54
7	$3,9Me_2Xan$	1	0.6	2.19
7	$1,9Me_2Xan$	3	0.5	2.08
7	$1,3Me_2Xan$	9	2.8	6.49
9	1MeHyp	7	1.2	4.7
exo-N	Ado	1	>1.2[a,b]	>8.9[a,b]
exo-N	Cyd	3	2.8[a,b]	9.3[a,b]
8	$1,3Me_2Xan$	9	<-0.72[b]	4.9[b]

[a]Relative to minimum value for deprotonation of exocyclic amine on free ligand.

[b]Note that hydrogen-binding sites are different in free and complexed ligands.

The Ru(III) exo-N bound complexes of adenosine and cytidine exhibit enormous increases in acidity over the free ligands. The similarity of the spectra of these complexes with those involving 1MeAdo and 3MeCyd indicate the proton to be bound to the pyrimidine ring nitrogen rather than the adjacent exo-N. Therefore, the changes in acidity relative to the free ligands are due not only to the proximity of the cationic metal to the deprotonation site, but also to the displacement of the favored protonation site from the exo-N to the adjacent pyrimidine ring nitrogen. In addition, H-bonding interactions of the type indicated in Fig. 4 further stabilized the deprotonated form [63].

Surprisingly, the C-8-coordinated Ru(III) xanthine complexes are less acidic than the N-bound isomers, in spite of the fact that the metal is situated more closely to the ionization site. In the case of the Ru(II) complexes this effect is understandable on the basis of increased retrodative bonding of the C-8 complex over that of the N-7-coordinated species, and so explains the negative ΔpK_a value entered for the $1,3Me_2Xan$ complex [66] in Table 4.

4.5.3. *Hydrolysis of the Sugar-Purine Bond*

Release of the sugar group from purine nucleosides can be strongly affected by the presence of a metal ion on the heterocyclic ring. In the only study of this phenomenon yet reported, $(NH_3)_5Ru(III)$ coordinated at the 7 position of guanosine and deoxyguanosine was utilized. With both of these complexes, the proton-catalyzed rate of sugar hydrolysis decreased by a factor of 30 relative to that for the free ligands at 56°C [65]. This effect is due to (1) metallization of the preferred protonation site forcing proton association at the less catalytically effective N-3 position and (2) decreasing the basicity of the purine ring so that protonation becomes more difficult. When the rate constant is corrected for the second effect, i.e., when the rates of hydrolysis of the protonated species are compared, it is evident that the protonated-metallized guanosine complex is more effective at liberating the sugar group than the protonated nucleoside, alone.

Early on, it had been suggested that coordination of a metal cation at the N-7 position of guanosine might lead to scission of the glycosidic link in a manner analogous to that of alkylation at this position [111]. This suggestion is supported by still preliminary studies with N-7-bound pentaammineruthenium(III) complexes of dGuo and 1MeAdo, which appear to hydrolyze several orders of magnitude faster than the corresponding free nucleosides at neutral pH [63,143,144].

4.6. Migration of the Metal Ion on Purines

Movement of metal ions on purine or pyrimidine rings represents means whereby metals can migrate to more or less biologically crucial positions or even between adjacent heterocyclic bases in polynucleotides. Relatively little has been done to study this phenomenon or how it may be affected by changes in the biochemical environment. However, recent work has shown that alterations in the oxidation state of the metal and the protonation state of the purine ligand can induce linkage isomerization in pentaammineruthenium complexes [63,66,67]. Figure 5 demonstrates the four such rearrangements which have been observed to date.

As mentioned above, N-7-bound xanthine complexes of $(NH_3)_5$-Ru(II) undergo an acid-catalyzed rearrangement to yield stable ylidene complexes at room temperature [63]. This isomerization reaction probably involves initial protonation on an octahedral face of the metal to yield a seven-coordinate intermediate. Such protonation would also labilize the adjacent C-8 proton and so facilitate attack of the metal ion at this site. Those purines which can undergo proton addition to form cationic ligands have not yielded C-8-bound compounds.

While $(NH_3)_5Ru(II)$ appears to coordinate firmly to the three available nitrogen sites on hypoxanthine, $(NH_3)_5Ru(III)$ is stable at the N-3 position only when the N-1 site is deprotonated [67]. The negative charge residing on the otherwise electron-deficient pyrimidine ring stabilizes binding of the harder metal ion at N-3.

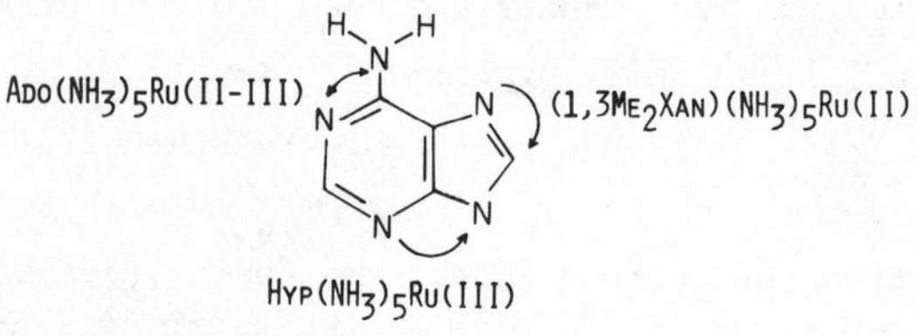

Ado$(NH_3)_5$Ru(II-III)

(1,3Me$_2$Xan)$(NH_3)_5$Ru(II)

Hyp$(NH_3)_5$Ru(III)

FIG. 5. Linkage isomerization reactions observed with purine-$(NH_3)_5Ru(III)$ complex ions.

However, at lower pH where the N-1 atom is protonated, the metal readily migrates to the adjacent N-9 site of the more electron-rich imidazole ring with retention of the five ammine ligands. The geometry of the ligand is well suited to facilitating this intramolecular shift so that the reaction proceeds much more rapidly ($t_{1/2}$ = 6 hr) than would otherwise be expected [67].

Electrochemical reduction of $(NH_3)_5Ru(III)$ ion coordinated at the exo-N of adenosine results in the movement of the resultant Ru(II) species to the adjacent N-1 position on the pyrimidine ring. The pyrimidine ring provides a better π-acceptor site for the metal in the lower oxidation state and N-1 coordination appears to be stable for at least short periods of time following reoxidation [63]. This rearrangement reaction is proton-assisted, probably due to the necessity of reprotonating the exo-N when the metal leaves this site. Subsequent oxidation of the N-1 species at neutral pH appears to result in linkage isomerization of the Ru(III) ion to the exo-N position [63].

4.7. Coenzyme Complexes

Since a number of coenzymes contain adenine residues, they can be labeled with pentaammineruthenium and similar species in the same manner as other nucleotides. Indeed, a number of coenzymes can be coordinated under the mildest of conditions via redox catalysis. The spectra of complexes so prepared involving NAD, NADH, NADP, and FAD indicate that $(NH_3)_5Ru(III)$ attaches firmly to the exo-N of adenine. All of these complexes are stable in aqueous solution for fairly long periods so that purification is possible by chromatographic methods. However, preliminary experiments with a range of enzymes and substrates indicate that these species are not enzyme-active [141].

Coenzyme residues other than adenine also offer potential binding sites for ruthenium ammines. Recently work has shown that cis-$(H_2O)_2(NH_3)_4Ru(II)$ will coordinate at the N-5 and O-4 positions of isoalloxazine derivatives to form a stable chelate [68]. Retrodative bonding in these complexes is considerable, resulting in the shortest Ru-N bond distance reported for Ru(II) bound to a neutral heterocyclic molecule and severely altering the electrochemical properties of the flavin ring. Subsequent work with FAD indicates that either the flavin or adenine residues can be preferentially labeled, depending on the reaction conditions.

5. COMPLEXES WITH NUCLEIC ACIDS

5.1. Binding Sites

Pentaammineruthenium(II) exhibits considerable selectivity for binding to heterocyclic base residues on DNA [64,116]. Subsequent oxidation of solutions containing $[(NH_3)_5Ru(II)]_n$-DNA allows the formation of the corresponding air-stable Ru(III) species. Since each heterocyclic base gives rise to a unique set of LMCT absorptions, ultraviolet-visible spectroscopy can be employed to identify the Ru(III) coordination sites. The Ru(III) coordination bond is sufficiently strong that acid hydrolysis of the metallated DNA yields the intact $(NH_3)_5Ru(III)$-purine complexes [65,116], which can be chromatographically separated and characterized individually. Experiments of this kind carried out with double-helical DNA clearly indicate that the N-7 of guanine, which is readily available in the major groove, is the preferred binding site. Coordination of the interior exocyclic-nitrogen sites of adenine and cytosine occurs with single-stranded DNA, so that localized, metal-induced melting of helical DNA may be responsible for an observed increase in exo-N binding with increasing metal concentration [116].

5.2. Coordination Under Equilibrium Conditions

While ammineruthenium(II) complexes bind nitrogenous ligands much more rapidly than ammineplatinum(II) species, the coordination reactions of both are largely under kinetic control. In a cleverly conceived series of experiments Brown, Sutton, and Taube [74] utilized tr-$[(H_2O)(SO_3)(NH_3)_4Ru(II)]$, which rapidly reaches equilibrium with purine ligands, to establish the relative affinities of Ru(II) for guanosine and adenosine. The strong discrimination between these two ligands ($K_G/K_A = 95$) coupled with the fact that the N-7 of guanine is much more readily accessible for coordination in helical DNA than the interior N-1-adenine and N-3- or exo-N cytosine sites clearly indicates that equilibrium binding with Ru(II) could be exploited to selectively label guanine residues on DNA. The metal can be fixed in place by peroxide oxidation to the sulfato-ruthenium(III) species and the coordinated SO_4^{2-} can then be selectively replaced by iodide via redox catalysis. This technique could provide a sufficiently dense area of electron density to be of use in defining guanine residues in nucleic acids by high-resolution electron microscopy [145].

5.3. Possible Effects of Ruthenium Binding to Nucleic Acids

Initial attack of ruthenium ammine complexes at the N-7 of guanine in coiled DNA could result in a number of possible interactions leading to the disruption of the DNA molecule. The hydrogen bonding of cis-aquo or ammine ligands to the adjacent carbonyl oxygen should weaken this group's ability to hydrogen bond with cytosine. In the case of Ru(III), the presence of the metal ion significantly increases the acidity of the N-1 proton ($pK_a = 7.4$) so that it might be lost in a manner further weakening the base-pair linkage. Finally, a relatively hard metal ion bound to a ring nitrogen of a purine or pyrimidine should polarize the π-electron cloud and so affect the

TABLE 5

Potential DNA Cross-linking Modes Open to Ruthenium Ammines

Bases involved	Probable coordination sites on respective bases[a]	
	Ru(II)	Ru(III)
A-A	N_1-N_1	N_6-N_6
G-G	N_7-N_7	N_7-N_7
C-C	N_4-N_4	N_4-N_4
A-G	N_1-N_7	N_6-N_7
A-C	N_1-N_4	N_6-N_4
G-C	N_7-N_4	N_7-N_4

[a] N_6 is exocyclic nitrogen on adenine, N_4 is exo-N on cytosine.

ligand's tendency toward ring stacking with adjacent bases. This latter effect should be pronounced for Ru(III) which may serve as a π-acceptor toward at least the imidazole ring on purines, but should be less so for Ru(II) [65]. The close proximity of a cationic metal complex to the phosphate backbone on the exterior of the DNA may well yield electrostatic and H-bonding interactions between these moieties which would also deform the DNA in coordinated regions.

Coordination of these metal ions to sites normally on the interior of the nucleic acid may be assisted by initial attack on the exterior of the macromolecule, resulting in localized denaturation [116]. Once binding to the H-bonding nitrogen sites of cytosine and adenine occurs, disorder of the polymer in this region is assured. More severe damage to the nucleic acid could be inflicted through metal-induced depurination of the polynucleotide [65,144].

While studies with chelatable ruthenium species such as cis-$[(H_2O)_2(NH_3)_4Ru]^{2+}$ are incomplete, it seems clear that the sites complexed by this and the corresponding Ru(III) species are identical with those previously discussed. Thus both intra- and interstrand cross-links in DNA similar to those suggested for the cis-diammine-platinum(II) complexes are possible [7,40,111]. Moreover, additional modes of cross-linking involving exocyclic nitrogens are also open, particularly in the case of Ru(III). The most probable cross-linking connections for ruthenium in both oxidation states are summarized in Table 5. In addition it should be noted that, in the special environment holding when an ammine complex ion is coordinated to a nucleic acid, the displacement of coordinated ammonia by an impinging nitrogen base should be possible. The situation is analogous to the presence of a pendant portion of a polydentate ligand, so that addition of the closely held atom is facilitated [50, 133]. Finally, at least in the case of $(NH_3)_5Ru(II)$ coordinated to a ring nitrogen, substitution in the cis position should be favored [50].

6. DISPOSITION OF RUTHENIUM COMPLEXES IN LIVING ORGANISMS

6.1. Modes of Ru(III) Reduction by by Cellular Components

While tumor cells are capable of producing energy by respiration, the low availability of oxygen in most tumor tissue prevents this [118-126]. Anoxic or hypoxic mitochondria might be expected to "back up" reducing power in the form of NADH, reduced dehydrogenases, and Krebs cycle intermediates so that transition metal ions could be readily reduced. In particular, reduction of Ru(III) complexes via single-electron transfer components such as flavoproteins or cytochromes [102-104] should be rapid. However, reduction by organic molecules such as NADH, succinate or malate, which prefer to transfer electrons in two-electron steps, is usually slow.

In a series of experiments utilizing succinate as the source of reducing power, semipurified preparations of rat-liver mitochondria as the redox catalyst $(Cl(NH_3)_5Ru(III)$ and isonictoinamide to indicate the presence of Ru(II), the net rate of Ru(II) production was significantly higher when the reaction was allowed to proceed under a nonoxidizing atmosphere rather than in air. Production of Ru(II) was also markedly decreased by the introduction of Krebs-cycle inhibitors such as malonate (for succinate dehydrogenase) and 2,4-dichlorophenol (for malate dehydrogenase). These studies clearly show that respiratory components, probably flavoproteins or low-potential cytochromes, can be quite effective in the redox activation of Ru(III) prodrugs toward coordinating heterocyclic ligands. While the degree of isonicotinamide binding decreased when the reaction was run under nitrogen rather than argon, it still remained at a high level, so that draining of Ru(II) in the form of the dinitrogen complex [39,44] appears to be a surmountable problem [146].

Microsomal redox enzymes are known to serve as effective catalysts for the NADH reduction of metal ions [31,147] and reducing proteins also exist in the soluble fraction of animal cells. To test the effectiveness of these components in reducing Ru(III) complexes, experiments analogous to those discussed for the mitochondrial fractions were performed, except that NADH served as the electron source [146]. Microsomal catalysis proved far superior to the mitochondrial in reducing the metal ion, since much higher concentrations had to be utilized in the latter case. In view of this, it may be that the mitochondrial outer membrane, which consists of microsomal-like material [148], may hold the actual reductant of the metal ion, since reducing power can be passed from the mitochondria to the surrounding cytosol via the malate-aspartate shuttle [149].

Sulfhydryl groups are also thermodynamically capable of reducing many Ru(III) complexes at neutral pH [150] and the work of Kuehn indicates that ruthenium(II) ammine complexes of RS^- are sufficiently strong as reductants to transfer electrons to acido complexes of ruthenium(III) [45]. It should also be pointed out that the lower pH of cancer tissue due to glycolytic lactic acid production will favor the reduction of those Ru(III) complexes whose reduction potentials are pH-dependent [63,118].

6.2. Tissue Distribution of Ruthenium Trichloride and Nitrosyl Complexes

Ruthenium chloride and nitrosyl complexes occur in the processing of reactor wastes so that their biodistribution has been repeatedly investigated. While the complexes used in these studies have usually not been well characterized, a few useful generalities can be drawn. When these materials are injested, the bulk of the metal absorbed appears to be through the small intestine within the first hour [8-10,21,22]. Radiotracer studies show that adherence to the intestinal mucosa is predominant with the highest concentrations in the body occurring in the epethelial cells of the villi shortly after ingestion [10]. Mixing with food prior to ingestion or with the contents of the gut converts most of the ruthenium to a nonabsorbable form and the bulk of the metal is excreted within 24 hr [151]. The nitrosyl form is more rapidly absorbed than the trichloro species [22], with one study indicating 13% of the amount injested to be absorbed for the former but only 3% for the latter [8]. It is likely that the trichloro species is converted to a hydrated oxide on reaching the intestine and little difference in the distribution of these two forms was noted at concentrations where no precipitate of the oxide formed. The ruthenium absorbed into the intestinal wall appears to be eliminated with a half-life of 2.3 days [21].

Once absorbed into the blood or when introduced intravenously, the metal becomes widely distributed [12,55] and is eliminated much more slowly [22]. Following oral ingestion, the concentration of ruthenium in the blood rises rapidly and then reaches a maximum within 12 hr. A substantial fraction (50-80%) is eliminated from the blood within 4-6 days, but the half-life for the elimination of the remaining ruthenium from the blood is 7-10 days [8,45]. The ruthenium which is not eliminated within the first month after ingestion lingers in the body for extended periods of time ($t_{1/2}$ = 4-7 weeks) [9,21]. Since most of the ruthenium contained in the blood is excreted through the urinary tract, high concentrations are evident in the kidneys [8-12]. The liver also shows concentrations comparable with those found in the kidneys.

Several studies indicate that the bulk of the ruthenium in the blood is bound to plasma proteins [9,12,24]. The introduction of chelating agents can promote excretion of the metal [12,23], but the efficacy of these agents is sharply diminished when the metal has remained in the blood long enough to allow binding [12]. Considering the high affinity of ruthenium for mercaptans [45] and the efficiency of sulfur sequestering agents in removing the metal from the blood, it appears likely that the metal is firmly held by sulfhydryl and nitrogen residues on plasma proteins. Subsequent to binding, the catabolic rate of the coordinating substance and elimination through the liver and kidneys probably limits the excretion of ruthenium from the body [8-12]. When administered intravenously in a colloidal form, higher concentrations occur in the liver and spleen [11]. When toxic doses are introduced, elevated levels are also found in the bone [9,11].

Studies involving ruthenium trichloride injected subcutaneously into rats indicate that the resulting concentration in the nuclei of liver cells reaches an equilibrium value 1.5 times that of the cytosol concentration and that the metal is exponentially lost from the cytosol with a half-life of 100 hr. The concentration in the nucleus builds up more slowly than that in the cytoplasm, but then reaches a higher value which decays at approximately the same rate. These authors suggested macromolecular binding of the metal, followed by a rate-limiting diffusion across the nuclear membrane, as most consistent with their data [19]. An in vitro investigation utilizing K_2RuCl_6 showed that addition of this compound inhibited rat liver mitochondrial respiration at a concentration of 2.5 μM [152]. An analogous study indicated that millimolar concentrations of $RuCl_3$ inhibited mitochondrial electron transport and all energy-linked functions, but was without effect at the micromolar level [153].

Several studies have utilized radioactive ruthenium trichloride to delineate tumors by γ-ray imaging. Investigations involving injections of ruthenium trichloride in acid media or neutral solutions, strongly buffered with citrate, indicated sufficient localization in subcutaneous Ehrlich's or AH-130 ascites tumors to allow imaging with a scintigraphic camera [26,28]. However, high concentrations in the lung, liver, and kidney would appear to interfere with tumor imaging in these areas. Elimination from the body was more rapid with the citrate solutions (75% after 1 day), whereas when the $RuCl_3$ was not accompanied by a chelating agent, the metal was eliminated in a biphasic manner with half-lives of 15 hr for the first phase and 18 days for the second. Studies in man involving 37 cancer patients showed that malignant tumors were well delineated in 54% and poorly delineated in 24% of the cases. Head, neck, and primary lung tumors were strongly imaged, while hepatomas were not identifiable [27].

6.3. Biological Studies With Ruthenium Red

Ruthenium red [154], $[(NH_3)_5Ru\text{-}O\text{-}(NH_3)_4Ru\text{-}O\text{-}Ru(NH_3)_5]^{6+}$, has long been used as a cytological stain for both visible and electron microscopic studies [155]. It selectively binds and precipitates animal acid mucopolysaccharides, probably by forming ion pairs with sulfate or carboxylate groups and hydrogen bonding through the ammine ligands to sugar hydroxyls. This characteristic interaction has led to its widespread utilization as a selective stain for cartilage matrix [155], epithelial cells [156], cell fine structure [157], and mast cell heparin granules [158]. Fusion of sarcoplasmic reticulum membranes in frog skeletal muscle cells has also been noted on exposure to ruthenium red, possibly by cross-linking anionic sites on opposite faces of the intermediate cisterns [159]. This type of mechanism may also be responsible for ruthenium-red-induced cell agglutination [160].

Ruthenium red serves as a selective stain for mitochondria in intact tissue and specifically inhibits the mitochondrial uptake of calcium [52]. At slightly higher concentrations (10-50 nM mg^{-1} of mitochondria) it also inhibits mitochondrial respiration and is capable of inhibiting both low and high affinity calcium binding, as well as potassium-linked and energy-driven calcium transport [153]. Interference with the high-affinity calcium-binding sites is probably due to its ability to bind anionic calcium specific sites on glycoproteins. It should be noted, however, that the commercially available ruthenium red preparations usually contain no more than 70% of the designated compound, even after recrystallization, and it may well be that a lower molecular weight ruthenium complex is responsible for the inhibition of mitochondrial calcium transport [162].

Ruthenium red can also penetrate peripheral nerve fibers and induce neurotoxicity [161,163]. Concentrations in the range of 2.5-10 μM reduces the synaptic potential to subthreshold levels [164]. Intracranial administration induces epilepsy while intraperitoneal injection produces flaccid paralysis. These effects are probably due to blocking calcium interaction with the presynaptic membrane and preventing the calcium-dependent release of neurotransmitter [165].

Anghileri has suggested that the selectivity ruthenium red exhibits for acid mucopolysaccharides may be exploited to produce tumor-specific radioscintigraphic agents employing ^{103}Ru or ^{97}Ru [166]. Agents of this type should be capable of selective interactions with the stromal ground substance of tumors, which is rich in mucopolysaccharides. Studies utilizing a solid subcutaneously transplanted Novikoff hepatoma in rats showed localization of intravenously ruthenium red in the kidney, liver, tumor, bone, spleen, and lungs with a maximum tumor-to-blood ratio of 4.0 occurring after 24 hr. These observations indicate that ruthenium red does tend to localize in tissues which contain a considerable amount of mucopolysaccharides and mucoproteins and that it may be useful not only for tumor imaging but also in other pathological states involving glycoprotein metabolism.

6.4. Oncological Properties of Ruthenium Coordination Complexes

The high antitumor activity of the cis-ammineplatinum(II) drugs and the activation by reduction hypothesis previously discussed suggested that ruthenium ammine complexes should exhibit the same type of biochemical and antitumor activity as the platinum compounds, but should have additional mechanisms open to them which might provide better tumor specificity. In an attempt to develop a convenient screen for the potential antitumor activity of ruthenium complexes, Kelman investigated their effect on DNA and protein synthesis in rabbit kidney cells in tissue culture [117,167]. While there eventually proved to be no correlation between the in vitro inhibition of DNA synthesis and later in vivo results (see Table 6), these studies show that a number of ruthenium complexes are capable of inhibiting cellular DNA synthesis. The inhibition of macromolecular synthesis has also been investigated as a function of metal concentration for a number of complexes [117]. Only about 10% of DNA synthesis is inhibited by a 1.0 mM concentration of $[(NH_3)_5RuCl]^{2+}$, but this increases exponentially to about 99% inhibition at 5.0 mM.

The results of collaborative efforts involving several laboratories on the oncological activity of ruthenium complexes are summarized in Table 6. The work of Yasbin, Miehl, and Matthews on the mutagenic activity of ruthenium indicates that many complexes of this metal are remarkably effective toward the induction of error-prone DNA repair [168]. This type of repair is likely to occur when the cell detects serious DNA lesions, so that these studies provide evidence that ruthenium complexes probably interact directly with chromatin DNA, presumably by one of the mechanisms indicated in Sec. 5. Further indications of DNA involvement are the filamentous growth of *E. coli* in the presence of a number of ruthenium complexes

TABLE 6

Antitumor and Mutagenic Activity of Ruthenium Complexes

Compound	Therapeutic dose[a] ($mg\ kg^{-1}$)	T/C[b] (%)	Toxic dose[c] ($mg\ kg^{-1}$)	Inhibition of DNA synthesis[d] (%)	Ames test[e]
$Cl_3(NH_3)Ru(III)$	50	189	100	86	
cis-$[Cl_2(NH_3)_4Ru(III)]Cl$	12.5	157	50	86	
cis-$[Cl_2(en)_2Ru(III)]Cl$	50	139	100	26	+++
cis-$[Br_2(NH_3)_4Ru(III)]Br$	25	118	100	45	
cis-$[(H_2O)_2(NH_3)_4Ru](TFMS)_3$	100	157	200	11	
cis-$[Br_2(dach)_2Ru(III)]Br$					++
$[Ox(NH_3)_4Ru(III)]Cl$	50	142	100	61	
$[Ox(en)_2Ru][(Ox)_2enRu]$	50	151	200	33	++
$[Ox(dach)_2Ru][(Ox)_2(dach)Ru]$	25	119	50	79	
tr-$[Cl_2(NH_3)_4Ru(III)]Cl$					++
tr-$[Cl_2(en)_2Ru(III)]Cl$					++
$[Cl(NH_3)_5Ru(III)]Cl_2$	1.5	116[f]	50		++
$[Br(NH_3)_5Ru(III)]Br_2$	12.5	120	50	40	
$[HCOO(NH_3)_5Ru(III)](ClO_4)_2$	12.5	128	50	69	
$[CH_3COO(NH_3)_5Ru](ClO_4)_2$	12.5	149	25	50	
$[CH_3CH_2COO(NH_3)_5Ru](ClO_4)_2$	12.5	163	50	0	

$[(Ino)(NH_3)_5Ru(III)]Cl_3$	25	125	100	96	+++
$[(1,3Me_2Xan)(NH_3)_5Ru]Cl_3$	50	90[g]	100	35[h]	+++
$[(Guo)(NH_3)_5Ru(III)]Cl_3$	25	102[g]	50		
$[(Ado)(NH_3)_5Ru(III)]Br_3$				40[h]	+++
$[(Hyp)(NH_3)_5Ru(III)]Cl_3$				25[h]	++
$[(1MeCyt^-)(NH_3)_5Ru](BF_4)_2$					+++
$[(His)(NH_3)_5Ru(III)]Cl_3$					±
$[(NH_3)_6Ru(III)]Cl_3$	3.1	113[f,g]	25		0
$K_3[(Ox)_3Ru(III)]$	12.5	122	150	99	
$K[Cl_4(phen)Ru(III)]$					±
$K[Cl_4(bipy)Ru(III)]$					+
$K_2[(H_2O)Cl_5Ru(III)]$					0
$K_3[Cl_6Ru(III)]$	25	138	100	6	
$(NH_4)_2[COCl_5Ru(III)]$	12.5	108	200		
$[Isn(NH_3)_5Ru(II)](PF_6)_2$	6.25	104[g]	200		
$[Ox(bipy)_2Ru(II)]$	3.13	101[g]	>200		+
$[Ox(phen)_2Ru(II)]$	6.25	96[g]	>200		
$[Cl_2(bipy)_2Ru(III)]Cl$	50	97[g]	100		
$[Cl_2(phen)_2Ru(III)]ClO_4$	6.25	90[g]	25		
$[(NO_2)_2(bipy)_2Ru(II)]$	6.25	95[g,i]	>200		
$[(NO)(NH_3)_5Ru(II)]$	3.12	127[f,g]	25	62	
cis-$[(NO)I(NH_3)_4Ru(II)]$	25	144[g]	200		

TABLE 6 (continued)

Compound	Therapeutic dose[a] ($mg\ kg^{-1}$)	T/C[b] (%)	Toxic dose[c] ($mg\ kg^{-1}$)	Inhibition of DNA synthesis[d] (%)	Ames test[e]
tr-$Na_2[(NO)OH(NO_2)_4Ru(II)]$	50	131[i]	100		
cis-$Cl_2(DMSO)_4Ru(II)$	565	125[g,j]	800[k]		++

[a]Dose of each of three intraperitoneal injections given on days 1, 5, and 9, at which the T/C value was obtained.

[b]Ratio of test versus control median survival times multiplied by 100. All values in first three columns are from NCI screening program for antitumor compounds utilizing P388 lymphocytic leukemia system of male CD_2F (CDF_1) mice, unless otherwise noted. T/C values reported are highest obtained which were judged to be reproducible. NCI criterion for activity is T/C ≥ 120 [167].

[c]Dose at which animal deaths appeared before day 6.

[d]Evaluation of uptake of [^{3}H]thy relative to controls after 27-hr incubation utilizing tissue cultures of transformed rabbit kidney cells. [Ru] = 100 μM. Adapted from Refs. 117 and 167.

[e]Ames test was negative in all cases for TA 1535, 1537, and 1538 strains. Results reported are for TA 98 and TA 100 strains, which are indicative of induction of error-prone DNA repair [168]. + indicates twice the number of colonies appearing relative to background.

[f]Compounds prepared and submitted from Boston University by J. N. Armor.

[g]Screen performed using L1210 lymphoid leukemia. NCI criterion for activity is T/C ≥ 125.

[h]KB cells in tissue culture. [Ru] = 100 μM [117].

[i]Compounds prepared and submitted by S. J. Pell, Boston University.

[j]Taken from Ref. 6.

[k]Taken from Ref. 35.

[4,5], and the inhibition cellular of DNA synthesis. These investigations also point out the likelihood that a variety of ruthenium complexes, particularly those involving purine or pyrimidine ligands, may be carcinogenic.

Studies on the toxicity of cis- and trans-$Cl_2(NH_3)_4Ru(III)$ on cell cultures (see Fig. 6) show that the cis complex is much more toxic than the trans. In a comparison involving an initial 1-hr incubation versus continuous incubation with these complexes, little difference in the toxicity resulted. Pitha has suggested that this indicates binding of the metal to the cell surface with the metal entering the cell through pinocytosis as the outer membrane is recycled [169]. Interestingly, nucleoprotein binding on cell surfaces had previously been cited as a possible step in the mechanism of action of the cis-diammineplatinum(II) drugs [170].

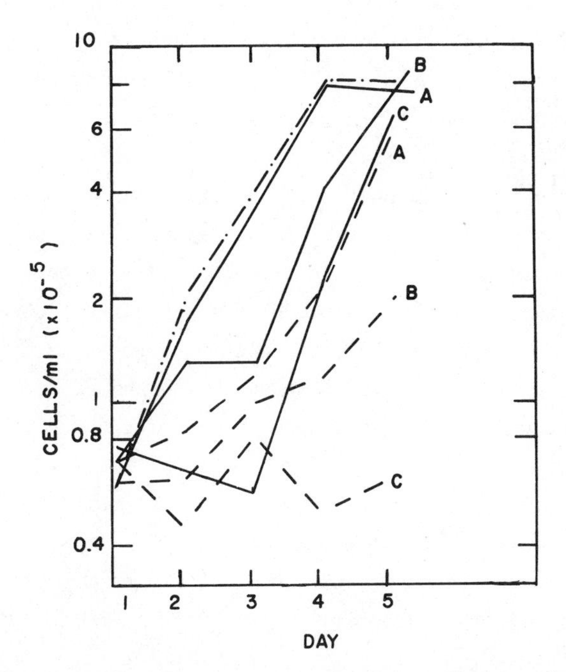

FIG. 6. Cell toxicity of erythryoleukemic cells as a function of the concentration of cis- and trans-$[Cl_2(NH_3)_4Ru(III)]^+$ after a 1-hr incubation with the metal complex: ——, trans-$[Cl_2(NH_3)_4Ru(III)]^+$; ---, cis-$[Cl_2(NH_3)_4Ru(III)]^+$; —•—•, control. Concentrations: A = 5 μg ml^{-1}, B = 50 μg ml^{-1}, C = 500 μg ml^{-1} [169].

While a possible pathway for metal binding in air could involve reaction between carbonyl groups on the cell surface and ammine ligands on Ru(III) to fix the metal by an imine link with simultaneous reduction to Ru(II) [129,130], the difference in toxicity between the cis and trans isomers suggest direct metal coordination. Degradation of these ruthenium-bound membrane components to monomeric or macromolecular fragments, which could diffuse into and possibly concentrate in the nucleus [19] and then translocate by redox activation, would appear to be a possible path to DNA binding.

The studies involving animal tumor systems given in Table 6 indicate that a number of ruthenium complexes possess antitumor activity. For comparison purposes, the average best T/C values for controls run with cis-$Cl_2(NH_3)_2Pt$ were 191 for the P388 system and 164 for the L1210, so that only the $(NH_3)RuCl_3$ compound approaches the activity of the platinum(II) drugs. Moreover, for many complexes the toxic doses are not greatly different from the dose at which the best T/C value occurs suggesting a low therapeutic index. On the other hand, many of the agents listed show antitumor activity and the compounds which exhibit the highest antitumor activity are indeed those which are expected to operate by an activation-by-reduction mechanism. The series of Ru(II) complexes stabilized by heterocyclic ligands [or closely related Ru(III) complexes with high reduction potentials which should be rapidly converted to the Ru(II) form] exhibit no activity and are relatively nontoxic. The $(NH_3)_3$-$RuCl_3$ and cis-$[Cl_2(NH_3)_4Ru]Cl$ compounds also showed activity against B-16 melanocarcinoma mice, with T/C values of 149 and 130, respectively [167].

Recent tissue distribution studies offer some insight into the in vivo activity of these complexes [170]. Those compounds which are expected to substitute readily on reduction exhibit tissue distributions that are similar to those of $RuCl_3$, suggesting that they also bind to plasma proteins, causing the metal to linger in the blood. Aside from the blood, the highest ruthenium concentrations for these complexes occur (in approximate order, 1 day after intravenous injection) in the kidney, liver, tumor, bone, and muscle. These results suggest an initial phase during which the metal binds with relatively rapid excretion of free metal ions, followed by slow elimination of ruthenium fixed in the tissues. The high and relatively constant concentration found in the kidneys (over the course of 1 day) suggests that kidney toxicity may present a problem in clinical use as it does for the platinum drugs. Nevertheless, complexes similar to cis-$Cl_2(NH_3)_4Ru(II)$ do exhibit a good degree tumor localization and the design of new agents along these lines, with substituents which might minimize the high blook and kidney concentrations, should be pursued.

Complexes of Ru(II) with tightly bound heterocyclic ligands such as phenanthroline derivatives are quickly eliminated from the blood and exhibit very high localization in the liver and bile [170]. This is not surprising in view of the inability of these complexes to form new coordination bonds and their high solubility in nonaqueous media [80]. It is likely that the fate of most of the Ru(II) complexes with heterocyclic ligands cited in Table 6 proceeds analogously.

6.5. Oncological Properties of Organoruthenium Complexes

The stability of the organometallic bonds in ruthenocene prevents the metal from binding other ligands directly; however, Wenzel has pointed out that derivatives of the aromatic ligand may be exploited to induce the metal to localize in selected tissues and so provide γ-ray imaging agents for diagnostic purposes [53]. Ruthenocene itself is readily eliminated from the body through the bile and urine following hydroxylation and formation of a glucuronide conjugate in the liver [54]. However, small quantities were detected in the liver and gastrointestinal system of rats weeks after injection, which may be due to decomposition of the complex to ionic ruthenium. The carboxylic acid derivative of ruthenocene is excreted rapidly by the kidneys ($t_{1/2}$ = 2 hr), and so shows promise as a renal-imaging agent. This compound also exhibited modest localization in Ehrlich's carcinoma [55]. The estradiol-17β and estrone esters of ruthenocenecarboxylic acid have been investigated as radioscintigraphic agents for the adrenals, ovary, and uterus, but exhibit their highest concentrations in the kidneys and liver [56].

Intraperitoneal injection of the cinnamoyl and 3-phenylpropen-1-one derivatives of ruthenocene reveals a remarkably high localization in the thymocytes, which is probably due to metabolism of the compound in the liver, yielding a product which is specific for the thymus tissue [57]. Acetylruthenocene undergoes oxidation of the terminal methyl group to yield a primary alcohol which is then conjugated with glucuronic acid in the liver. Studies involving injections of acetylruthenocene containing ^{103}Ru into rats indicated appreciable localization in the adrenals; however, high concentrations in the lungs, liver, and kidneys would interfere with its utility as an adrenal imaging radiopharmaceutical [58].

Investigations of ten ruthenocene derivatives in mice bearing Ehrlich's carcinoma revealed only two with good tumor-to-muscle ratios. These two compounds, vinylbenzoyl- and di(methylcarboxylate)ruthenocene, exhibited favorable tumor-organ ratios relative to the most widely used tumor radioscintigraphic agent, ^{67}Ga-citrate; however, their absolute concentrations in the tumor were much lower. Surprisingly, these ruthenocene derivatives showed higher concentrations in smaller tumors, indicating some potential for imaging micrometastases [59].

An interdisciplinary group at the University of Trieste has been systematically investigating the antitumor properties of cis-$Cl_2(DMSO)_4Ru$. This compound exhibited marginal activity against the mouse L1210 leukemia (see Table 6), but did effect cures at the highest dosage employed with Ehrlich ascites carcinomas [6]. Doses above therapeutic also resulted in inhibition of DNA, RNA, and protein synthesis. Histological studies indicated that the ruthenium complex caused a pattern of cell damage similar to that of cis-$Cl_2(NH_3)_2Pt$, but that its effects were slightly less on the intestinal mucosa and spleen. Mutagenicity studies indicated activity comparable to the platinum compound in the Ames test, but relatively much less when the pKM101 plasmid was added to enhance the sensitivity of the hisG46 strain. Thus it was concluded that both compounds interact with the DNA of intact bacteria, but that their mechanism of action is probably different [35]. Recent work suggests preferential binding of the ruthenium compound to adenine rather than guanine residues of DNA [171].

7. CONCLUSION

While no ruthenium compounds have been developed that appear superior to already existing anticancer drugs, several rational and promising approaches have been made toward the utilization of this metal in chemotherapeutic and radiodiagnostic agents. Ruthenium(III) compounds which require in vivo activation by reduction have not yet attained the desired level of selective tumor toxicity. However, soluble tridentate analogs of $(NH_3)_3RuCl_3$, the most active ruthenium complex tested thus far, should provide drugs of significantly higher activity. Derivatives of the most active Ru(II) complex, cis-$Cl_2(DMSO)_4Ru$, should also be pursued, since it is likely that the therapeutic value of this compound can be improved by substitution of ligands which could form hydrogen bonds with the carbonyl group of guanine. Despite the initial studies of Dwyer [2,3], at the present time it appears unlikely that Ru(II) compounds stabilized solely by aromatic nitrogen heterocycles will prove effective against many cancers. Moreover, the neurotoxicity noted for derivatives of $(phen)_3Ru(II)$ would further limit the utilization of this line of complexes [172]. This series of compounds should also be rapidly metabolized by the liver to yield glucuronide conjugates. More speculative approaches to attaining chemotherapeutic agents include the use of binuclear ruthenium complexes [173], which should behave analogously to the dialkylating antitumor drugs [1,72], and employing intercalating ligands attached to the metal by flexible aliphatic chains, which could insert into the DNA helix and so enhance the specificity of the metal for nucleic acid interactions.

The strong possibility of commercially available ^{97}Ru in the fairly near future and its excellent radiophysical properties (γ-ray energies of 215 keV (90%) and 324 keV (8%)) which would enable its use in presently equipped nuclear medicine facilities, argues strongly for its incorporation into tumor-specific radiodiagnostic agents. Ruthenium red, ruthenocene derivatives, and various Ru(III) coordination compounds already show promise in this regard. Ruthenium complexes, which should tend to localize in tumors by virtue of the properties of coordinated biological ligands such as amino acids, nucleotides, coenzymes, antibiotics, porphyrins [174,175], proteins, and nucleic acids, should also be investigated. Bleomycin, an anticancer antibiotic known to concentrate in tumors and interact with DNA and which also strongly binds several metal ions, is particularly attractive. The conveniently long half-life (2.9 days) of ^{97}Ru and the high affinity of ruthenium for nitrogen ligands should make the synthesis and quality control of compounds of this type relatively straightforward, representing a distinct advantage over most presently used radioscintigraphic agents.

Finally, it should be cautioned that, as in all other families of antitumor compounds, many ruthenium complexes are mutagenic and some may be carcinogenic, so that some care should be exercised in their handling. However, relative to many presently used anticancer drugs, this probably represents no particular disadvantage, and the concentrations necessary for radiodiagnostic study are so low that their effects should be negligible.

ACKNOWLEDGMENTS

The persevering efforts of the author's associates in the Bio-Inorganic Chemistry Laboratory at Boston College previously cited in the text are gratefully acknowledged. The author is also indebted to Prof. Ronald Yasbin of the Pennsylvania State University, Dr. Josef Pitha of the Gerontology Research Institute of the National Institutes of Health, Prof. Asher D. Kelman of the Boston University Medical School, P. Richards and Dr. Suresh Srivastava of Brookhaven National Laboratory, and Prof. Derek Hodgson of the University of North Carolina for their collaboration in providing many of the results reported here.

Support for work carried out by the author has been provided by grants from the NIH, CA-18522 (Boston College and Wheaton College), GM-13638 (Stanford University), and the American Cancer Society, Massachusetts Division.

ABBREVIATIONS

Ado	Adenosine
bipy	Bipyridine
Cyd	Cytidine
dach	1,2-diaminocyclohexane
dGuo	Deoxyguanosine
DMSO	Dimethylsulfoxide
DNA	Deoxyribonucleic acid
en	Ethylenediamine
Guo	Guanosine
His	Histidine
Ino	Inosine
Isn	Isonicotinamide
LMCT	Ligand-to-metal charge transfer
1MeAdo	1-methyladenosine
3MeCyd	3-methylcytidine
1MeCyt	1-methylcytosine
1,3Me_2Xan	1,3-dimethylxanthine
NADH	Nicotinamide-adenine dinucleotide (reduced)
Ox	Oxalate
phen	O-phenanthroline
pyr	Pyridine
TFMS	Triflouromethylsulfonate

REFERENCES

1. A. Korolkovas and J. H. Burckhalter, in Essentials of Medicinal Chemistry, Wiley-Interscience, New York, 1976, p. 366.
2. F. P. Dwyer, E. Mayhew, E. M. F. Roe, and A. Shulman, *Brit. J. Cancer, 19,* 195 (1965).
3. D. O. White, A. W. Harris, and A. Shulman, *J. Exp. Biol., 41,* 527 (1963).
4. B. Rosenberg, E. Renshaw, L. Van Camp, J. Hartwick, and J. Drobnik, *J. Bacteriol., 93,* 716 (1967).
5. J. R. Durig, J. Danneman, W. D. Behnke, and E. E. Mercer, *Chem. Biol. Interact., 13,* 287 (1976).
6. T. Giraldi, G. Save, G. Bertoli, G. Mestroni, and G. Zassinovich, *Canc. Res., 37,* 2662 (1977).
7. M. J. Cleare, *Coord. Chem. Rev., 12,* 349 (1974).
8. R. S. Bruce and T. E. F. Carr, *React. Sci. Technol., J. Nucl. Energy, 14,* 9 (1961).
9. R. S. Bruce and T. E. F. Carr, *React. Sci. Technol., J. Nucl. Energy, 14,* 145 (1961).
10. R. S. Bruce, T. E. F. Carr, and M. E. Collins, *Health Phys., 8,* 397 (1962).
11. L. A. Buldakov, in Raspredelenie, Biol. Deistvie i Migratsiya Radiat. Izotopov, Gos. Izd. Med. Lit, Sb., Moscow, 1961, p. 164.
12. D. Seidel, A. Catsch, and K. Schweer, *Strahlenther., 122,* 595 (1963).
13. G. G. Berg and E. Ginsberg, *Health Phys., 30,* 329 (1976).
14. C. E. Webber and J. W. Harvey, *Health Phys., 30,* 352 (1976).
15. J. F. Stara, N. S. Nelson, D. Rosa, J. Rocco, and L. K. Bustad, *Health Phys., 20,* 113 (1971).
16. J. E. Furchner, C. R. Richmond, and G. A. Drake, *Health Phys. 21,* 355 (1971).
17. W. M. Pusch, *Health Phys., 15,* 515 (1968).
18. B. V. Sastry, *Toxicol. Appl. Pharmacol., 9,* 419 (1966).
19. I. G. Gilbert, *Biochim. Biophys. Acta, 79,* 568 (1964).
20. R. C. Thompson, M. H. Weeks, O. L. Holles, and J. E. Ballou, *Amer. J. Roentgenol., 79,* 1026 (1958).
21. N. Tamagata, K. Iwashima, T. Hnuma, K. Watari, and T. Nagai, *Health Phys., 16,* 159 (1969).

22. Y. Enomoto, K. Watari, and R. Ichikawa, *J. Radiat. Res., 13,* 193 (1972).

23. A. Catsch and B. Kawin, in Radioactive Metal Mobilization in Medicine, C. Thomas Publications, Springfield, Ill., 1964.

24. G. Subramanian and J. G. McAfee, *Nucl. Med., 11,* 365 (1970).

25. D. Comor and C. Crouzel, *Radiochem. Radioanal. Lett., 27,* 307 (1976).

26. K. Mizukawa, G. Yamamoto, T. Tamai, M. Tanabe, and M. Yamato, *Radioisotopes, 27,* 19 (1978).

27. M. Tanabe, *Radioisotopes, 25,* 44 (1976).

28. M. Tanabe and G. Yamamoto, *Acta Medica Okayama, 29,* 431 (1975).

29. L. J. Anghileri, *Strahlenther., 129,* 173 (1975).

30. R. W. Yatscoff and J. E. Cummins, *Nature, 257,* 422 (1975).

31. J. E. Gruber and K. W. Jennette, *Biophys. Biochem. Res. Comm., 82,* 700 (1978).

32. D. R. Williams, *Chem. Rev., 72,* 203 (1977).

33. F. W. Sunderman, in Metal Toxicity in Mammals (T. D. Luckey and B. Venugopal, eds.), Plenum Press, New York, 1977.

34. NIOSH Carcinogen List, *Chem. Eng. News, 56,* No. 31, 20 (1978).

35. C. Monti-Bragadin, M. Tamaro, and E. Banfi, *Chem. Biol. Interact., 11,* 469 (1975).

36. H. Sigel, ed., Carcinogenicity and Metal Ions, Metal Ions in Biological Systems, Vol. 10, Marcel Dekker, New York, 1979. (See B. Rosenberg, Chap. 3 of this volume.)

37. W. P. Griffith, in The Chemistry of the Rarer Platinum Metals, Interscience, New York, 1968, p. 126

38. B. A. Moyer and T. J. Meyer, *J. Amer. Chem. Soc., 100,* 3601 (1978).

39. H. Taube, *Sur. Prog. Chem., 6,* 1 (1973).

40. H. Taube, *Coord. Chem. Rev., 26,* 33 (1978).

41. J. E. Huheey, in Inorganic Chemistry: Principles of Structure and Reactivity, 2nd ed., Harper and Row, New York, 1978, p. 276.

42. P. C. Ford, *Coord. Chem. Rev., 5,* 75 (1970).

43. P. C. Ford, D. F. P. Rudd, R. Gaunder, and H. Taube, *J. Amer. Chem. Soc., 90,* 1187 (1968).

44. G. D. Watt, *J. Amer. Chem. Soc., 94,* 7351 (1972).

45. C. G. Kuehn and H. Taube, *J. Amer. Chem. Soc., 98,* 689 (1976).

46. F. Bottomley, *Coord. Chem. Rev., 26,* 7 (1978).

47. H. J. Peresie and J. A. Stanko, *Chem. Commun.,* 1674 (1970).

48. S. E. Diamond and H. Taube, *J. Amer. Chem. Soc.*, *97*, 5921 (1975).

49. D. M. Klassen, C. W. Hudson, and E. L. Shaddix, *Inorg. Chem.*, *14*, 2733 (1975).

50. T. Matsubara and C. Creutz, *J. Amer. Chem. Soc.*, *100*, 6255 (1978).

51. E. E. Mercer and P. E. Dumas, *J. Amer. Chem. Soc.*, *94*, 6426 (1972).

52. (a) K. Rieder and H. Taube, *J. Amer. Chem. Soc.*, *99*, 7891 (1977).
(b) H. Fischer, G. M. Tom, and H. Taube, *J. Amer. Chem. Soc.*, *98*, 5512 (1976).

53. E. A. Stadlbauer, E. Nipper, and M. Wenzel, *Radiopharmaceuticals*, *13*, 491 (1977).

54. A. J. Taylor and M. Wenzel, *Xenobiotica*, *8*, 107 (1978).

55. M. Wenzel, E. Nipper, and W. Klose, *J. Nucl. Med.*, *18*, 367 (1977).

56. B. Riesselmann and M. Wenzel, *Physiol. Chem.*, *358*, 1353 (1977).

57. M. Wenzel, R. Herken, and W. Klose, *Z. Naturforsch.*, *32*, 473 (1977). M. Wenzel and W. Klose, *Klin. Wochensch.*, *55*, 559 (1977).

58. M. Wenzel, N. Subrumanian, and E. Nipper, *Naturwissenschaften* *63*, 341 (1976). A. J. Taylor and M. Wenzel, *Biochem. J.*, *172*, 77 (1978).

59. M. Wenzel, *Strahlenther.*, *154*, 506 (1978).

60. A. D. Allen, T. Eliades, R. O. Harris, and P. Reinsalu, *Can. J. Chem.*, *47*, 1605 (1969).

61. S. D. Pell and J. N. Armor, *Inorg. Chem.*, *12*, 873 (1973).

62. D. W. Raichart and H. Taube, *Inorg. Chem.*, *11*, 999 (1972).

63. M. J. Clarke, *J. Amer. Chem. Soc.*, *100*, 5068 (1978).

64. M. J. Clarke, Doctoral dissertation, Stanford University, Stanford, Ca., 1974.

65. M. J. Clarke and H. Taube, *J. Amer. Chem. Soc.*, *96*, 5413 (1974).

66. M. J. Clarke and H. Taube, *J. Amer. Chem. Soc.*, *97*, 1397 (1975).

67. M. J. Clarke, *Inorg. Chem.*, *16*, 738 (1977).

68. M. J. Clarke, M. Dowling, A. Garafalo, and T. F. Brennan, *J. Amer. Chem. Soc.*, *101*, 223 (1979).

69. A. M. Zwickel and C. Creutz, *Inorg. Chem.*, *10*, 2395 (1971).

70. R. E. Shepherd and H. Taube, *Inorg. Chem.*, *12*, 1392 (1973).

71. M. E. Gress, C. Creutz, and C. O. Quicksall, unpublished results.

72. D. Richardson, J. Walker, J. E. Sutton, K. Hodgson, and H. Taube, *Inorg. Chem.*, *18*, 2216 (1980).

73. H. Stynes and J. A. Ibers, *Inorg. Chem.*, *10*, 2304 (1971).

74. G. M. Brown, J. E. Sutton, and H. Taube, *J. Amer. Chem. Soc.*, *100*, 2767 (1978).

75. D. Franco and H. Taube, *Inorg. Chem.*, *17*, 572 (1978).

76. (a) J. A. Stritar and H. Taube, *Inorg. Chem.*, *8*, 2281 (1969). (b) J. F. Endicott and H. Taube, *Inorg. Chem.*, *4*, 437 (1965).

77. P. C. Ford, J. D. Petersen, and R. E. Hintze, *Coord. Chem. Rev.*, *14*, 67 (1974).

78. D. Waysbort, M. Evenor, and G. Navon, *Inorg. Chem.*, *14*, 514 (1975).

79. M. Swarc, in Ions and Ion Pairs in Organic Reactions, Wiley-Interscience, New York, 1972, p. 1.

80. T. Takamatsu, *Bull. Chem. Soc. Jap.*, *47*, 118 (1974).

81. R. J. Allen and P. C. Ford, *Inorg. Chem.*, *11*, 679 (1972).

82. R. J. Allen and P. C. Ford, *Inorg. Chem.*, *13*, 237 (1974).

83. S. S. Isied and H. Taube, *Inorg. Chem.*, *15*, 3030 (1976).

84. J. A. Marchant, T. Matsubara, and P. C. Ford, *Inorg. Chem.*, *16*, 2160 (1977).

85. J. H. Baxendale, R. A. J. Rogers, and M. D. Ward, *J. Chem. Soc.*, *A*, 1246 (1970).

86. C. M. Elson, I. J. Itzkovitch, J. McKenney, and J. A. Page, *Can. J. Chem.*, *53*, 2922 (1975).

87. G. M. Coleman, J. W. Gesler, F. A. Shirley, and J. R. Kuempel, *Inorg. Chem.*, *12*, 1036 (1973).

88. H. S. Lim, P. J. Barclay, and F. C. Anson, *Inorg. Chem.*, *11*, 460 (1972).

89. P. C. Ford, J. R. Kuempel, and H. Taube, *Inorg. Chem.*, *7*, 1976 (1968).

90. P. C. Ford and C. Sutton, *Inorg. Chem.*, *8*, 1545 (1969).

91. S. S. Isied and H. Taube, *Inorg. Chem.*, *15*, 3070 (1976).

92. S. S. Isied and H. Taube, *Inorg. Chem.*, *13*, 1545 (1974).

93. R. J. Sundberg, R. B. Bryan, I. F. Taylor, and H. Taube, *J. Amer. Chem. Soc.*, *96*, 381 (1974).

94. J. A. Broomhead, F. Basolo, and R. G. Pearson, *Inorg. Chem.*, *3*, 826 (1964).

95. L. A. P. Kane-Maguire and G. Thomas, *J.C.S. Dalton*, 1324 (1975).

96. J. N. Armor and H. Taube, *J. Amer. Chem. Soc.*, *92*, 6170 (1970).

97. (a) J. N. Armor, Doctoral dissertation, Stanford University, Stanford, Ca., 1970.
(b) D. Waysbort and G. Navon, *Inorg. Chem.*, *18*, 9 (1979).

98. A. Ohyoshi, T. Shinohara, Y. Hosoyamada, T. Yamada, and Y. Hiroshima, *Bull. Chem. Soc. Jap.*, *46*, 2133 (1973).

99. K. Ohkubo, H. Sakamoto, and A. Ohyoshi, *Chem. Lett.*, *969* (1973).

100. T. Matsubara and P. C. Ford, *Inorg. Chem.*, *15*, 1107 (1976).

101. V. Gutman, *Struc. Bond.*, *15*, 141 (1973).

102. H. Taube, Electron-Transfer Reactions of Complex Ions in Solution, Academic Press, New York, 1970.

103. L. E. Bennett, *Prog. Inorg. Chem.*, *18*, 1 (1973).

104. R. X. Ewalt and L. E. Bennett, *J. Amer. Chem. Soc.*, *96*, 940 (1974).

105. A. J. Miralles, R. E. Armstrong, and A. Haim, *J. Amer. Chem. Soc.*, *99*, 1416 (1977).

106. A. Ohyoshi, K. Toshinkuni, H. Ohtsuama, T. Yamashita, and S. Sakaki, *Bull. Chem. Soc. Jap.*, *50*, 666 (1977).

107. R. W. Craft and R. G. Gaunder, *Inorg. Chem.*, *14*, 1283 (1975).

108. M. Chou, C. Creutz, and N. Sutin, *J. Amer. Chem. Soc.*, *99*, 5615 (1977).

109. T. J. Meyer, *Acc. Chem. Res.*, *11*, 94 (1978).

110. D. M. Stanbury, O. Haas, and H. Taube, *Inorg. Chem.*, *19*, 518 (1980).

111. A. J. Thomson, R. J. P. Williams, and S. Reslova, *Struc. Bond.*, (Berlin), *11*, 1 (1972).

112. C. Monti-Bragadin, L. Ramani, L. Samer, G. Mestroni, and G. Zassinovich, *Antimicr. Ag. Chemother.*, *7*, 825 (1975).

113. (a) S. Mansy, B. Rosenberg, and A. J. Thomson, *J. Amer. Chem. Soc.*, *95*, 1663 (1973).
(b) W. M. Scovell and R. S. Reaoch, *J. Amer. Chem. Soc.*, *101*, 174 (1979).

114. See Ref. 64. The biological activity of this compound was kindly tested by Dr. B. Rosenberg's group at the Michigan State University, 1974.

115. M. J. Clarke, unpublished results.

116. M. J. Clarke, M. Buchbinder, and A. D. Kelman, *Inorg. Chim. Acta*, *27*, L87 (1978).

117. A. D. Kelman, M. J. Clarke, S. D. Edmonds, and H. J. Peresie, *J. Clin. Hematol. Oncol.*, *7*, 274 (1977).

118. P. M. Gullino, *Adv. Exp. Biol. Med.*, *75*, 521 (1975).

119. U. Del Monte, *Proc. Soc. Exp. Biol. Med.*, *125*, 853 (1967).

120. P. Vaupel and G. Thews, *Adv. Exp. Med. Biol.*, *75*, 547 (1976).

121. V. S. Shapot, *Adv. Exp. Med. Biol.*, *75*, 581 (1976).

122. P. Vaupel, *Microvasc. Res.*, *13*, 399 (1977).

123. G. Thews and P. Vaupel, *Adv. Exp. Med. Biol.*, *75*, 537 (1976).

124. (a) C. T. Gregg, in Growth Nutrition and Metabolism of Cells in Culture, Vol. I (G. H. Rothblat and V. J. Cristofalo, eds.), Academic Press, New York, 1972, p. 83.
(b) C. E. Wenner, *Adv. Enzymol.*, *29*, 321-390 (1967).
(c) S. Weinhaus, *Z. Krebsforsch.*, *87*, 115 (1976).

125. R. W. McKee, A. Dickey, and M. E. Parks, *Arch. Biochem. Biophys.*, *126*, 760 (1968).

126. D. B. Cater and A. F. Phillips, *Nature*, *174*, 121 (1954).

127. S. Diamond, Doctoral dissertation, Stanford University, Stanford, Ca., 1975.

128. S. E. Diamond, G. M. Tom, and H. Taube, *J. Amer. Chem. Soc.*, *97*, 2661 (1975).

129. I. P. Evans, G. W. Everett, and A. M. Sargeson, *J. Amer. Chem. Soc.*, *98*, 8041 (1976).

130. K. Schug and C. P. Guengerich, *J. Amer. Chem. Soc.*, *97*, 4136 (1975).

131. S. D. Pell and J. N. Armor, *J. Amer. Chem. Soc.*, 5012 (1975).

132. (a) K. Gleu and W. Breuel, *Z. Anorg. Allgem. Chem.*, *237*, 355 (1938).
(b) K. Gleu and I. Buddeckar, *Z. Anorg. Allgem. Chem.*, *268*, 202 (1952).
(c) H. Hartman and C. Buschbeck, *Z. Phys. Chem.*, *11*, 120 (1957).

133. V. E. Alvarez, R. J. Allen, T. Matsubara, and P. C. Ford, *J. Amer. Chem. Soc.*, *96*, 7686 (1974).

134. R. Sundberg and G. Gupta, *Bioinorg. Chem.*, *3*, 39 (1973).

135. C. R. Matthews, P. M. Erickson, D. L. Van Vliet, and M. Petersheim, *J. Amer. Chem. Soc.*, *100*, 2261 (1978).

136. S. S. Isied, *J. Amer. Chem. Soc.*, *100*, 6754 (1978).

137. A. Zanella and H. Taube, *J. Amer. Chem. Soc.*, *93*, 7166 (1971).

138. H. Krentzien, M. J. Clarke, and H. Taube, *Bioinorg. Chem.*, *4*, 143 (1975).

139. M. E. Kastner, S. Edmonds, K. Eriks, and M. J. Clarke, unpublished results, 1979.

140. B. J. Graves and D. J. Hodgson, *J. Amer. Chem. Soc.*, *101*, 5608 (1979).

141. M. J. Clarke, K. Coffey, and G. Kirvan, unpublished results, 1979.

142. (a) C. Yu, S. Peng, I. Akiyama, J. Lin, and P. R. LeBreton, *J. Amer. Chem. Soc.*, *100*, 2279 (1978).
(b) S. Peng, A. Padva, and P. R. LeBreton, *Proc. Nat. Acad. Sci.*, *73*, 2966 (1976).
(c) D. Dougherty and S. P. McGlynn, *J. Chem. Phys.*, *67*, 1289 (1977).

143. M. J. Clarke and K. Coffey, work in progress.

144. J. A. Zoltewicz, D. F. Clark, T. W. Sharpless, and G. Grahe, *J. Amer. Chem. Soc.*, *92*, 1741 (1970). L. Hevesi, E. Wolfson-Davidson, J. B. Nagy, O. B. Nagy, and A. Bruylants, *J. Amer. Chem. Soc.*, *94*, 4715 (1972).

145. (a) M. Beer and I. Moudrianakis, *Proc. Nat. Acad. Sci. U.S.*, *48*, 409 (1962).
(b) A. M. Fiskin and M. Beer, *J. Mol. Biol.*, *86*, 865 (1974).

146. M. J. Clarke, D. Rennert, M. Buchbinder, A. D. Kelman, and S. Bitler, unpublished results.

147. (a) T. Okamoto, A. Ohno, and S. Oka, *J.C.S. Chem. Comm.*, *181* (1977).
(b) G. Blankenhorn, *Eur. J. Biochem.*, *50*, 351 (1975).

148. (a) D. F. Parsons, G. R. Williams, and B. Chance, *Acad. Sci.*, *137*, 643 (1966).
(b) G. C. Shore and J. R. Tata, *J. Cell. Biol.*, *72*, 714 (1977).
(c) G. L. Sottocasa, B. Kuylenstierna, L. Ernster, and A. Bergstrand, *J. Cell. Biol.*, *32*, 415 (1967).

149. A. L. Lehninger, in Biochemistry, Worth, New York, 1975, p. 535.

150. D. E. Metzler, in Biochemistry, Academic Press, New York, 1977, p. 421.

151. In view of these results and the high affinity of ruthenium for sulfur and nitrogen ligands, an antidote for ruthenium ingestion might be a meal high in sulfur and protein, such as eggs, cheese or peanuts, administered as quickly as possible to bind the metal and facilitate its excretion.

152. F. D. Vasington, P. Gazzotti, R. Tiozzo, and E. Carofoli, *Biochim. Biophys. Acta*, *256*, 43 (1972).

153. C. L. Moore, *Biochem. Biophys. Res. Commun.*, *42*, 298 (1971).

154. P. M. Smith, T. Fealey, J. E. Earley, and J. V. Silverton, *Inorg. Chem.*, *10*, 1943 (1971).

155. J. H. Luft, *J. Cell. Biol.*, *27*, 61A (1965).

156. J. H. Luft, *J. Cell. Biol.*, *23*, 54A (1964).

157. J. H. Luft, *Anat. Rec.*, *171*, 369 (1971).

158. G. T. Gustafson and E. Pihl, *Acta Path. Microbiol. Scand.*, *68*, 393 (1967).

159. J. N. Howell, *J. Cell. Biol.*, *62*, 242 (1974).

160. K. Utsumi, Y. Matsunage, and T. Oda, *Symp. Cell Biol.*, *24*, 19 (1973).

161. W. Bondareff, *J. Neurosurg.*, *32*, 145 (1970).

162. K. C. Reed and F. L. Bygrave, *FEBS Lett.*, *46*, 109 (1974).

163. M. Singer, N. Krishnan, and D. A. Fyfe, *Anat. Rec.*, *173*, 375 (1972).

164. R. Rahamimoff and E. Alnaes, *Proc. Nat. Acad. Sci.*, *70*, 3613 (1973).

165. R. Tapia, M. Graciela, L. Duran, and R. Drucker-Colin, *Brain Res.*, *116*, 101 (1976).

166. L. J. Anghileri, *J. Nucl. Med.*, *16*, 795 (1975).

167. M. J. Clarke, A. D. Kelman, and S. J. Edmonds, unpublished results.

168. R. Yasbin, C. R. Matthews, and M. J. Clarke, *Chem. Biol. Interact.*, in press, 1980.

169. Private communication, J. Pitha; see *Eur. J. Biochem.*, *82*, 285 (1978) for experimental details.

170. (a) B. Rosenberg, *Cancer Chemother. Rep.*, Part 1, *59*, 589 (1975).
(b) S. Srivastava, P. Richards, G. Neinken, P. Som, H. Atkins, S. Larson, Z. Grunbaum, J. Rasey, M. Dowling, and M. J. Clarke, Abstr. 2d Int. Symp. Radiopharm., Seattle, Washington, March, 1979.

171. B. T. Khan and A. Mehmood, *J. Inorg. Nucl. Chem.*, *40*, 1938 (1978).

172. F. P. Dwyer, E. C. Gyarfas, W. P. Rogers, and J. H. Koch, *Nature*, *170*, 490 (1952).

173. D. Richardson, Stanford University, private communication, 1979.

174. G. M. Brown, F. R. Hopf, J. A. Ferguson, T. J. Meyer, and D. G. Whitten, *J. Amer. Chem. Soc.*, *95*, 5939 (1973).

175. M. Tsutsui, D. Ostfeld, L. M. Hoffman, *J. Amer. Chem. Soc.*, *95*, 1820 (1971). M. Tsuitsui, D. Ostfeld, J. N. Francis, and L. M. Hoffman, *J. Coord. Chem.*, *1*, 115 (1971).

Chapter 6

METAL COMPLEXES OF ALKYLATING AGENTS AS POTENTIAL ANTICANCER AGENTS

Melvin D. Joesten
Department of Chemistry
Vanderbilt University
Nashville, Tennessee

1. INTRODUCTION 285
2. ALKYLATING AGENTS USED IN CHEMOTHERAPY 287
   2.1. Major Classes 287
   2.2. Mechanism of Action 287
   2.3. Evaluation of Clinical Effectiveness 294
3. METAL COMPLEXES OF ALKYLATING AGENTS 295
   3.1. Complexes of Aziridinyl Phosphine Oxides 296
   3.2. Complexes of Aziridinyl Phosphine Sulfides 298
   3.3. Complexes of Cyclophosphamide 298
4. NEW ANTICANCER DRUG DESIGN 299
5. RECENT DEVELOPMENTS 301
   ABBREVIATIONS 301
   REFERENCES 302

1. INTRODUCTION

The author and his associates first became interested in metal complexes of alkylating agents during their investigation of the influence of various organic substituents on the coordinating ability of the phosphoryl group in

$$\underset{\mathrm{I}}{R_3P{=}O} \quad \text{and} \quad \underset{\mathrm{II}}{R_2P(O){-}O{-}P(O)R_2}$$

Two of the most versatile phosphoryl ligands are hexamethylphosphoramide [1] [R=N$(CH_3)_2$ in Struct. I] and octamethylpyrophosphoramide [R=N$(CH_3)_2$ in Struct. II] [2]. They chose tris-(1-aziridinyl)-phosphate oxide [R=N◁ in Struct. I] as a ligand to determine whether the reduced steric requirements of the aziridine ring would enhance the coordinating properties of the phosphoryl oxygen atom. Although the ligand is not as versatile as hexamethylphosphoramide, the resulting complexes were screened in the hope that such complexes would have a higher therapeutic index (toxic dose/effective dose) than the parent ligand (see Sec. 3.1.). The results of this study encouraged the author and his coworkers to consider the potential of metal complexes of alkylating agents as anticancer drugs.

During the past 40 years many different types of metal complexes have been tested for antitumor activity (see Chap. 1). However, no systematic study of metal complexes of alkylating agents has been carried out even though Furst, in a book entitled *Chemistry of Chelation in Cancer*, proposed that alkylating agents formed chelating agents in situ [3]. Furst also proposed that the metal complex

$$\underset{\mathrm{III}}{[(ClCH_2CH_2)_2S]_2PtCl_2}$$

might be a useful antitumor agent [4]. Such predictions are fascinating in view of the recent success of cis-$[Pt(NH_3)_2Cl_2]$ as an anticancer drug.

Alkylating agents have been a major class of anticancer drugs since research in the 1940s indicated that nitrogen mustards have antitumor action [5]. After the initial use of the nitrogen mustard, methylbis(2-chloroethyl)amine (HN2), a wide variety of alkylating agents have been synthesized and tested. Several recent reviews provide an excellent overview of cancer research on alkylating agents [6-14].

2. ALKYLATING AGENTS USED IN CHEMOTHERAPY

2.1. Major Classes

Biological alkylating agents are compounds capable of transferring alkyl groups to nucleophilic sites under physiological conditions. Several, such as cyclophosphamide, are not alkylating agents in vitro but react in vivo to give metabolites which are alkylating agents. All effective biological alkylating agents either have two or more alkylating groups or combine alkylation with some other inhibiting action. Compounds which have alkylating capacity as all or part of their anticancer activity fall into six major classes: mustards, aziridine derivatives, epoxides, nitrosoureas, sulfonates, and triazenes. Compounds representative of each class are shown in Figs. 1-3. All except the epoxides are used as anticancer drugs, and Table 1 summarizes the therapeutic uses of compounds from the other classes.

2.2. Mechanism of Action

A general representation of biological alkylation is

$$RCH_2Y + X^- \rightarrow RCH_2X + Y^-$$

IV

where the alkylating agent contributes an alkyl group, $RCH2^-$ to a nucleophilic center on a biological substrate, X^- (X and Y may be neutral). Although alkylating agents are capable of reacting with a variety of O, N, and S nucleophilic sites in cellular compounds,

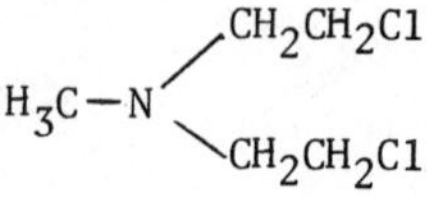

Nitrogen mustard, mechlorethamine (HN2)

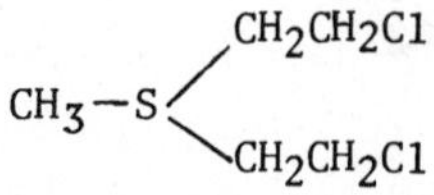

Sulfur mustard

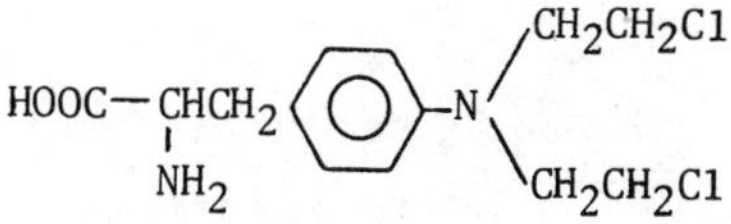

L-phenylalanine mustard, L-sarcolysin, Melphalan

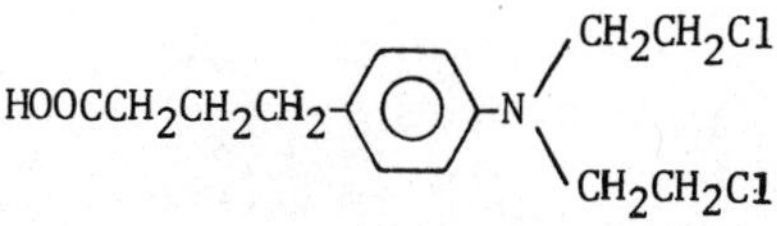

Chlorambucil

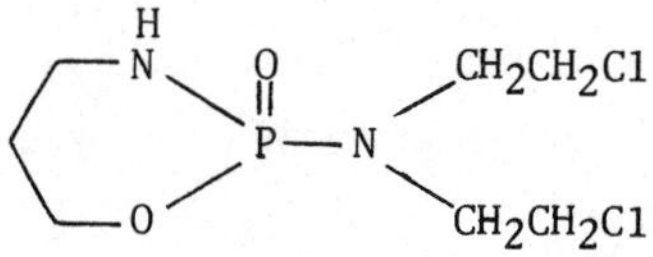

Cyclophosphamide, cytoxan

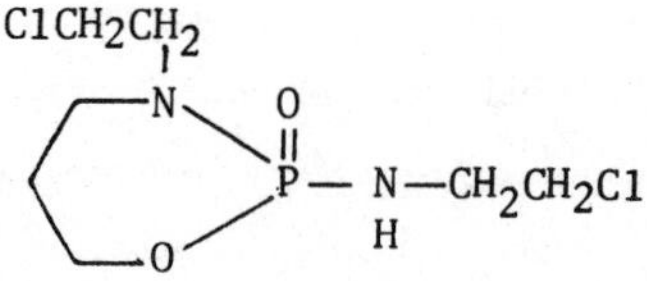

Isophosphamide

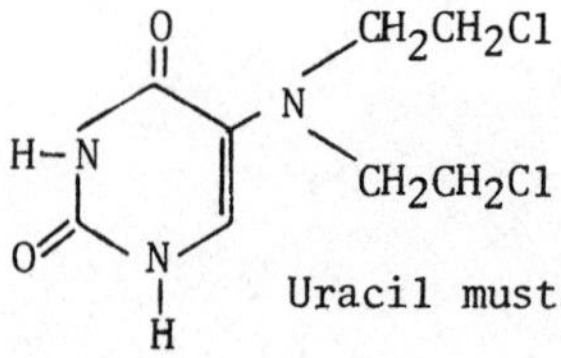

Uracil mustard

FIG. 1. Nitrogen and sulfur mustards.

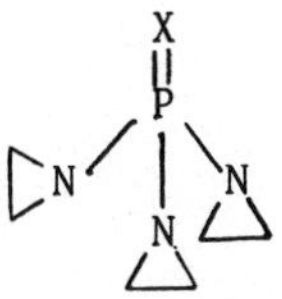

X = O TEPA (APO)

X = S thio-TEPA (APS)

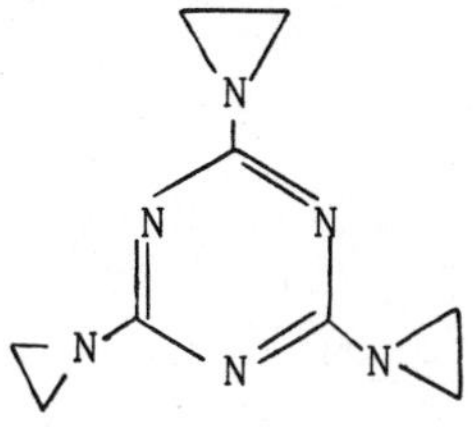

Triethylenemelamine

(TEM)

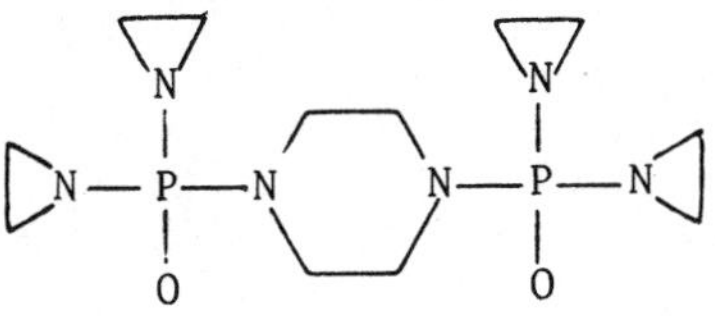

Dipine

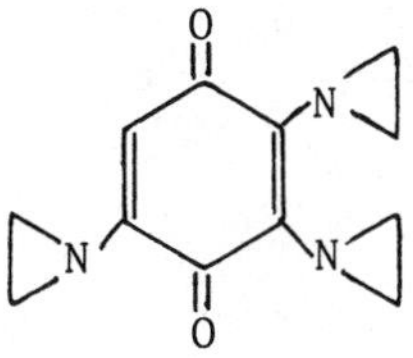

2,3,S-tris(aziridino)-1,4-

benzoquinone; Trenimon

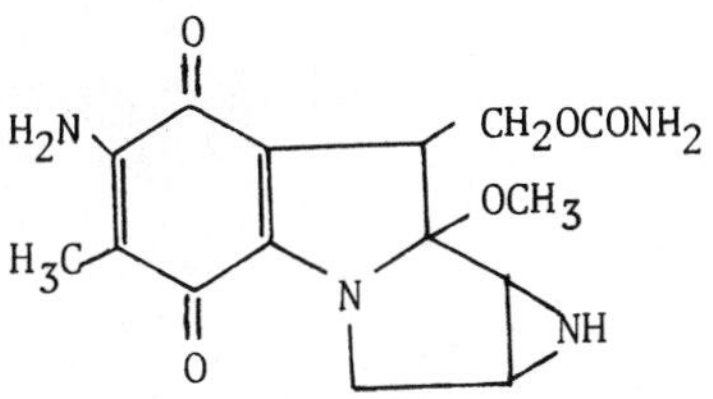

Mitomycin C

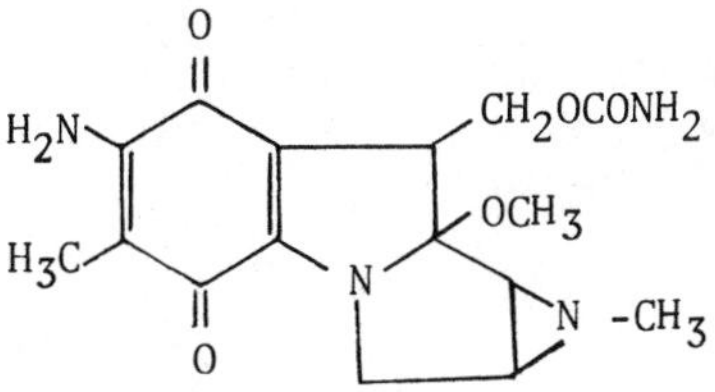

Porfiromycin

FIG. 2. Aziridine derivatives.

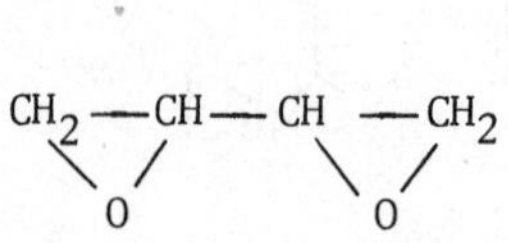

$$\underset{\diagdown O \diagup}{CH_2-CH}-\underset{\diagdown O \diagup}{CH-CH_2}$$

1,2:3,4-diepoxybutane

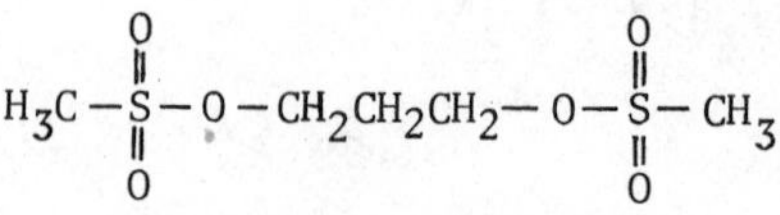

$$H_3C-\overset{O}{\underset{O}{\overset{\|}{\underset{\|}{S}}}}-O-CH_2CH_2CH_2-O-\overset{O}{\underset{O}{\overset{\|}{\underset{\|}{S}}}}-CH_3$$

Busulfan, Myleran

$$ClCH_2CH_2-\underset{NO}{N}-\overset{O}{\overset{\|}{C}}-\underset{H}{N}-CH_2CH_2Cl$$

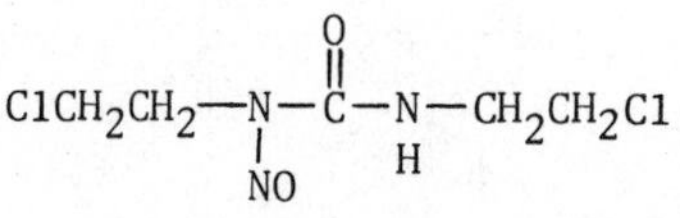

bishchloroethylnitrosourea

(BCNU)

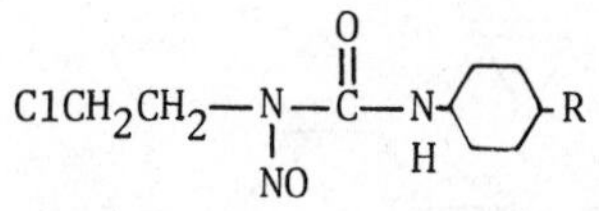

$$ClCH_2CH_2-\underset{NO}{N}-\overset{O}{\overset{\|}{C}}-\underset{H}{N}-C_6H_{10}-R$$

R = H, chloroethylcyclohexyl-

nitrosourea (CCNU)

R = CH_3, methyl CCNU

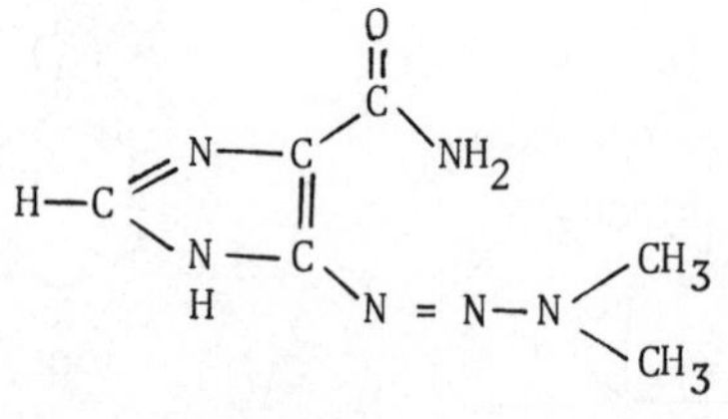

5-(3,3-dimethyl-1-triazeno)imidazole-4-carboxamide (DTIC)

FIG. 3. Epoxides, sulfonates, nitrosoureas, and triazenes.

TABLE 1

Chemotherapeutic Uses of Alkylating Agents[a]

Agent	Cancers
Mechlorethamine (used as hydrochloride salt)	Lymphomas;[b] breast and ovary carcinomas
Melphalan	Multiple myeloma; breast and ovary carcinomas; polycythemia vera
Chlorambucil	Chronic lymphocytic leukemia; lymphomas; breast, ovary, and testis carcinomas
Cyclophosphamide	Acute and chronic lymphocytic leukemias, lymphomas; multiple myeloma; neuroblastoma; Wilms' tumor; rhabdomyosarcoma; breast, ovary, and lung carcinomas
Isophosphamide	Hodgkin Disease, bronchogenic, breast, and ovary carcinomas
Uracil mustard	Chronic lymphocytic leukemia; lymphomas; primary thrombocytosis; ovary carcinoma
TEM	Chronic leukemias; lymphomas; retinoblastoma; breast and ovary carcinomas
Thio-TEPA	Lymphomas; retinoblastoma; breast and ovary carcinomas
Mitomycin C	Carcinomas; lymphomas; melanoma; sarcomas
Busulfan	Chronic granulocytic leukemia; polycythemia vera; primary thrombocytosis
BCNU and CCNU	Lymphomas; primary and metastatic brain tumors
DTIC	Hodgkin Disease; melanoma; sarcomas

[a]See Ref. 12, p. 1250.

[b]In this table, lymphomas include Hodgkin Disease.

the alkylation of the N-7 atom of guanine in DNA is considered the primary site of alkylation by biological alkylating agents [6,15-20]. Figure 4 illustrates how the bifunctionality of nitrogen mustard results in (top) inter- or intrastrand cross-linking of DNA; and/or (bottom) cross-linking of DNA and protein. Although other reactions take place, these are the most important since the reactions are irreversible [6,12,16].

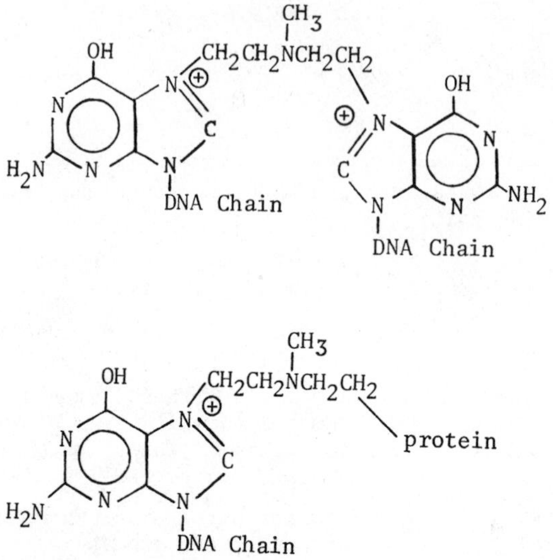

FIG. 4. Irreversible damage to DNA by (top) cross-linking DNA strands and (bottom) cross-linking DNA and proteins.

The mechanism for the biological alkylation step is generally S_N2. For aliphatic nitrogen mustards, a cyclic ethylenimmonium ion is rapidly formed followed by the bimolecular reaction with the nucleophilic site. The three-membered rings in aziridine derivatives are unstable and undergo bimolecular reaction with X^-.

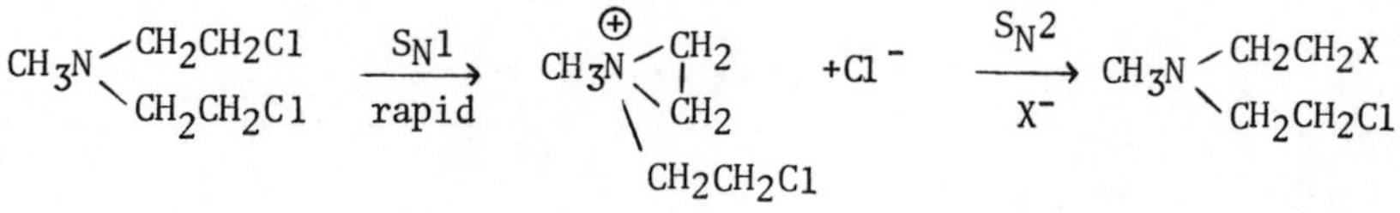

V

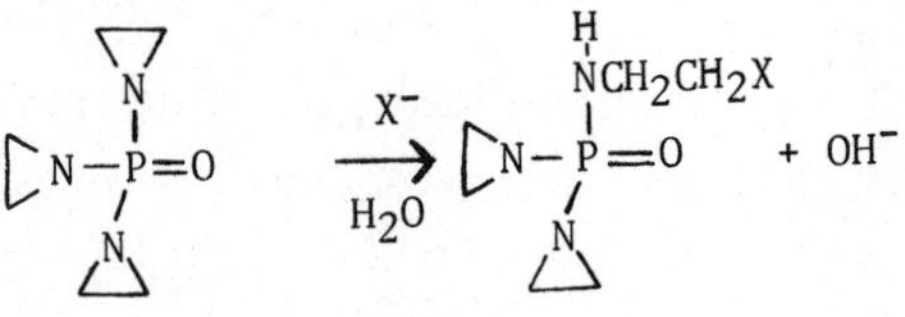

VI

Nitrosoureas hydrolyze rapidly under physiological conditions to give species which are either alkylating agents or carbamylating agents [12,13,20,21]. BCNU is a bifunctional alkylating agent due to the generation of vinyl carbonium ion and 2-chloroethyl carbonium ion. The carbamylating species generated by BCNU is 2-chloroethylisocyanate. CCNU generates the alkylating agent, vinyl carbonium ion, along with cyclohexylisocyanate, a carbamylating species. The isocyanates generated by nitrosoureas react with proteins while the carbonium ions cross-link DNA strands or DNA and protein. Alkylating activity appears to be the most important function of nitrosoureas [22], and recent research on DNA-protein cross-linking and DNA interstrand cross-linking indicates that chloroethylnitrosoureas with one alkylating group are also capable of cross-linking [23].

Cyclophosphamide is converted to an active alkylating agent by the liver. Although a number of different metabolites have been identified (see Fig. 5), the phosphoramide mustard appears to be

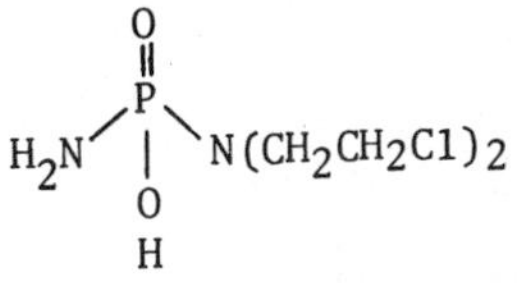

$$H_2N-P(=O)(OH)-N(CH_2CH_2Cl)_2$$

VII

the therapeutically active metabolite [24,25]. The clinical success of cyclophosphamide and the extensive research on the identification of its metabolites [26] has stimulated interest in both the design of new latent precursors for phosphoramide mustard derivatives like Struct. VII and the synthesis of a variety of new phosphamide mustards, both active and latent [27,28].

Mechanisms for the other biological alkylating agents shown in Figs. 1-3 will not be discussed here. Wheeler [13] describes mechanisms for the alkylating activity of busulfan and mitomycin C, and Oliverio [29] reviews the mechanisms in support of alkylating activity for the triazene DTIC.

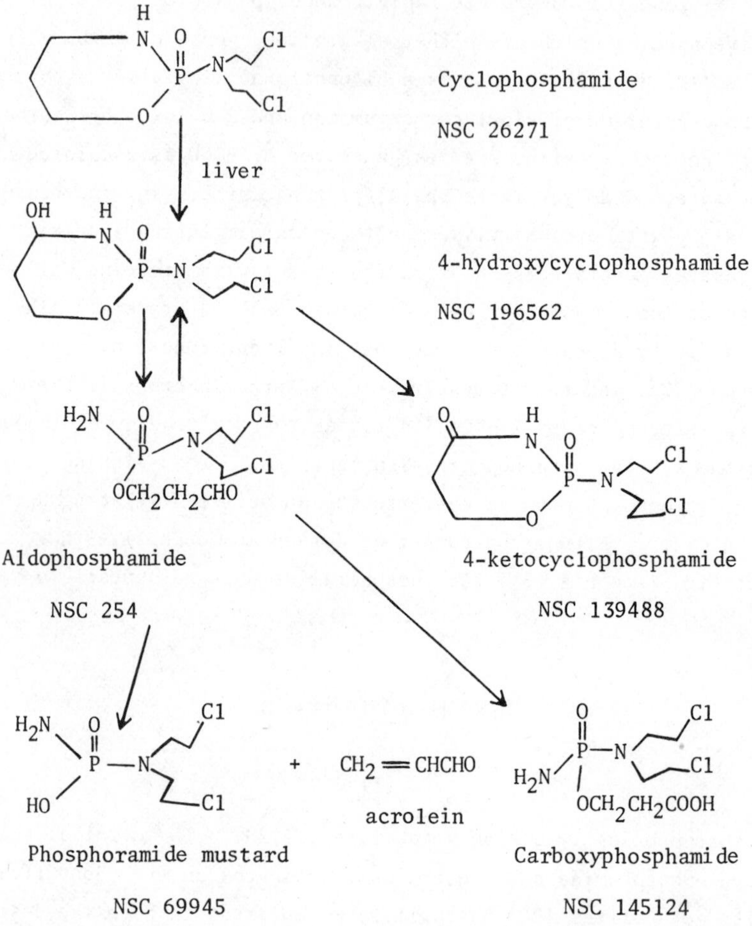

Cyclophosphamide

NSC 26271

liver

4-hydroxycyclophosphamide

NSC 196562

Aldophosphamide

NSC 254

4-ketocyclophosphamide

NSC 139488

Phosphoramide mustard

NSC 69945

+ $CH_2{=}CHCHO$

acrolein

Carboxyphosphamide

NSC 145124

FIG. 5. Metabolism of cyclophosphamide.

2.3. Evaluation of Clinical Effectiveness

2.3.1. Single Drug

Table 1 lists a number of therapeutic uses for biological alkylating agents. Sufficient data are not available for a precise comparison of the clinical effectiveness of these anticancer drugs. However, useful comparisons have been made by looking at the relative clinical effectiveness of anticancer drugs against a variety of cancers. For example, Apple [30] has proposed a list of the 13 "best" anticancer drugs, which includes the 6 alkylating agents, cyclophosphamide, melphalan, mechlorethamine, hexamethylamine, BCNU, and mitomycin C. Mitomycin C behaves like a bifunctional alkylating agent after chemical or enzymatic reduction of the quinone ring. However, mitomycin derivatives which do not contain the aziridine ring are still biologically active, and this has been attributed to intercalative binding with DNA [19].

2.3.2. *Combination of Drugs*

Single-agent chemotherapy has largely given way to combination chemotherapy because of the success in finding a therapeutic synergistic effect with drug combinations. For example, most of the cancers listed in Table 1 for treatment by cyclophosphamide are actually treated by the combination of cyclophosphamide with other anticancer drugs [10]. Since alkylating agents are toxic and cause harmful side effects, combinations of drugs were sought which would decrease the toxicity to the host without decreasing cytotoxicity to tumor cells. Success in finding combinations which are more effective than single agents has maintained interest in the use of alkylating agents in spite of their low therapeutic index. Several reviews have been published on both research and clinical applications of combination chemotherapy [30-32].

3. METAL COMPLEXES OF ALKYLATING AGENTS

The objective of the research in this area is to determine what effect metals have on the anticancer activity of alkylating agents. It is the author's hope that metal complexes of alkylating agents might increase the therapeutic index through a synergistic effect. The metal complexes might also serve as latent precursors to biologically active metabolites. Although the idea of using metal complexes as anticancer agents is not new [3,33], the discovery of Rosenberg and his coworkers [34-36] that platinum complexes have antitumor activity has resulted in renewed interest in metal complexes as a class of potential anticancer drugs [37,38].

3.1. Complexes of Aziridinyl Phosphine Oxides

In 1975 the author and his associates reported [39] the preparation and characterization of metal complexes of tris-(1-aziridino) phosphine oxide (APO), tris[1-(2-methyl) aziridino]-phosphine oxide (MAPO), bis-(1-aziridino)-N,N'-dimethylaminophosphine oxide (DO), and 1-aziridinobis (N,N'-dimethylamino)phosphine oxide (MO). A magnesium complex of P,P-bis-(1-aziridinyl)-N-phenylphosphinic amide (BAPPA) has also been prepared and tested [40]. The general structure of these ligands is

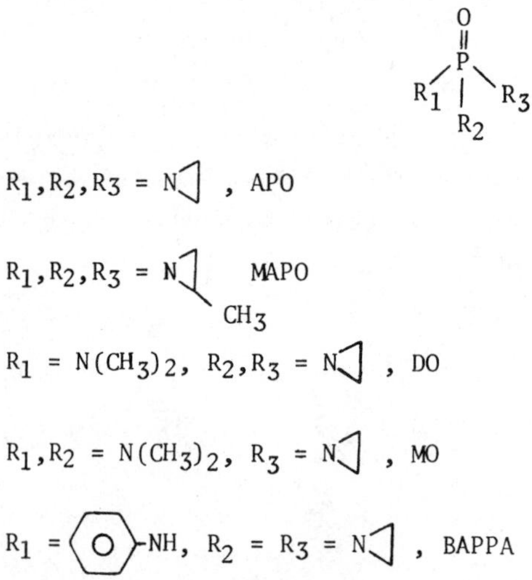

$$O{=}P(R_1)(R_2)(R_3)$$

R_1,R_2,R_3 = N◁ , APO VIII

R_1,R_2,R_3 = N◁(CH_3) MAPO IX

$R_1 = N(CH_3)_2$, R_2,R_3 = N◁ , DO X

$R_1,R_2 = N(CH_3)_2$, R_3 = N◁ , MO XI

R_1 = C_6H_5-NH, $R_2 = R_3$ = N◁ , BAPPA XII

The number of aziridine rings was varied from one to three to determine what effect this would have on both the stability and toxicity of the metal complexes. Despite the similarity of these ligands to hexamethylphosphoramide (HMPA), the replacement of the dimethylamino group with the aziridine ring reduces the coordinating ability of the phosphoryl oxygen.

Screening for all the compounds listed in Table 2 was done by the Drug Evaluation Branch of the Drug Research and Development Department of the National Cancer Institute. Most of the compounds were screened against L1210 leukemia, and a few against P388 leukemia, in mice. The T/C value in Table 2 represents the percentage of increase in mean survival time of the treated mice over that of the untreated controls. In general, T/C values must be greater

TABLE 2

Screening Results

Compound	NSC no.	Maximum nontoxic dose ($mg\ kg^{-1}$)	Effective dose ($mg\ kg^{-1}$)	Maximum T/C%	Tumor
$Mg(MO)_4(ClO_4)_2 \cdot 2H_2O$	169819	400	20	111	L1210
$Ba(MO)_4(ClO_4)_2$	169818	200	25	111	L1210
$Mg(DO)_4(ClO_4)_2 \cdot 2H_2O$	224968	200	50	96	P388
$Mg(MAPO)_4(ClO_4)_2 \cdot 2H_2O$	169773	400	300	150[a]	L1210
$Ca(MAPO)_3Cl_2$	169772	400	150	153	L1210
$Fe(MAPO)_6(ClO_4)_3 \cdot 5H_2O$	191282	400	400	115	L1210
$Cu(MAPO)_4(ClO_4)_2 \cdot 7H_2O$	189789	100	100	119	L1210
$Cu(MAPO)_4(BF_4)_2 \cdot 7H_2O$	189791	400	40	115	L1210
$Mg(BAPPA)_4(ClO_4)_2$	220465	12.5	6.25	172	P388
$Ag(APS)_2ClO_4$	224969	200	200	90	P388
$Co(MAPS)_6Cl_2$	187749	100	50	112	L1210
$Pt(MAPS)_2Cl_2$	218488	200	100	103	L1210
$Pt(MAPS)_2Cl_4$	186036	25	25	114	L1210
$[Pd(MAPS)Cl_2]_2$	237016	200	100	98	L1210
$[Pt(PAPS)Cl_2]_2$	237017	200	100	100	L1210
$Rh_2(propionate)_4 \cdot 2CP$	301994	4	4	107	L1210

[a]See Sec. 3.1.

than 125% for L1210 leukemia and 130% for P388 leukemia for a compound to be considered for second-stage screening. (T/C values in Table 2 which satisfy this requirement are underlined.) Two of the complexes, $Mg(MAPO)_4(ClO_4)_2 \cdot 2H_2O$ and $Ca(MAPO)_3Cl_2$, have undergone second-stage screening, but the results were not sufficiently promising to justify further testing.

3.2. Complexes of Aziridinyl Phosphine Sulfides

Metal complexes of tris(1-aziridino)phosphine sulfide (APS), bis-(1-aziridino)phenylphosphine sulfide (PAPS), and bis[1-(2-methyl)-aziridino]phenylphosphine sulfide (MPAPS) have also been prepared [41]. Since platinum readily coordinates to phosphine sulfides [42], the author and coworkers were interested in determining whether platinum complexes of aziridinyl phosphine sulfides would be more effective anticancer agents than the ligands by themselves. Table 2 lists the complexes and screening data. The screening data indicate that the metal complexes of aziridinyl phosphine sulfides have no activity. Since these complexes are insoluble in a wide range of solvents, biological activity is probably prevented by their inability to reach the appropriate sites.

3.3. Complexes of Cyclophosphamide

Since cyclophosphamide (CP) is widely used as an anticancer drug (see Secs. 2.3.1 and 2.3.2), the author and his associates were interested in determining whether metal complexes of cyclophosphamide would offer any advantage as anticancer agents. The only complexes which we have been able to isolate and characterize are $Rh_2(acetate)_4 \cdot 2CP$ and $Rh_2(propionate)_4 \cdot 2CP$ [43]. The parent rhodium(II) carboxylates have been shown to exhibit antitumor activity against L1210 leukemia and Ehrlich ascites tumors in mice [44,45]. The structure of the dimeric rhodium carboxylates is

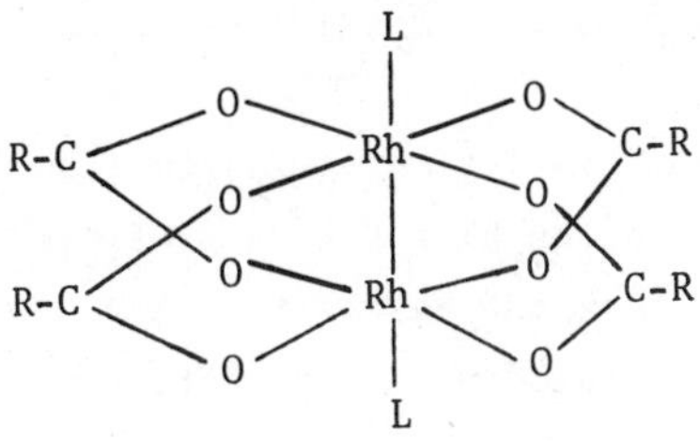

XIII

Adducts have been synthesized with L = water, alcohols, imidazole, diethyl sulfide, dimethyl sulfoxide, pyridine, and triphenyl phosphine [46-49]. The screening results listed in Table 2 for the $Rh_2(propionate)_2 \cdot 2CP$ adduct indicate the compound is not active.

Although these results are discouraging, the therapeutic synergism observed for the combination of cyclophosphamide and cis-$[Pt(NH_3)_2Cl_2]$ [50] leaves open the possibility that other coordination compounds of cyclophosphamide might be effective.

4. NEW ANTICANCER DRUG DESIGN

Although none of the compounds described in Sec. 3 have sufficient activity to warrant additional testing, a variety of complexes of alkylating agents remain to be synthesized and tested.

The anticancer drug effectiveness of alkylating agents can be improved by (1) reducing toxicity and harmful side effects; (2) enhancing membrane transport, and (3) increasing alkylating efficiency of DNA in tumor cells. Two types of metal complexes of alkylating agents which might prove to be effective anticancer drugs are (1) complexes which release active alkylating agents in vivo (latent precursor); and (2) complexes which dissociate in vivo to give an active alkylating species along with a metal complex species capable of inhibitory action. The remainder of this section will be devoted to a discussion of compounds which the author proposes as examples of these two types of metal complexes. None of these compounds have been synthesized, but hopefully the arguments presented here will stimulate interest in the synthesis, characterization, and testing of these compounds.

The ability of cis-[Pt$(NH_3)_2Cl_2$] to form intrastrand cross-links in DNA, because the bite distance of cis-Pt is similar to the separation between adjacent base sites on DNA (3.3 Å versus 3.4 Å), could be utilized in building a more effective blocking agent by incorporating alkylating functions and cis-Pt in the same molecule. Both interstrand and intrastrand cross-linking could lead to a more efficient inhibition of DNA synthesis, which might result in a higher therapeutic index. The sulfur mustard complex (Struc. III) proposed by Furst [4] is an example of a compound of this type. Other compounds which are possibilities are

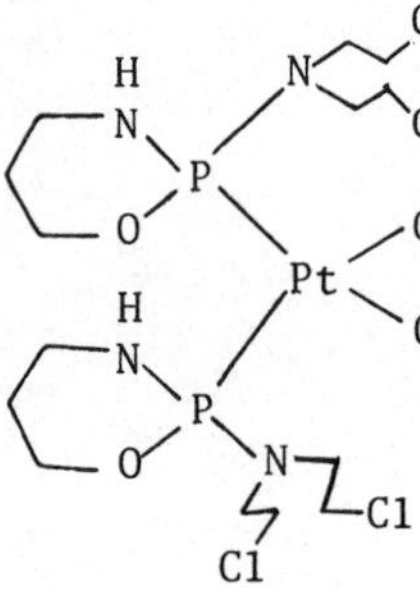

XIV 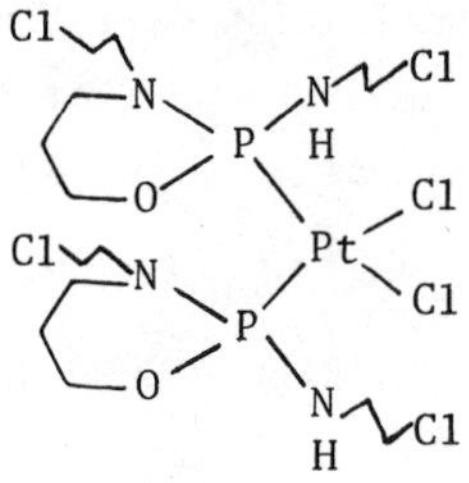 XV 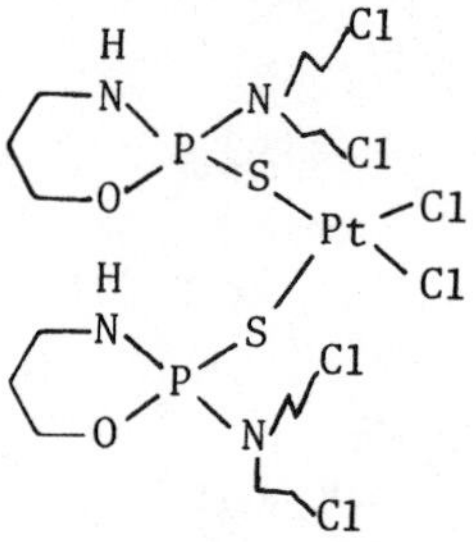 XVI

The phosphine analogs of cyclophosphamide and isocyclophosphamide are proposed because of the greater stability of Pt-P bonds. The isophosphamide derivative (Struct. XV) may be more effective because of its steric arrangement of 2-chloroethyl nitrogen groups. The thio analog of cyclophosphamide (Struct. XVI) is also proposed as a ligand because Pt-S bonds are stronger than Pt-O bonds.

Another series utilizes a partially or fully substituted ethylenediamine complex of cis-[$PtCl_2$] (Struct. XVII). A Cu(II) complex of an ethylenediamine mustard was reported in 1961 [51]. Although the complex was quite toxic, recent developments in the use of platinum complexes as antitumor agents suggests that compounds such as Struct. XVII might be useful.

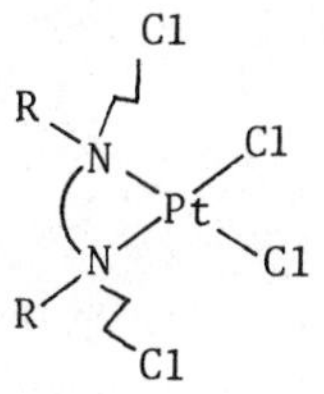

$R = H, CH_3, CH_2CH_2Cl$

XVII

The antitumor activity of bis(thiosemicarbazonato)copper(II) complexes (see Chap. 4) illustrates how the metal complex can improve membrane transport and/or enhance antitumor activity. Metal complexes of alkylating agents, particularly those of cobalt(III), copper(II), platinum(II), rhodium(III), ruthenium(II), palladium(II), gold(III), iridium(III), gallium(III), and aluminum(III), should be synthesized and tested to determine whether these complexes have more favorable therapeutic indices than the parent ligand.

5. RECENT DEVELOPMENTS

Compounds XIV and XV have been synthesized and tested [A. Okruszek and J. G. Verkade, *Phosphorus and Sulfur*, 7, 235 (1979)]. These compounds have T/C values between 87 and 104 for L1210 mouse tumors. The low T/C values eliminate these compounds from further consideration as effective antitumor agents.

ABBREVIATIONS

APO	tris(1-aziridino)phosphine oxide
APS	tris(1-aziridino)phosphine sulfide
BAPPA	P,P-bis(1-aziridinyl)-N-phenylphosphinic amide
BCNU	bis-chloroethylnitrosourea
CCNU	chloroethyl cyclohexyl nitrosourea

CP	cyclophosphamide
DO	bis(1-aziridino)-N,N'-dimethylaminophosphine oxide
DTIC	5-(3,3-dimethyl-1-triazeno)imidazole-4-carboxamide
HN2	methyl-bis(2-chloroethyl)amine = mechlorethamine
MAPO	tris[1-(2-methyl)aziridino]phosphine oxide
MO	1-aziridinobis(N,N'-dimethylamino)phosphine oxide
MPAPS	bis[1-(2-methyl)aziridino]phenylphosphine sulfide
PAPS	bis(1-aziridino)phenylphosphine sulfide

REFERENCES

1. M. W. G. DeBolster and W. L. Groenweld, in Topics in Phosphorus Chemistry, Vol. 8 (E. J. Griffith and M. Grayson, eds.), John Wiley and Sons, New York, 1976, p. 273.
2. (a) M. D. Joesten, *Inorg. Chem.*, *6*, 1598 (1967), and references cited therein.
   (b) M. F. Prysak and M. D. Joesten, *Inorg. Chim. Acta*, *4*, 383 (1970).
3. A. Furst, Chemistry of Chelation in Cancer, C. C. Thomas, Springfield, Ill., 1963, p. 46.
4. Ibid., p. 101.
5. A. Gilman and F. S. Phillips, *Science*, *103*, 409 (1946).
6. D. B. Ludlum, in Cancer, A Comprehensive Treatise, Vol. 5 (F. F. Becker, ed.), Plenum Press, New York, 1977, p. 285.
7. W. C. J. Ross, in Handbook of Experimental Pharmacology, Vol. 38/1 (A. C. Sartorelli and D. G. Johns, eds.), Springer-Verlag, New York, 1974, p. 33.
8. L. H. Schmidt, R. Fradkin, R. Sullivan, and A. Flowers, *Cancer Chemother. Rep. Suppl.*, *2*, 1 (1965).
9. L. L. Bennett, Jr. and J. A. Montgomery, in Methods in Cancer Research, Vol. 3 (H. Busch, ed.), Academic Press, New York, 1967, p. 549.
10. D. L. Hill, A Review of Cyclophosphamide, Charles C. Thomas, Springfield, Ill., 1975.
11. J. A. Stock, in Drug Design, Vol. 2 (E. J. Ariens, ed.), Academic Press, New York, 1971, p. 532.
12. P. Calabresi and R. E. Parks, Jr., in The Pharmacological Basis of Therapeutics (L. S. Goodman and A. Gilman, eds.), 5th ed., MacMillan, New York, 1975, p. 1254.
13. G. P. Wheeler, in Cancer Medicine (E. Frei, III and J. Holland, eds.), Lea and Febiger, Philadelphia, 1973, p. 791.

14. W. C. J. Ross, Biological Alkylating Agents, Butterworth, London, 1962.

15. C. C. Price, in Handbook of Experimental Pharmacology, Vol. 38/2 (A. C. Sartorelli and D. G. Johns, eds.), Springer-Verlag, New York, 1975, p. 1.

16. D. B. Ludlum, ibid., p. 6.

17. T. A. Connors, ibid., p. 18.

18. B. W. Fox, ibid., p. 35.

19. H. Kersten, ibid., p. 47.

20. G. P. Wheeler, ibid., p. 65.

21. Proceedings of the 7th New Drug Seminar: Nitrosoureas, *Cancer Treat. Rept.*, *60*, 645 (1976).

22. H. E. Kann, Jr., *Cancer Res.*, *38*, 2363 (1978).

23. R. A. G. Ewig and K. W. Kohn, *Cancer Res.*, *38*, 3197 (1978).

24. M. Colvin, R. B. Brundrett, M. N. Kan, I. Jardine, and C. Fenselau, *Cancer Res.*, *36*, 1121 (1976).

25. O. M. Friedman, I. Wodinsky, and A. Myles, *Cancer Treat. Rep.*, *60*, 337 (1976).

26. Proceedings of the Symposium on the Metabolism and Mechanism of Action of Cyclophosphamide, *Cancer Treat. Rep.*, *60*, 299 (1976).

27. J. A. Montgomery and R. F. Struck, *Cancer Treat. Rep.*, *60*, 381 (1976).

28. M. Szekerke, *Cancer Treat. Rep.*, *60*, 347 (1976).

29. V. T. Oliverio, in Cancer Medicine (E. Frei, III and J. Holland, eds.), Lea and Febiger, Philadelphia, 1973, p. 806.

30. M. A. Apple, in Cancer, A Comprehensive Treatise, Vol. 5 (F. F. Becker, ed.), Plenum Press, New York, 1977, p. 599.

31. A. Goldin, J. M. Venditti, and N. Mantel, in Handbook of Experimental Pharmacology, Vol. 38/1 (A. C. Sartorelli and D. G. Johns, eds.), Springer-Verlag, New York, 1974, p. 411.

32. E. Frei, III and J. A. Gottlieb, ibid., p. 449.

33. M. J. Cleare, *Coord. Chem. Rev.*, *12*, 349 (1974).

34. B. Rosenberg, L. Van Camp, and T. Krigas, *Nature* (London), *205*, 698 (1965).

35. B. Rosenberg, L. Van Camp, J. E. Trosko, and V. H. Mansour, *Nature* (London), *222*, 385 (1969).

36. B. Rosenberg and L. Van Camp, *Cancer Res.*, *30*, 1799 (1970).

37. T. A. Connors and J. J. Roberts, eds., Platinum Coordination Complexes in Cancer Chemotherapy, Springer-Verlag, Berlin, 1974.

38. A. P. Zipp and S. G. Zipp, *J. Chem. Educ.*, *54*, 739 (1977).

39. R. O. Inlow and M. D. Joesten, *J. Inorg. Nucl. Chem.*, *37*, 2353 (1975).

40. C. White and M. D. Joesten, unpublished results, 1975.

41. R. O. Inlow and M. D. Joesten, *J. Inorg. Nucl. Chem.*, *38*, 359 (1976).

42. N. M. Karayannis, L. M. Mikulski, and L. L. Pytlewski, *Inorg. Chim. Acta Rev.*, *5*, 69 (1971).

43. G. Hebrank and M. D. Joesten, unpublished results, 1978.

44. A. Erck, L. Rainen, J. Whileyman, I. M. Chang, A. P. Kimball, and J. L. Bear, *Proc. Soc. Exp. Biol. Med.*, *145*, 1278 (1974).

45. J. L. Bear, H. B. Gray, Jr., L. Rainen, I. M. Chang, R. Howard, G. Serio, and A. P. Kimball, *Cancer Chemother. Rep.*, Part I, *59*, 611 (1975).

46. S. A. Johnson, H. R. Hunt, and H. M. Neumann, *Inorg. Chem.*, *2*, 960 (1963).

47. J. Kitchens and J. L. Bear, *J. Inorg. Nucl. Chem.*, *31*, 2415 (1969).

48. T. A. Stephenson, S. M. Morehouse, A. R. Powell, J. P. Heffer, and G. Wilkinson, *J. Chem. Soc.*, 3632 (1965).

49. Y. B. Koh and G. G. Christoph, *Inorg. Chem.*, *17*, 2590 (1978).

50. R. J. Woodman, A. E. Sirica, M. Gang, I. Kline, and J. M. Venditti, *Chemother.*, *18*, 169 (1973).

51. M. Ishidate, Y. Sakurai, and K. Sawatari, *Chem. Pharm. Bull. (Tokyo)*, *9*, 679 (1961).

Chapter 7

METAL BINDING TO ANTITUMOR ANTIBIOTICS

James C. Dabrowiak
Department of Chemistry
Syracuse University
Syracuse, New York

1. INTRODUCTION 306
2. BLEOMYCIN 307
   2.1. Structure 307
   2.2. Mechanism of Action 309
   2.3. Interaction with DNA 310
   2.4. Metal Binding Properties 311
3. TALLYSOMYCIN 319
   3.1. Structure 319
   3.2. Mechanism of Action 320
   3.3. Interaction with DNA 320
   3.4. Metal Binding Properties 321
4. PHLEOMYCIN 322
   4.1. Structure 323
   4.2. Mechanism of Action 323
   4.3. Interaction with DNA 324
   4.4. Metal Binding Properties 324
5. STREPTONIGRIN 324
   5.1. Structure 324
   5.2. Mechanism of Action 325
   5.3. Interaction with DNA 325
   5.4. Metal Binding Properties 326

6. DAUNOMYCIN AND ADRIAMYCIN 326
6.1. Structure 326
6.2. Mechanism of Action 327
6.3. Interaction with DNA 327
6.4. Metal Binding Properties 327
7. CHROMOMYCIN, OLIVOMYCIN, AND MITHRAMYCIN 328
7.1. Structure 328
7.2. Mechanism of Action 328
7.3. Interaction with DNA 329
7.4. Metal Binding Properties 330
8. CONCLUSIONS 330
REFERENCES 331

1. INTRODUCTION

The use of naturally occurring substances for therapeutic purposes has its origins in antiquity. Perhaps the earliest documentation of useful remedies can be found in the Thebes papyrus, which dates back to 3300 B.C. That ancient work is a compendium of medicinals, mainly organic in nature, which can be derived from the leaves and roots of plants. On the list are compounds such as morphine and quinine, which are still in use today.

In some ways, the modern-day Thebes papyrus is the *Merck Index*. Although the *Index* does not solely contain therapeutic agents, it does collect the physical and chemical characteristics of a large number of active drugs. A perusal of the *Index* will reveal that many drugs contain heteroatoms (O, N, S) at points in their structure that allow the formation of stable five- and six-membered chelate rings with a metal ion. Since biological systems are replete with metal ions, the prospect of altering a drug's activity via chelation in vivo is significant. Chelation in biological fluids is known to be a highly competitive phenomenon and ascertaining whether or not a drug is complexed or uncomplexed under physiological conditions is generally difficult [1]. Nonetheless, a growing group of cancer drugs appear to require a biologically available metal ion for their activity. This chapter focuses on the structure, mechanism, and binding properties (toward metal ions and DNA) of antitumor antibiotics for which chelation may play an important mechanistic role.

2. BLEOMYCIN

Bleomycin (BLM) provides the best, and perhaps the clearest, example of how a metal ion can participate in the mechanism of action of an antitumor agent. The drug is produced by a fungus, *Streptomyces verticillus*. The antibiotic is initially isolated as a mixture of blue Cu(II) complexes [2]. Subjection of the copper complexes to hydrogen sulfide results in the destruction of the coordination compounds and the liberation of 11 antitumor-active glycolpeptides. The peptide mixture is widely used, under the tradename Blenoxane, for the treatment of squamous cell carcinomas, lymphomas, and testicular carcinomas [2,3].

2.1. Structure

The structure of the antibiotic was established by Umezawa et al. in the period 1968-1973 [2]. However, the originally proposed structure has been recently revised [4]. Bleomycin A_2, the most abundant component in the clinical mixture of bleomycins, is shown in Fig. 1. The revised structure does not contain a β-lactam moiety, but possesses in its place a primary amide group and a secondary amine function. By the addition of various amines to the culture medium, or by the application of certain synthetic approaches, it is possible to make a variety of bleomycins, and about 200 different analogs are known [2,5,6]. With three exceptions, all of the variations have the same basic framework as bleomycin A_2 and differ only in the amine substituent (R) attached to the

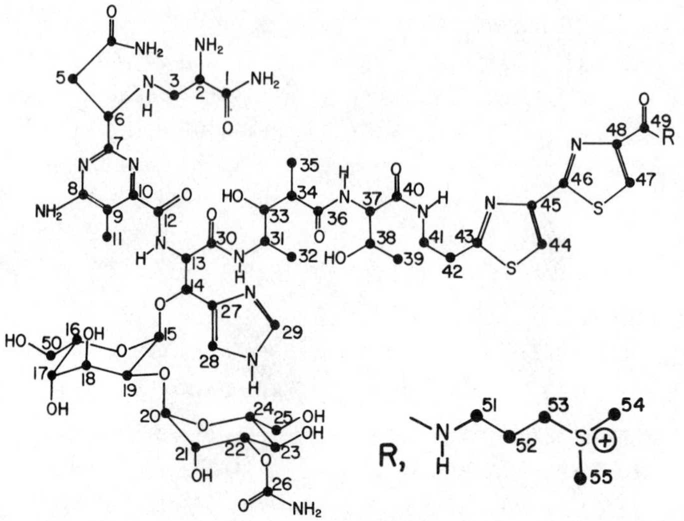

FIG. 1. Structure of Bleomycin A_2 [2,4].

2,4'-bithiazole moiety of the drug. One of the exceptions, epi-bleomycin, has an inverted arrangement of groups (the R configuration) about carbon atom 6 (Fig. 1) [7]. Isobleomycin, on the other hand, has the bleomycin structure but possesses a carbamoyl group which has migrated from the 3 (carbon 22) to the 2 hydroxyl position (carbon 21) of the D-mannose sugar of bleomycin [8]. Finally, bleomycin hydrolase, which is widely distributed in animal tissues [9], reacts with the bleomycin to yield desamidobleomycin. Desamidobleomycin has the amide group of the α-amino carboxamide moiety of the drug hydrolyzed to an acid function. With the exception of bleomycinic acid, which has an acid function attached to the bithiazole moiety, isobleomycin and desamidobleomycin, all others are active against DNA breakage [5,10].

2.2. Mechanism of Action

The mechanism of action of bleomycin is not fully understood. Most of the early efforts, as well as those currently in progress, have focused on the interaction of bleomycin (most often, bleomycin A_2) with DNA. The glycolpeptide is capable of causing single- and double-strand breaks in DNA [2,11]. Recent studies have shown that DNA breakage by bleomycin occurs primarily at GT and GC sites [12, 13]. The in vitro degrading process, which leads to the production of free bases [14-17], is strongly metal-ion-dependent. Addition of Cu(II), Zn(II), and Co(II) salts [18,19] to the reaction mixture prevents DNA breakage. It is known that these metal ions complex to bleomycin, and in so doing, apparently inactivate the drug. The addition of Fe(II) salts to the reaction mixture, on the other hand, under conditions which lead to the formation of the Fe(II)-BLM complex, greatly enhances the DNA degrading ability of bleomycin [20-23]. The observed increase in activity of the drug in the presence of iron salts has led to the postulate that the antibiotic requires an iron ion for its in vitro, as well as for its in vivo, activity.

Horwitz and coworkers [21-23] have shown that Fe(II)-BLM is air-sensitive and that it is readily oxidized to Fe(III)-BLM in the atmosphere [Eq. (1)]. These workers speculated that a radical is formed in the oxidation process and that is the radical that leads

$$\text{Fe(II)-BLM} \underset{\text{R-SH}}{\overset{O_2}{\rightleftarrows}} \text{Fe(III)-BLM} + \text{radical} \tag{1}$$

to DNA breakage. The existence of radicals has been confirmed via esr measurements [24-27]. The nature of the radical(s) and the effects of various radical scavengers (e.g., superoxide dimutase) on the DNA-degrading process by bleomycin is under active investigation. Sulfhydryl groups present in the cell [Eq. (1)] would offer a convenient means of reducing the trivalent iron-bleomycin complex to Fe(II)-BLM, thus allowing the production of additional radicals from the iron-bleomycin complex. In experiments where divalent iron has not been added to the reaction mixture, DNA breakage by bleomycin occurs, but at a reduced rate [20-23]. The reason that it occurs at all has been attributed to the presence of trace amounts of iron contributed by DNA, bleomycin, or both. It should be pointed out that the degrading process by bleomycin can be quenched by the strong metal-chelating agent EDTA [14,28,29]. The fact that the reaction exhibits an oxygen requirement is also supportive of the metal hypothesis [20-23].

Recently Umezawa and coworkers [30] have drawn attention to the possible role of the copper(II)-BLM complex in the DNA degrading process. It is known that both copper-free BLM and its metal complex inhibit DNA synthesis in cells [31], as well as preventing growth of animal and bacterial cells [31-35]. However, as was pointed out earlier, Cu(II)-BLM does not cause scission of DNA in vitro [19]. In the search for a cellular mechanism to explain the activity of Cu(II)-BLM, Umezawa et al. [30] identified a protein which is capable of reducing Cu(II) to Cu(I) in Cu(II)-BLM. They suggested that a sulfhydryl group of a protein carries out the reduction and that the resulting Cu(I) ion binds to a second protein in the cell [Eqs. (2) and (3)]. The liberated BLM [Eq. (3)] presumably binds to cellular iron and forms the iron-BLM complex which causes DNA damage.

$$\text{Cu(II)-BLM} \xrightarrow{\text{protein-SH}} \text{Cu(I)-BLM} \tag{2}$$

$$\text{Cu(I)-BLM} + \text{protein} \rightarrow \text{Cu(I)-protein} + \text{BLM} \tag{3}$$

2.3. Interaction with DNA

The functional groups utilized by bleomycin in binding to DNA have been determined. Using ^{1}H nmr and fluorescence spectroscopy, Horwitz and coworkers [36] were the first to show that the bithiazole moiety of bleomycin binds to DNA. Dichroism experiments on small double-stranded DNAs in the presence of an applied magnetic field have further suggested that the bithiazole is involved in binding and that intercalation may be the primary mode of interaction between the heterocycle and DNA [37]. Recent investigations by Umezawa et al. [38] have indicated that the bithiazole moiety preferentially binds to guanine bases in nucleic acids, in agreement with other studies [12,13], and that the positive charge at the terminal amine of bleomycin facilitates binding. It is known that the proton resonances of the dimethyl sulfonium cation, the terminal group of BLM-A_2, broadened in the presence of DNA--an observation which is supportive of a substantial sulfonium-DNA interaction [36]. No evidence for the participation of the pyrimidine-imidazole portion of the bleomycin molecule, the metal-binding area, in DNA binding has been advanced. A variety of techniques have shown [36,37,39,40] that each drug molecule utilizes three to five base pairs in binding to DNA.

2.4. Metal Binding Properties

Bleomycin is a potent binder of transition, as well as nontransition, metal ions. Most of the complexes with nontransition metal ions are γ emitters and they are widely used for tumor imaging [41-44]. However, since transition metal ions play a central role in the mechanism of action of the pharmaceutical, attention shall be focused on what is known about the compounds formed between the antibiotic and the nonradioactive cations Fe-Zn.

2.4.1. Fe(II,III)-BLM

Bleomycin is capable of binding a single Fe(II) ion [21,23,25]. The pinkish-orange Fe(II)-BLM complex is exceedingly air-sensitive and physical measurements on it must be done under rigorously anaerobic conditions. Neither the spin state nor the coordination number of the bleomycin-bound iron ion are known. However, the Fe(II)-BLM complex is esr-quiet. This behavior is expected for the S = 0 state, which is diamagnetic, and of the remaining spin possibilities, S = 1 and S = 2, only the former is a reasonable esr candidate [45]. The visible and near-ultraviolet absorption spectrum of the complex exhibits two broad bands (at 370 and 475 nm), both having molar extinction coefficients of less than ~400 [21, 23,25].

Potentiometric titration studies have shown that the pK_a's at 7.7, 5.0, and 2.9 assigned to the deprotonation of the amino group of the α amino carboxamide moiety [Eq. (4)], the deprotonation of the imidazole function [Eq. (5)], and the deprotonation of the secondary amine group [Eq. (6)], respectively (the latter being an

$$-NH_3^+ \longrightarrow -NH_2 + H^+ \qquad pK_a = 7.7 \tag{4}$$

$$ImH^+ \longrightarrow Im + H^+ \qquad pK_a = 5.0 \tag{5}$$

$$>NH_2^+ \longrightarrow >NH + H^+ \qquad pK_a = 2.9 \tag{6}$$

anomalously low value for a secondary amine), are missing from the titration curve of Fe(II)-BLM [24,46]. Thus, these three functionalities appear to be metal-binding sites. Titration studies have also shown that the drug releases a total of four protons upon binding Fe(II) [46]. In addition to the three sites discussed above [Eqs. (4)-(6)], the amide nitrogen between C-12 and 13 (Fig. 1) has been postulated as the fourth deprotonation site [Eq. (7)]. In addition, polarographic investigations using a dropping mercury

$$M\text{—}N(\text{—}C{=}O)\text{—}H \longrightarrow M\text{—}N^{\ominus}(\text{—}C{=}O) + H^+ \tag{7}$$

electrode have indicated that the 4-amino pyrimidine group, is also a metal binding site [47]. The two-electron reduction process at -1.22 V, associated with the reduction of the pyrimidine moiety to a tetrahydropyrimidine in BLM, is missing in the dc polarogram of

Fe(II)-BLM. Complexation of the pyrimidine residue to the metal ion shifts the reduction potential for the heterocycle out of the polarographic window, so that it cannot be observed.

Exposing an aqueous buffered solution of Fe(II)-BLM to dioxygen or potassium ferricyanide results in a rapid oxidation to Fe(III)-BLM. Extensive esr investigations have shown that the number and types of Fe(III)-BLMs which are produced by oxidation are sensitive to pH, buffer composition, and the oxidant used [25]. It is important to note that the degrading of DNA by bleomycin is also sensitive to the first two variables [23]. When Fe(II)-BLM is oxidized by dioxygen in pH 8.0 phosphate buffer solutions, at least three different Fe(III)-BLM complexes form. Two low-spin species ($S = 1/2$), having parameters $g_x = 2.41$, $g_y = 2.18$, and $g_z = 1.89$ [Fe(III)-BLM-1/2a] (stable) and $g_x = 2.25$, $g_y = 2.17$, and $g_z = 1.94$ [Fe(III)-BLM-1/2b] (unstable), as well as at least one high-spin compound ($S = 5/2$) with a single transition at $g = 4.3$ [Fe(III)-BLM-5/2], can be observed [24,25,27]. The esr data indicate that the iron ion in Fe(III)-BLM-5/2 resides in a completely rhombic crystal field. Allowing the esr solutions to stand at room temperature for 30 min results in the complete disappearance of Fe(III)-BLM-1/2b. This species does not appear when the oxidation is carried out anaerobically, using potassium ferricyanide as an oxidant.

Esr studies have shown that free radicals are produced in the air oxidation of Fe(II)-BLM to Fe(III)-BLM [24-27]. Using the spin trap α-phenyl-N-tert-butyl nitrone (PBN), Sugiura and Kikuchi [27] have shown that both $\cdot OH$ and $O_2^-\cdot$ are produced in the reaction and that the radical type is dependent on the concentration of Fe(II)-BLM present in solution. At high concentration (~1.0 M), hydroxyl radicals are produced, while at low concentration (0.02 M) $O_2^-\cdot$ or $\cdot O_2H$ adducts with the spin trap were observed.

Little is known about the Fe(III)-binding site of bleomycin. However, electrochemical studies have shown that the pyrimidine moiety of bleomycin binds to Fe(III) [24,25,47]. The reduction wave for the heterocycle is missing from the polarograms of the Fe(III)-BLM complexes produced by air or by ferricyanide oxidation of Fe(II)-BLM. It is also missing from the polarograms of the Fe(III)-BLMs formed by the reaction of Fe(III) $(ClO_4)_3$ with the drug. It is interesting to note that forming Fe(III)-BLM from BLM and Fe(III)-$(ClO_4)_3$ in a phosphate buffer does not give a complex containing a bound-pyrimidine function [24,47]. In view of the complicated nature of the iron system, further studies with a variety of techniques will be necessary before a clear picture of the role of the iron ion in the degrading process evolves.

2.4.2. Co(II,III)-BLM

Addition of Co(II) salts to aqueous solutions (pH 7.0) of BLM under nitrogen yields Co(II)-BLM. Esr studies clearly show that the compound is low-spin (S = 1/2) and that it has a dz^2 ground state [48, 49]. The measured esr parameters for the complex were $g_{\parallel} = 2.025$ and $g_{\perp} = 2.272$, with $|A_{\parallel}(Co)|$, 92.5G, and $|A_{\parallel}(N)|$, 13G. The fact that a nitrogen superhyperfine interaction can be observed on the parallel transition indicates that one of the axial ligating atoms in Co(II)-BLM is a nitrogen atom. Polarographic studies have shown that at least one metal ligating site is the 4-amino pyrimidine moiety of the drug [47]. Similar to Fe(II), the drug releases four protons upon binding Co(II) and the metal ion appears to utilize the primary amine, imidazole, and secondary amine moieties of the drug [46].

Oxidation of Co(II)-BLM with dioxygen proceeds through the intermediate superoxide complex, Co(III)-BLM($O_2^-\cdot$), which has esr parameters of $g_{\perp} = 2.007$, $g_{\parallel} = 2.098$, and $|A_{\perp}(Co)|$ of 12.4G and $|A_{\parallel}(Co)|$ of 20.2G [48,49]. This complex slowly decomposes to yield Co(III)-BLM and superoxide radical, the latter of which has been detected using the spin-trapping agent PBN [49]. The fact that the superoxide complex, Co(III)-BLM($O_2^-\cdot$), exhibited slightly different esr parameters in the presence of calf thymus DNA, $g_{\parallel}$ = 2.106 and $g_{\perp}$ = 2.004, $|A_{\parallel}(Co)|$, 18.9G and $|A_{\parallel}(N)|$, 11.5G, than it did in its absence, was attributed to a difference in orientation of the superoxide radical about the cobalt ion [48].

The Co(III)-BLM complex has been reported [18]. The compound can be synthesized by the addition of Co(II) salts to BLM in the presence of the atmosphere. As expected, the complex is exchange-inert and the metal ion cannot be removed by addition of the strong metal chelating agent EDTA. Visible absorption and C.D. spectra indicate that the metal ion in Co(III)-BLM resides in primarily a nitrogen environment. Polarographic studies have confirmed the fact that the pyrimidine moiety of the drug is a ligating site for Co(III) [47].

2.4.3. Ni(II)-BLM

Relatively little information about Ni(II)-BLM is known. Visible absorption studies with the clinical mixture of bleomycins, Blenoxane, has shown that the nickel ion in Ni(II)-BLM is high-spin (S = 1) and that it must reside in a primarily six-coordinate nitrogen environment [50]. Two very weak d-d electronic absorption bands for the complex can be observed. By comparison with the spectrum of $[Ni(II)(CH_3NH_2)_6]^{2+}$ [51], the absorption bands have been assigned to the ^3T_2g (926 nm, εM, 7.5) and ^3T_1g (575 nm, εM, 3.7) transitions of the six-coordinate metal ion.

Although ^{13}C nmr studies on Ni(II)-BLM have not been conducted, nuclear magnetic resonance experiments on a structurally related antibiotic, tallysomycin [52], have suggested that Ni(II) binds to the pyrimidine-imidazole portion of BLM. The drug, similar to Fe(II)-BLM and Co(II)-BLM, releases four protons on binding Ni(II) [46]. Electrochemical studies have shown that the metal ion in Ni(II)-BLM is not easily oxidized or reduced and that one of the metal-ligating sites must be the 4-amino pyrimidine moiety of the drug [47].

2.4.4. Cu(I,II)-BLM

Umezawa et al. [2] were the first to report Cu(II)-BLM. They also stated, but provided little evidence, that the pyrimidine-imidazole portion of the drug is utilized for metal binding. Recent ^{13}C nmr studies on Cu(II)-BLM have clearly shown that the pyrimidine-imidazole region of the antibiotic is involved in binding Cu(II) [53]. Complexation of the drug to Cu(II) eliminated 23 resonances from the carbon spectrum of the antibiotic. All of the affected carbon atoms were located in the pyrimidine-imidazole region of the drug. The fact that the carbonyl carbon atom of the carbamoyl moiety, C-26, (Fig. 1) was also missing from the spectrum of Cu(II)-BLM, is indicative of the carbamoyl group being close to, and probably bound to, the copper ion. Titration [24,46], visible absorption [54,55], and electrochemical studies [47] have confirmed the fact that the primary amine, the pyrimidine, the imidazole, and the secondary amine groups are metal ligating functions. The log K for Cu(II)-BLM of 12.6 is the highest of the first-row divalent transition metal ion complexes with BLM [46].

Significant progress toward establishing the binding geometry which exists in Cu(II)-BLM has been made through the x-ray structural analysis of Cu(II) P-3A [56], a copper complex containing a metabolite of bleomycin. Although the structure possesses considerable disorder (R, 0.131) the analysis showed that the Cu(II) ion in Cu(II) P-3A resides in a square pyramidal crystal field (Fig. 2). In addition to the four bleomycin functional groups assigned via chemical and spectroscopic techniques [2,24,46,53-55], the secondary amine moiety and a deprotonated amide function (very likely the source of the fourth proton release in the titration studies on metallobleomycins) complete the primary coordination sphere of the metal.

On the basis of both the structural analysis of Cu(II) P-3A and the structural revision in the antibiotic itself [4], the orig-

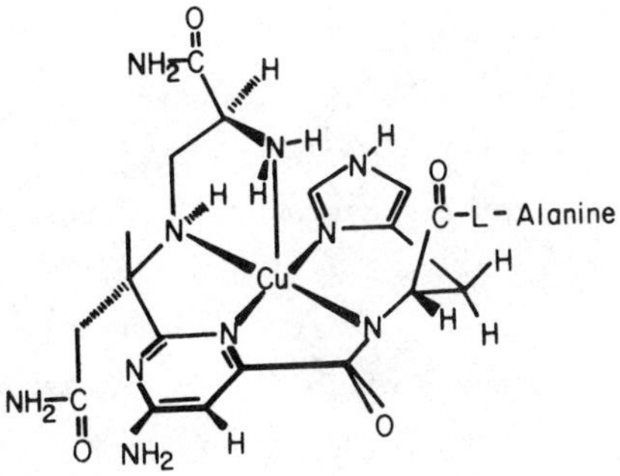

FIG. 2. Structure of Cu(II) P-3A [56].

inally proposed binding-site geometry of Cu(II)-BLM will have to be revised [53-54]. The substitution of the secondary amine function for the relatively rigid β-lactam moiety in bleomycin complicates the stereochemical analysis of the metallobleomycins. However, with the revised structure it is clear that coordination of N-1 of the pyrimidine moiety [24,53] restricts the deprotonated amide nitrogen atom and the secondary amine nitrogen atom to metal coordination sites which are coplanar with the pyrimidine ring. The positions of the primary amine function, N-1 of the imidazole residue, and the sugar carbamoyl group on the coordination polyhedron of Cu(II)-BLM are less certain.

Very little is known about Cu(I)-BLM. The compound can be made by the addition of Cu(I) salts, e.g., Cu(I)Cl, to aqueous solutions of BLM or by the reduction of Cu(II)-BLM using mercaptans [30,47]. Similar to the Co(II) and Fe(II)-BLM complexes, the Cu(I) analog is oxygen-sensitive and it is readily oxidized in the atmosphere to Cu(II)-BLM. The Cu(I) ion in Cu(I)-BLM apparently does not bind to the 4-amino pyrimidine moiety of BLM. Polarographic studies on Cu(I)-BLM have shown that the $E_{1/2}$ of the two-electron pyrimidine-centered reduction process is unaffected by the presence of Cu(I). These results are in general agreement with the proposed mechanism explaining the observed cytotoxicity of Cu(II)-BLM [30].

2.4.5. Zn(II)-BLM

Because it is diamagnetic, the zinc-bleomycin complex has been extensively studied using nmr. Employing that technique, Dabrowiak et al. [54] reported that bleomycin utilized both the 4-amino pyrimidine and imidazole moieties for binding Zn(II). These findings have been recently confirmed on the clinical mixture of bleomycins using nmr [57]. The binding of Zn(II) to bleomycin caused the hydrogens, H-28 and H-29, the imidazole hydrogens, to shift to lower field. Since the system is in slow exchange on the nmr time scale, the assignment of the shifted resonances is not unequivocal. However, the most reasonable assignment indicates that the shift of H-29 is three times that of H-28, implying that the zinc ion is bound to N-1 and not to N-3 of the imidazole fragment. The use of this bleomycin atom, N-1, as a metal ligating site has been confirmed through the structural analysis of Cu(II) P-3A [56]. The shift in the pyrimidine methyl resonance, from δ 2.06 to 2.47 ppm, provides evidence that the pyrimidine is a metal ligating site for Zn(II).

^{13}C nmr studies on the drug and its Zn(II) complex have shown that 38 carbon resonances shift when Zn(II) binds to the antibiotic [53]. The carbon atoms, which are most affected by the presence of the zinc ion, lie in the pyrimidine-imidazole-sugar region of the drug. These nmr observations, along with visible absorption [54], polarographic [47], and titration [24,46] results on both Zn(II)-BLM and Cu(II)-BLM suggest that both ions bind to the same drug binding site. However, since Zn(II) does not possess crystal field stabilization energy, the stereochemistry of Zn(II)-BLM is more difficult to define than that of the copper analog.

3. TALLYSOMYCIN

Tallysomycin (TLM) is a bleomycin analog which was isolated from *Streptoalloteichus hindustanus* [58]. The compound has been found to be more potent than bleomycin against certain types of tumors, bacteria, and fungi [40,59-61].

3.1. Structure

Tallysomycin can be isolated in two forms, TLM-A and TLM-B [4,62]. The structures of both compounds are shown in Fig. 3. TLM has four features which structurally differentiate it from BLM. Unlike bleomycin, the new antibiotic does not possess a methyl group on carbon 34. In addition, the compound has an L-talose moiety (attached to C-41), a hydroxyl group on C-42, and, in the case of TLM-A, a

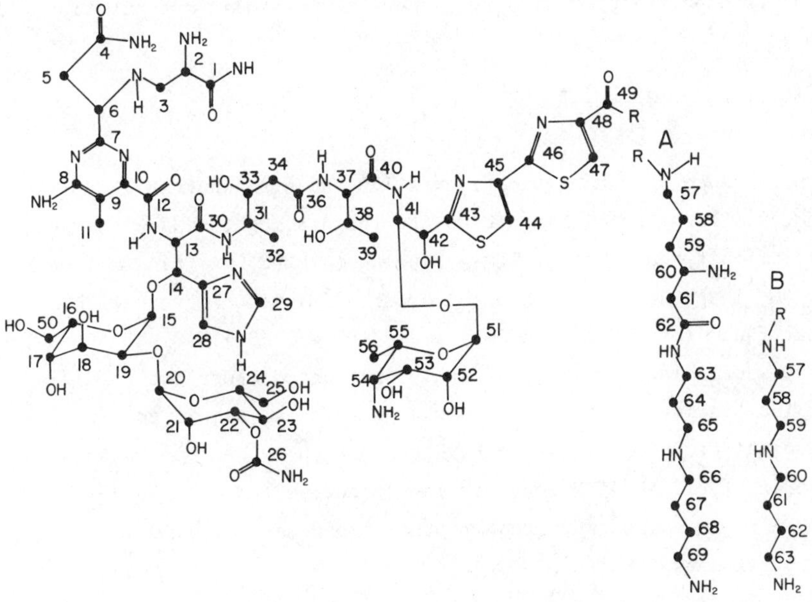

FIG. 3. Structures of tallysomycin A and B [4,60].

β-lysine-spermidine side chain attached to the bithiazole function of the framework. Tallysomycin B has a spermidine moiety attached to the bithiazole function of the framework.

3.2. Mechanism of Action

Because of the structural similarity of TLM and BLM, the receptor site of TLM is believed to be the DNA within a tumor cell. Since the antibiotic is capable of complexing to metal ions [52,59], it presumably utilizes the same mechanism for DNA degradation as that proposed for BLM. Recent work by Strong and Crooke [40] has shown that TLM-A is capable of inflicting both single- and double-strand breaks on covalently closed, circular PM-2 DNA. Although it is more active in vivo than is BLM, it is less efficient at DNA breakage (in vitro) than is bleomycin. The in vivo effectiveness of the drug has been attributed to pharmacological drug disposition, cellular drug transport and/or drug resistance to metabolic inactivation [40].

3.3. Interaction with DNA

Fluorescence quenching experiments [40] have shown that TLM, like BLM, utilizes the bithiazole moiety for binding to DNA. However, the affinity constant of TLM-A binding to PM-2 DNA ($8.47 \pm 1.06 \times 10^5\ M^{-1}$) appears to be about an order of magnitude greater than that found for BLM toward DNA, $1.2 \times 10^5\ M^{-1}$ [36]. Since the β-lysine-spermidine terminus is highly protonated under the conditions of the binding experiment, and charge effects associated with the terminal group of bleomycin have been shown to be important in DNA binding [38], the TLM behavior is not unexpected. The binding of TLM-A to DNA appears to require the same number of base pairs (~5), as does BLM [36,37,39,40].

3.4. Metal Binding Properties

The study of the metal binding properties of TLM are just beginning. In addition to the "bleomycin" binding site in the pyrimidine-imidazole portion of the molecule, TLM-A has been shown to have a second metal ligating site [52]. However, the stability of the complexes which are formed with the second site appears to be lower than those associated with the "bleomycin" site and as such dinuclear complexes with TLM will probably not figure prominently in the pharmacology of the antibiotic.

3.4.1. Ni(II)-TLM-A and $(Ni(II))_2$TLM-A

These complexes have been briefly examined. Electrochemical studies have shown that the 4-amino pyrimidine group of TLM-A is a metal-ligating site in Ni(II)-TLM-A [47]. ^{13}C nmr experiments indicate that the "bleomycin" site of TLM-A is occupied by Ni(II) in Ni(II)-TLM-A. Carbon resonances associated with the pyrimidine-imidazole region of Ni(II)-TLM-A are shifted from their original positions and are significantly broadened [52]. Nmr experiments also show that the drug can weakly bind a second nickel ion to form the dinuclear complex $(Ni(II))_2$-TLM-A. The carbon atoms of the drug which are affected by the binding of the second metal ion to Ni(II)-TLM-A lie in the β-lysine-spermidine region of Ni(II)-TLM-A. Since the resonances associated with the second site are only slightly shifted and broadened, (at pH = 8.0) it appears that the equilibrium Ni(II) + Ni(II)-TLM-A $\rightleftharpoons$ $(Ni(II))_2$-TLM-A lies well to the left.

3.4.2. Cu(II)-TLM-A and $(Cu(II))_2$-TLM-A

The copper complex Cu(II)-TLM-A was described in the isolation of tallysomycin [59]. Visible absorption, electrochemical and esr studies have shown that Cu(II)-TLM-A is very similar in structure, if not identical to that of Cu(II)-BLM [52]. However, unlike BLM, TLM-A is capable of forming the dinuclear complexes $(Cu(II))_2$-TLM-A.

^{13}C Nmr and esr studies have shown that the $(Cu(II))_2$-TLM-A complex is stable within the pH range 6-10. The measured esr parameters of the copper ion occupying the second site, $g_{\parallel} = 2.258$, $|A_{\parallel}(Cu)| = 0.0178\ cm^{-1}$ and $g_{\perp} = 2.06$, obtained by making the mixed complex Zn(II)-Cu(II)-TLM-A, indicated that the second site possessed a primarily nitrogen environment [52]. This information, combined with the observation that certain carbon resonance emanating from carbon atoms located in the β-lysine-spermidine region and the talose moiety of the drug broaden or disappear from the spectrum of $(Cu(II))_2$-TLM-A, suggests that the primary amine functions attached to carbon atoms 54, 60, and 69 and the amide nitrogen atom between C-62 and C-63 of TLM-A are metal ligating sites. Since the second copper complex exhibits fast exchange on the nmr time scale and it is stable over a limited pH range, the stability of complexes formed at the second site appear to be less than the mononuclear TLM-A complexes and the metallobleomycins.

3.4.3. Zn(II)-TLM-A

This complex was briefly studied in connection with the ^{13}C nmr assignments of the tallysomycins [63]. The binding of Zn(II) to tallysomycin affects the carbon spectrum of the drug in a manner which is nearly identical to that observed for Zn(II)-BLM [24,52, 53]. Although Zn(II)-TLM-A binds a second copper ion to form Zn(II)-Cu(II)-TLM-A the dinuclear complex, $(Zn(II))_2$-TLM-A, does not appear to form.

4. PHLEOMYCIN

Phleomycin (PHM) was the first antitumor agent to be isolated by Umezawa [2]. Although the antibiotic showed strong inhibition of several animal tumors, the compound exhibited irreversible renal toxicity in dogs, and as a result, the planned clinical study on the drug had to be abandoned. In comparison to bleomycin, only a limited amount of chemical and biochemical information on phleomycin is known.

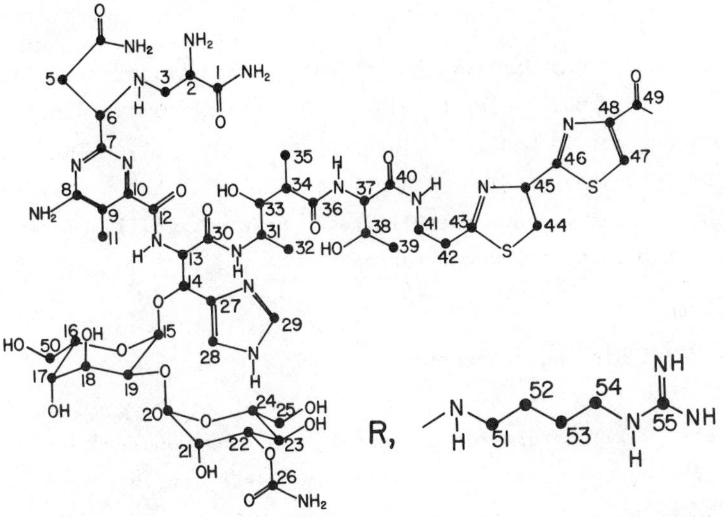

FIG. 4. Structure of phleomycin D_1 [4,64].

4.1. Structure

Structurally, except for a reduced double bond in one of the thiazole rings of the 2,4' bithiazole moiety, the phleomycins are identical to the bleomycins (Fig. 4) [4,64]. In the presence of oxidizing agents such as MnO_2, it is possible to oxidize the bithiazole moiety of a phleomycin and convert the phleomycin to the corresponding bleomycin [64].

4.2. Mechanism of Action

Since phleomycin has been shown to induce DNA breakage in vitro [5,65] using essentially the same condition as employed for bleomycin and tallysomycin, all three antibiotics appear to have the same mechanism of action.

4.3. Interaction with DNA

As was pointed out earlier, both bleomycin and tallysomycin utilize the bithiazole moiety for binding to DNA. Although it would appear to be important to understanding how these antibiotics interact with DNA, binding studies with phleomycin, which has a modified bithiazole moiety, have been limited. However, Cu(II)-PHM was found to affect the thermal denaturation of DNA [19].

4.4. Metal Binding Properties

Only the Cu(II) complex of phleomycin has been reported [66]. In view of the structural similarities of phleomycin, bleomycin and tallysomycin in the pyrimidine-imidazole region, phleomycin would be expected to have metal ligating properties which are identical to those of the other two glycolpeptides [55].

5. STREPTONIGRIN

Streptonigrin is a quinoid antibiotic which was isolated from *Streptomyces flocculus* [67]. The compound is active against a variety of animal tumors [68].

5.1. Structure

The structure of streptonigrin (Fig. 5) was established both by chemical [67] and by single-crystal x-ray analysis [69]. The antibiotic is composed of aminoquinone, pyridyl, and phenyl fragments. Several streptonigrin analogs are known [70,71], some of which are active against DNA breakage [71,72].

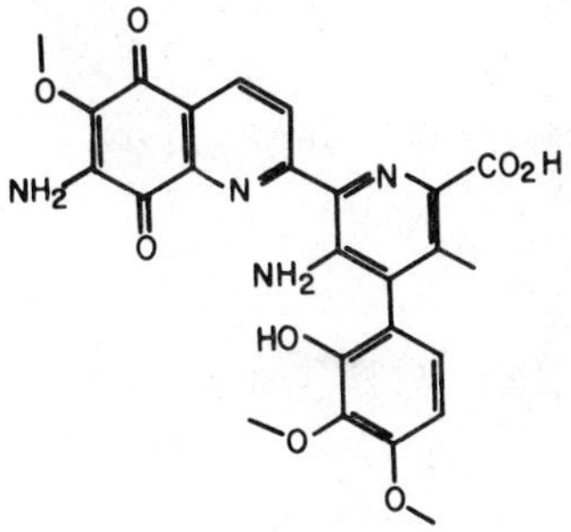

FIG. 5. Structure of streptonigrin [67,69].

5.2. Mechanism of Action

The mechanism of action of streptonigrin appears to be strikingly similar to that of bleomycin [73]. The compound is capable of causing single-strand breaks in various types of DNAs [74,75], and the breakage process appears to be metal-ion- and oxygen-dependent and to proceed via a free-radical mechanism. Cone et al. [75] have shown that Fe(II), and to a lesser extent, Cu(II) enhances the DNA-degrading ability of streptonigrin. Addition of Co(II) salts to the reaction medium had no effect on the degradation of DNA by streptonigrin. It was postulated that the species responsible for DNA degrading by the antibiotic is a hydroxy radical [75].

5.3. Interaction with DNA

Streptonigrin binds both reversibly and irreversibly to DNA [76]. One antibiotic molecule appears to bind per 2000 base pairs of the biopolymer. The type of interaction between the drug and DNA has not yet been established.

5.4. Metal Binding Properties

Studies on the metal complexes formed by streptonigrin have not been reported.

6. DAUNOMYCIN AND ADRIAMYCIN

The fungus-derived antibiotics daunomycin [77,78] and adriamycin [79] are the earliest members of a growing class of anthracyclines [80] which exhibit high cytotoxic activity against normal and neoplastic cells.

6.1. Structure

Both daunomycin and adriamycin possess an amino sugar, 2,3,6-trideoxy-3-amino-L-lyxohexose, daunosamine, and an anthracycline moiety (Fig. 6). The structure and stereochemistry of daunomycin have been confirmed by x-ray analysis of N-bromoacetyldaunomycin [81].

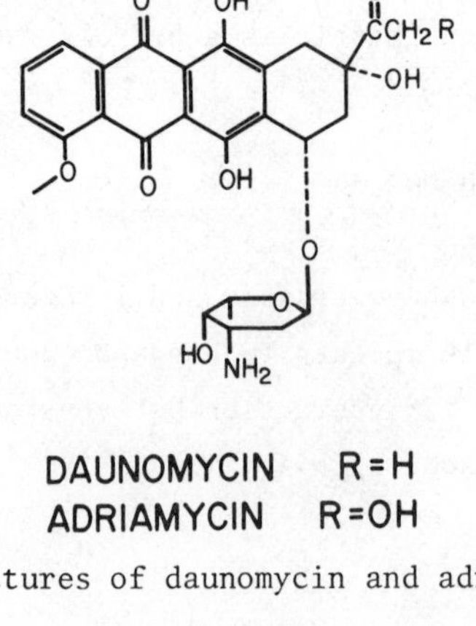

O OH O CCH_2R OH O O OH O O HO NH_2

DAUNOMYCIN R=H
ADRIAMYCIN R=OH

FIG. 6. Structures of daunomycin and adriamycin [81].

6.2. Mechanism of Action

Both daunomycin and adriamycin are believed to bind to DNA and to interfere with its template function [82]. The binding process may have a metal requirement and Co(II), Ni(II), and Cu(II), have been shown to facilitate the interaction [83]. More recent work [84] indicates that these antibiotics may operate at several biochemical levels.

6.3. Interaction with DNA

Two modes of binding between the anthracyclines and DNA have been recognized [85]. "Strongly bound" daunomycin appears to intercalate into DNA with a binding constant of $\sim 3 \times 10^6\ M^{-1}$. Molecular model studies have shown that the intercalation of daunomycin is sterically possible [86]. "Weakly bound" antibiotic molecules are presumed to be attached to DNA via electrostatic forces between the phosphate groups of DNA and the protonated daunosamine amino group. Metal ions such as Mg(II), Ag(I), Hg(II), and Cu(II) are known to affect the binding of daunomycin to DNA [81,83,87-89].

6.4. Metal Binding Properties

Compared to what is known about the metal binding properties of the glycolpeptides bleomycin and tallysomycin, the ligating properties of daunomycin and adriamycin have been little studied. Calendi et al. [87] examined the changes which occurred in the visible absorption spectrum of daunomycin when the drug binds to Al(III), Fe(II, III), Cu(II), Mg(II), and Na(I). No attempt was made to elucidate the structure of the metal complexes. In a later study, Yesair et al. [90] reported that Cu(II), Mn(II), Ni(II), Co(II), and Mg(II) are capable of forming 1:1 complexes with both adriamycin and daunomycin in ethanol solution. It was also shown that complexation of adriamycin to Cu(II) markedly inhibited the reductive glycosidic cleavage of the drug to the reduced aglycone and daunosamine. Although the glycosidic cleavage was totally inhibited by metal complexation, metal chelation did not affect the reduction of the 13-carbonyl function of the antibiotic. No evidence on the structure of the metal complexes was forwarded.

7. CHROMOMYCIN, OLIVOMYCIN, AND MITHRAMYCIN

Chromomycin, olivomycin, and mithramycin are three chemically related antitumor antibiotics which appear to function by the same biochemical mechanism. Chromomycin was isolated in Japan [91], olivomycin in the USSR [92], and mithramycin in the United States [93]. Two reviews on the properties of these antibiotics can be found [94,95].

7.1. Structure

The structures of the most abundant forms of the three antibiotics are shown in Fig. 7 [96,97]. Chromomycin A_3 and mithramycin possess the same aglycone, chromomycinone, but have different di- and trisaccharides. Chromomycin A_3 and olivomycin A contain identical disaccharide moieties, and except for an ester function located on the terminal sugar residue of the trisaccharide, they also have identical trisaccharide groups. The absence of a methyl group on the aglycone of olivomycin, olivin, further differentiates the two antibiotics.

7.2. Mechanism of Action

All three antibiotics preferentially inhibit RNA synthesis by binding to DNA [94].

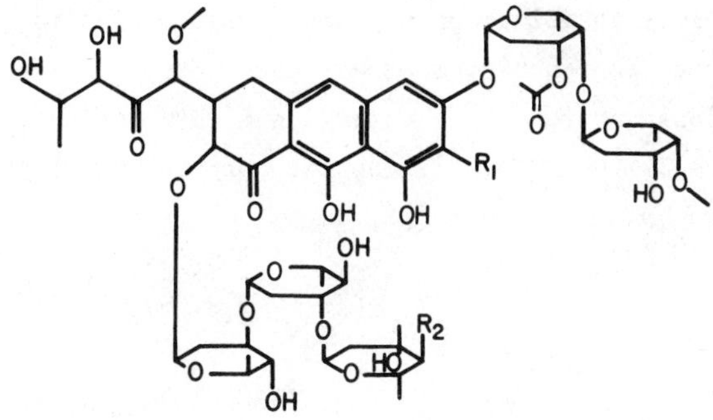

CHROMOMYCIN A_3 $R_1 = CH_3$, $R_2 = CH_3CO_2$
OLIVOMYCIN A $R_1 = H$, $R_2 = (CH_3)_2CHCO_2$

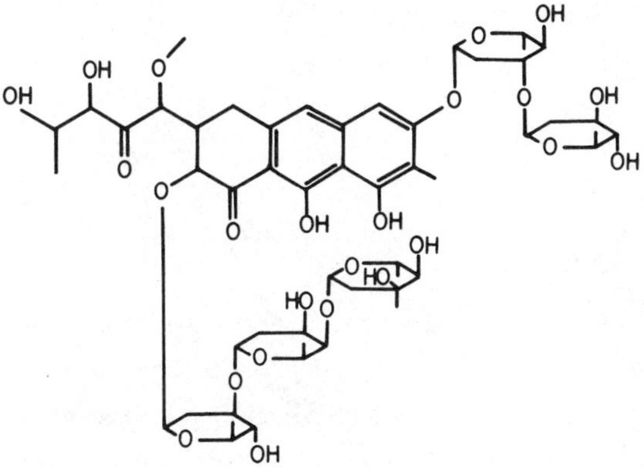

MITHRAMYCIN

FIG. 7. Structures of chromomycin A_3, olivomycin A, and mithramycin [96,97].

7.3. Interaction with DNA

The binding interaction between these antibiotics and DNA has been little explored. However, it is known that Mg(II) is required for the interaction [98] and that the binding primarily takes place at guanine-rich segments of the biopolymer [99]. One study has shown that the limit of association is reached when one antibiotic molecule is bound to four nucleotide base pairs [99]. While the apparent binding constant for the chromomycin A_3-DNA interaction has been determined, $\sim 10^6\ M^{-1}$ [99], the mode of binding of any of the antibiotics to DNA is unknown [95].

7.4. Metal Binding Properties

The three antibiotics appear to form complexes with a variety of metal ions and complexation studies with Mg(II), Ca(II), Al(III), Mn(II), Fe(III), Co(II), Ni(II), Cu(II), and Zn(II) have been reported [98-100]. Aside from recording the changes which occur in the spectra of the drugs upon metal binding, no detailed investigations of the metal complexes formed with the antibiotics have been published. It would appear such investigations would be a necessary prerequisite to understanding how the drugs interact with DNA.

8. CONCLUSIONS

It should be evident from this review that the study of metal interactions with an important class of pharmaceutical agents, antitumor antibiotics, is just beginning. While chelation in pharmacology is certainly not a new theme, bleomycin and its cogners demonstrate that it has a new, subtle but nonetheless important, permutation. Namely, that it is possible for metal-centered oxidation and reduction processes to be associated with the activity of an anticancer drug. The extent to which this observation will become important in pharmacology is at this point difficult to assess. However, anticancer agents such as streptonigrin, and possibly neocarzinostatin [101], may operate by metal-mediated redox mechanisms. It is hoped that by summarizing the known metal binding properties of a group of antitumor antibiotics that new interest in this important but largely overlooked area of pharmacology will be stimulated.

REFERENCES

1. D. D. Perrin and R. P. Agarwal, in Metal Ions in Biological Systems, Vol. 2 (H. Sigel, ed.), Marcel Dekker, New York, 1973, p. 167.
2. H. Umezawa, *Biomed.*, *18*, 459 (1973).
3. R. H. Blum, S. K. Carter, and K. A. Agre, *Cancer*, *31*, 903 (1973).
4. T. Takita, Y. Muraoka, T. Nakatani, A. Fujii, Y. Umezawa, H. Naganawa, and H. Umezawa, *J. Antibiot.*, *31*, 801 (1978).
5. H. Umezawa, in Antibiotics, Vol. III, Mechanism of Action of Antimicrobial and Antitumor Agents (J. W. Corcoran and F. E. Hahn, eds.), Springer-Verlag, New York, 1975, p. 21.
6. T. Takita, A. Fujii, T. Fukuoka, and H. Umezawa, *J. Antibiot.*, *26*, 252 (1973).
7. M. Kunishima, T. Fujii, Y. Nakayama, T. Takita, and H. Umezawa, *J. Antibiot.*, *29*, 853 (1976).
8. Y. Nakayama, M. Kunishima, S. Omoto, T. Takita, and H. Umezawa, *J. Antibiot.*, *26*, 400 (1973).
9. H. Umezawa, T. Takeuchi, S. Hori, T. Sawa, M. Ishizuka, T. Ichikawa, and T. Komai, *J. Antibiot.*, *25*, 409 (1972).
10. H. Asakura, M. Hori, and H. Umezawa, *J. Antibiot.*, *28*, 537 (1975).
11. L. F. Povirk, W. Wübker, W. Kohnlein, and F. Hutchinson, *Nucl. Acid Res.*, *4*, 3573 (1977).
12. A. D. D'Andrea and W. A. Haseltine, *Proc. Nat. Acad. Sci. U.S.*, *75*, 3608 (1978).
13. A. Takeshita, A. Grollman, E. Ohtsubo, and H. Ohtsubo, *Proc. Nat. Sci. U.S.*, *75*, 5983 (1978).
14. C. W. Haidle, *Mol. Pharmacol.*, *7*, 645 (1971).
15. C. W. Haidle, K. K. Weiss, and M. T. Kuo, *Mol. Pharmacol.*, *8*, 531 (1972).
16. M. T. Kuo and C. W. Haidle, *Biochim. Biophys. Acta*, *335*, 109 (1974).
17. W. E. G. Müller, Z. Yamazaki, H. J. Breter, and R. K. Zahn, *Eur. J. Biochem.*, *31*, 518 (1972).
18. A. Kono, Y. Matsushima, M. Kojima, and T. Maeda, *Chem. Pharm. Bull. Jap.*, *25*, 1725 (1977).
19. K. Nagai, H. Yamaki, H. Suzuki, N. Tanaka, and H. Umezawa, *Biochim. Biophys. Acta*, *179*, 165 (1969).

20. J. W. Lown and S. Sim, *Biochem. Biophys. Res. Commun.*, *77*, 1150 (1977).

21. E. A. Sausville, J. Peisach, and S. B. Horwitz, *Biochem. Biophys. Res. Commun.*, *73*, 814 (1976).

22. E. A. Sausville, J. Peisach, and S. B. Horwitz, *Biochem.*, *17*, 2740 (1978).

23. E. A. Sausville, R. W. Stein, J. Peisach, and S. B. Horwitz, *Biochem.*, *17*, 2746 (1978).

24. J. C. Dabrowiak, F. T. Greenaway, and F. S. Santillo, in Bleomycin: Chemical Biochemical and Biological Aspects (S. Hecht, ed.), Springer-Verlag, New York, 1979, p. 137.

25. J. C. Dabrowiak, F. T. Greenaway, F. S. Santillo, and S. T. Crooke, *Biochem. Biophys. Res. Commun.*, *91*, 721 (1979).

26. J. W. Lown, in Bleomycin: Chemical Biochemical and Biological Aspects (S. Hecht, ed.), Springer-Verlag, New York, 1979, p. 184.

27. Y. Sugiura and T. Kikuchi, *J. Antibiot.*, *31*, 1310 (1978).

28. I. Shirakawa, M. Azegami, S. Ishii, and H. Umezawa, *J. Antibiot.*, *24*, 761 (1971).

29. M. Takeshita, A. P. Grollman, and S. B. Horwitz, *Virol.*, *69*, 453 (1976).

30. K. Takahashi, O. Yoshioka, A. Matsuda, and H. Umezawa, *J. Antibiot.*, *30*, 861 (1977).

31. H. Suzuki, K. Nagai, H. Yamaki, N. Tanaka, and H. Umezawa, *J. Antibiot.*, *21*, 379 (1968).

32. T. Ichikawa, A. Matsuda, K. Miyamoto, M. Tsubosaki, T. Kaihara, K. Sakamoto, and H. Umezawa, *J. Antibiot.*, *20*, 149 (1967).

33. M. Ishizuka, H. Takayama, T. Takeuchi, and H. Umezawa, *J. Antibiot.*, *20*, 15 (1967).

34. H. Umezawa, M. Ishizuka, K. Kimura, J. Iwanaga, and T. Takeuchi, *J. Antibiot.*, *21*, 592 (1968).

35. H. Umezawa, K. Maeda, T. Takeuchi, and Y. Akami, *J. Antibiot.*, *Ser. A.*, *19*, 200 (1969).

36. M. Chien, A. P. Grollman, and S. B. Horwitz, *Biochem.*, *16*, 3641 (1977).

37. L. F. Povirk, M. Hogan, and N. Dattagupta, *Biochem.*, *18*, 96 (1979).

38. H. Kashi, H. Naganawa, T. Takita, and H. Umezawa, *J. Antibiot.*, *31*, 1316 (1978).

39. K. S. R. Sastry, J. G. Hallee, M. E. Ottlinger, and E. W. Westhead, Abstr., 4th International Conference on Hyperfine Interactions, Madison, N. J., June 13-17, 1977.

40. J. E. Strong and S. T. Crooke, *Cancer Res.*, *38*, 3322 (1978).

41. A. D. Nunn, *J. Antibiot.*, *29*, 1102 (1976).

42. J. P. Novel, H. Renault, J. Robert, C. Jeanne, and L. Wicart, *Nouv. Presse Méd.*, *1*, 25 (1971).

43. M. V. Merrick, S. W. Gunasekera, J. P. Lavender, A. D. Nunn, M. L. Thakur, and E. D. Williams, *Med. Radioiso. Scint.*, *2*, 721 (1972).

44. M. S. Lin, D. A. Goodwin, and S. L. Kruse, *J. Nucl. Med.*, *15*, 338 (1974).

45. B. R. McGarvey, *Trans. Met. Chem.*, *3*, 90 (1966).

46. Y. Sugiura, K. Ishizu, and K. Myoshi, *J. Antibiot.*, *32*, 453 (1979).

47. J. C. Dabrowiak and F. S. Santillo, *J. Electrochem. Soc.*, *126*, 2091 (1979).

48. Y. Sugiura, *J. Antibiot.*, *31*, 1206 (1978).

49. J. C. Dabrowiak and F. T. Greenaway, unpublished results, 1978.

50. J. C. Dabrowiak, W. Longo, M. Van Husen, F. T. Greenaway, and S. T. Crooke, Abstr. 179th National Meeting of the American Chemical Society, Chicago, Ill., September, 1977, Abstr. MEDI 51.

51. R. S. Drago, D. W. Meek, R. Longhi, and M. D. Joesten, *Inorg. Chem.*, *2*, 1056 (1963).

52. F. T. Greenaway, J. C. Dabrowiak, M. Van Husen, R. Grulich, and S. T. Crooke, *Biochem. Biophys. Res. Commun.*, *85*, 1407 (1978).

53. J. C. Dabrowiak, F. T. Greenaway, and R. Grulich, *Biochem.*, *17*, 4090 (1978).

54. J. C. Dabrowiak, F. T. Greenaway, W. E. Longo, M. Van Husen, and S. T. Crooke, *Biochim. Biophys. Acta*, *517*, 517 (1978).

55. T. Takita, Y. Muraoka, T. Nakatani, A. Fujii, Y. Iitaka, and H. Umezawa, *J. Antibiot.*, *31*, 1073 (1978).

56. Y. Iitaka, H. Nakamura, T. Nakatani, Y. Muraoka, A. Fujii, T. Takita, and H. Umezawa, *J. Antibiot.*, *31*, 1070 (1978).

57. A. E. G. Cass, A. Galdes, H. A. O. Hill, and C. E. McClelland, *FEBS Lett.*, *89*, 187 (1978).

58. K. Tomita, Y. Uenoyama, K. Numata, T. Sasahira, Y. Hoshino, K. Fujisawa, H. Tsukiura, and H. Kawaguchi, *J. Antibiot.*, *31*, 497 (1978).

59. H. Kawaguchi, H. Tsuriura, K. Tomita, M. Konishi, K. Saito, S. Kobaru, K. Namata, K. Fujisawa, T. Miyaki, M. Hatori, and H. Koshiyama, *J. Antibiot.*, *30*, 779 (1977).

60. H. Imanishi, M. Ohbayashi, Y. Nishyama, and H. Kawaguchi, *J. Antibiot.*, *31*, 667 (1978).

61. W. T. Bradner, in The Bleomycins: Current Status and New Developments (S. Carter, H. Umezawa, and S. T. Crooke, eds.), Academic Press, New York, 1978, p. 333.

62. M. Konishi, K. Saito, K. Numata, T. Tsuno, K. Asama, H. Tsukiura, T. Naito, and H. Kawaguchi, *J. Antibiot.*, *30*, 789 (1977).

63. F. T. Greenaway, J. C. Dabrowiak, R. Grulich, and S. T. Crooke, *Org. Mag. Res.*, in press.

64. T. Takita, Y. Muraoka, A. Fujii, H. Itoh, K. Maeda, and H. Umezawa, *J. Antibiot.*, *25*, 197 (1972).

65. M. J. Sleigh, *Nucl. Acid Res.*, *3*, 891 (1976).

66. T. Takita, K. Maeda, and H. Umezawa, *J. Antibiot.*, *Ser. A*, *12*, 111 (1959).

67. K. V. Rao, K. Biemann, R. B. Woodward, *J. Amer. Chem. Soc.*, *85*, 2532 (1963).

68. J. J. Oleson, L. A. Calderella, K. J. Mjos, A. R. Reith, R. S. Thie, and I. Toplin, *Antibiot. Chemother.*, *11*, 158 (1961).

69. Y. H. Chiu and W. N. Lipscomb, *J. Amer. Chem. Soc.*, *97*, 2525 (1975).

70. W. B. Kremer and J. Laszio, *Cancer Chemother. Rep.*, *51*, 19 (1967).

71. J. W. Lown and S. Sim, *Can. J. Chem.*, *54*, 2563 (1976), and references therein.

72. J. W. Lown and S. Sim, *Can. J. Chem.*, *54*, 445 (1976).

73. E. F. Gale, E. Cundliffe, P. E. Reynolds, M. H. Richmond, and M. J. Waring, in The Molecular Basis of Antibiotic Action, Wiley Interscience, New York, 1972, p. 255.

74. H. L. White and J. R. White, *Biochim. Biophys. Acta*, *123*, 648 (1966).

75. R. Cone, S. K. Hasan, J. W. Lown, and A. R. Morgan, *Can. J. Chem.*, *54*, 219 (1976).

76. N. S. Mizuno and D. P. Gilboe, *Biochim. Biophys. Acta*, *224*, 319 (1970).

77. A. DiMarco, M. Gaetani, P. Orezzi, B. Scarpinato, R. Silvertrini, M. Soldati, T. Dasdia, and L. Valentini, *Nature*, *201*, 706 (1964).

78. A. DiMarco, M. Gaetani, and B. Scarpinato, *Cancer Chemother. Rep.*, *53*, 33 (1969).

79. F. Arcamone, G. Cassinelli, G. Fantini, A. Grein, P. Orezzi, C. Pol, and C. Spalla, *Biotechnol. Bioeng.*, *11*, 1101 (1969).

80. D. E. Nettleton, Jr., W. T. Bradner, J. A. Bush, A. B. Coon, J. E. Moseley, R. W. Myllymaki, F. A. O'Herron, R. H. Schreiber, and A. L. Vulcano, *J. Antibiot.*, *30*, 525 (1977).

81. R. Angiuli, E. Foresti, L. Riva DiSansenerino, N. W. Isaacs, O. Kennard, W. D. S. Motherwell, D. L. Wampler, and F. Arcamone, *Nature New Biol.*, *234*, 78 (1971).

82. A. DiMarco, F. Arcamone, and F. Zunino, in Antibiotics: Mechanism of Action of Antimicrobial and Antitumor Agents, Vol. 3 (J. W. Corcoran and F. E. Hahn, eds.), Springer-Verlag, New York, 1975, p. 101.

83. P. Midelens and W. Levinson, *Bioinorg. Chem.*, *9*, 441 (1978).

84. N. R. Bachur, S. L. Gordon, and M. V. Gee, *Mol. Pharmacol.*, *13*, 901 (1977), and references therein.

85. F. Zunino, R. A. Gambetta, A. DiMarco, and A. Zaccara, *Biochim. Biophys. Acta*, *277*, 489 (1972).

86. W. J. Pigram, W. Fuller, and L. D. Hamilton, *Nature New Biol.*, *235*, 17 (1972).

87. E. Calendi, A. DiMarco, M. Reggiani, B. M. Scarpinato, and L. Valentini, *Biochim. Biophys. Acta*, *103*, 25 (1965).

88. M. M. Fishman and I. Schwartz, *Biochem. Pharmacol.*, *23*, 2147 (1974).

89. A. Rusconi, *Biochim. Biophys. Acta*, *123*, 627 (1966).

90. D. W. Yesair, S. McNitt, and L. Bittman, in Proceedings of the 3rd International Symposium on Microsomes and Drug Oxidation (V. Ullrich, ed.), Pergamon Press, New York, 1977, p. 688.

91. S. Tatsuoka, A. Miyake, and K. Mizuno, *J. Antibiot., Ser. B*, *13*, 332 (1960).

92. M. G. Brazhnikova, E. B. Kruglyak, I. N. Kovsharova, N. V. Konstantinova, and V. V. Proshlyakova, *Antibiotiki*, *7*, 39 (1962).

93. K. V. Rao, W. R. Cullen, and B. O. Sobin, *Antibiot. Chemother.*, *12*, 182 (1962).

94. G. F. Gause, in Antibiotics: Mechanism of Action, Vol. 1 (D. Gottlieb and P. D. Shaw, eds.), Springer-Verlag, New York, 1967, p. 246.

95. G. F. Gause, in Antibiotics: Mechanism of Action of Antimicrobial and Antitumor Agents, Vol. 3 (J. W. Corcoran and F. E. Hahn, eds.), Springer-Verlag, New York, 1975, p. 197.

96. K. A. Sedov, I. V. Sorokina, Y. A. Berlin, and M. N. Kolosov, *Antibiotiki*, *14*, 721 (1969).

97. Y. A. Berlin, O. A. Kiseleva, M. N. Kolosov, M. M. Shemyakin, V. S. Soifer, I. V. Vasina, and I. V. Yartseva, *Nature*, *218*, 193 (1968).

98. D. C. Ward, E. Reich, and I. H. Goldberg, *Science*, *149*, 1259 (1965).

99. W. Behr, K. Honikel, and G. Hartman, *Eur. J. Biochem.*, *9*, 82 (1969).

100. M. G. Brazhnikova, E. B. Kruglyak, V. N. Borisova, and G. B. Fedorova, *Antibiotiki*, *9*, 141 (1964).

101. R. M. Burger, J. Peisach, and S. B. Horwitz, *J. Biol. Chem.*, *14*, 4830 (1978).

Chapter 8

INTERACTIONS OF ANTICANCER DRUGS WITH ENZYMES

John L. Aull, Harlow H. Daron[†],
Michael E. Friedman, and Paul Melius
Department of Chemistry and
[†]Department of Animal and Dairy Sciences
Auburn University
Auburn, Alabama

1. INTRODUCTION 338

2. INTERACTIONS OF ENZYMES WITH METAL-CONTAINING ANTICANCER COMPOUNDS 339

   2.1. General Background 339

   2.2. Effects of Platinum Compounds on Enzymes and Proteins 341

   2.3. Effects of Other Metallocompounds on Enzymes 351

3. INTERACTIONS OF METALLOENZYMES AND METAL-ION-ACTIVATED ENZYMES WITH ANTICANCER COMPOUNDS 353

   3.1. Target Enzymes 353

   3.2. Drug Activation (Kinases) 362

   3.3. Drug Inactivation (Xanthine Oxidase) 364

ABBREVIATIONS 365

REFERENCES 366

1. INTRODUCTION

Enzymes are frequently the primary targets for chemotherapeutic compounds. In addition, these compounds are usually metabolized after entering the cell. In some cases enzyme-catalyzed processes are required for the administered drug to become therapeutically active, while in other cases metabolic processes destroy the chemotherapeutic properties of the compound. A number of enzymes that are inhibited by chemotherapeutic agents, and many enzymes that are involved in drug metabolism, contain metals or are activated by metal ions. The role of the metal in most drug-enzyme interactions has not been clearly defined. A large amount of interest in the potential role of metals in chemotherapy has been generated by the recent discovery that certain metal complexes have antineoplastic activity. Although there is evidence that the primary site of action of these compounds is on DNA, they also interact with and inhibit a number of enzymes.

The present chapter deals with two general areas, both of which involve interactions between enzymes, metals, and antineoplastic agents. The first area is concerned with the interactions of enzymes with metal-containing compounds, some of which have antitumor activity. A large portion of this section is devoted to describing the inhibition of enzymes by complexes of platinum; however, compounds containing other metals are also considered. The second general area concerns metalloenzymes and metal-ion-activated enzymes that are either inhibited by antitumor compounds or involved in the metabolism of these compounds. Specific examples of enzymes or classes of enzymes that are involved in these types of interactions are presented, with particular emphasis on the metal ion requirements of the enzymes and the possible roles of the metal in the catalytic mechanisms.

2. INTERACTIONS OF ENZYMES WITH METAL-CONTAINING ANTICANCER COMPOUNDS

2.1. General Background

Since Rosenberg's [1] discovery of the biological activity of the cis-dichlorodiamminoplatinum(II) (cis-PDD) complex, a voluminous literature has developed supporting the molecular basis of the biological interaction at the DNA level. Whether this effect is on DNA alone, or a combined effect on nucleic acid and protein, is yet to be defined. It has been known at least since 1967 [2] that platinum complexes inhibit certain enzymes. Also, the binding of platinum complexes was used to study the structures of chymotrypsin, ribonuclease, carboxypeptidase, subtilisin, and cytochrome c by x-ray crystallography [3]. The amino acid residues involved in binding were methionine, cysteine, histidine, and N-terminal amino acids. Morris and Gale [4] studied the binding of cis-dichlorodi-([^{3}H]pyridine)Pt(II) to DNA by equilibrium dialysis. The retention of the platinum complex by DNA was decreased by previous exposure to amino acids. The greatest inhibition of binding occurred when the platinum compound was exposed first to glutamic acid, aspartic acid, lysine, arginine, and cystine. Histones decreased the binding of platinum complexes to DNA by 50%. The platinum complex was also bound by polyglutamate and polyaspartate. Robins [5] found that the effectiveness of dichloroethylenediamine Pt(II) as an inhibitor of cell growth was lost within a short time because the drug bound to serum proteins in the culture medium. The binding was quite rapid with a half-life of 50 min, but the complex was acid-labile. The amount of platinum used was equivalent to 1 platinum atom per 5-10 intact protein molecules, and this treatment resulted in a change in the electrophoretic mobility of the protein.

Bannister et al. [6], using an ultrafiltration process and x-ray fluorescence to measure platinum, found that when 3.98 mg of cis-PDD was incubated in one liter of plasma at 37°C for 7 hr, no filterable platinum was detectable. The half-life for platinum-protein binding was found to be 156 min. These results agree with the work of Litterst et al. [7], who used atomic absorption spectroscopy to measure the platinum. Balice and Theophanides [8] used infrared spectroscopy to study the binding of platinum complexes to various amino acids. Melius and Friedman [2] have recently reviewed the interaction of platinum complexes with various amino acids and peptides.

Metal complexes and chelates as cancer chemotherapeutic agents were reviewed by Furst [9]. Wherever positive results were found, usually transition elements were involved. A carcinolytic effect for Yoshida carcinoma was reported for cobalt complexes [10], and zinc salts were reported to have some effect on induced tumors [11]. Fowler's solution (suspension of lead arsenate in benzene), colloidal lead or lead phosphate, small amounts of copper, lead, bismuth, ruthenium, and selenium have all been used in treatment of tumors [9]. Complexes of cobalt, copper, mercury, lead, and cadmium have also been reported to have some antitumor effects. On the contrary, iron compounds seem to be carcinogenic or to stimulate tumor growth.

On the very last page of his book, Furst [9], at the end of his section on speculations and conclusions, made the statement, "Perhaps platinum or palladium derivatives should be made. Since thioethers have a strong tendency to unite with these metals, sulfur mustard complexes and chelates may prove useful. A reaction between K_2PtCl_4 and sulfur mustard would produce three isomers." This has turned out to be quite a prophetic statement, as a variety of platinum complexes have been shown to have excellent antitumor activities and the cis-PDD compound has been studied in a large number of clinical experiments with humans. Carcinomas of the head and neck, ovaries and testes, as well as some sarcomas and lymphomas, have responded to treatment with cis-PDD [12]. One serious drawback is the toxicity incurred, especially in the kidneys and hearing; this has catalyzed a search for compounds with lower toxicity or administration techniques which will reduce the toxicity of the platinum compounds.

Some results have been particularly encouraging, such as those of the work by Krakoff [13], who used PDD in combination with actinomycin, cyclophosphamide, and vinblastine to obtain 95% observable regressions of cancer of the testis (30 patients) with two-thirds of the patients showing complete remission.

Rosenberg's first report of the antitumor activity of cis-PDD indicated that a single injection of the complex at 8 mg kg^{-1} would almost completely inhibit the growth of a transplanted sarcoma-180 tumor. The LD_{50} level was 14 mg kg^{-1}. The PDD complex could be injected 8 days after sarcoma 180 transplant and still produce an almost complete regression of large tumors in close to 100% of the mice. The effect lasted at least 11 months after the injection [14, 15]. Kociba [16] demonstrated a 100% cure in rats having the Walker 256 carcinoma and the Dunning ascitic leukemia using cis-PDD.

2.2. Effects of Platinum Compounds on Enzymes and Proteins

2.2.1. In Vivo Studies

First, the effect of platinum compounds on proteins and enzymes, beginning with in vivo systems will be discussed. An example of an effect on an enzyme system is the work of Kessel et al. [17], where it was found that cis-PDD treatment of a patient with carcinoma of the testis resulted in nearly a 75% decrease in plasma sialyltransferase, an enzyme reported to be elevated in breast and colon tumors [18]. The decrease in sialyltransferase activity accompanied a decrease in the size of the tumor. Apparently this was not a direct effect on the enzyme itself as 100 μg ml^{-1} (3×10^{-4} M) of the cis-PDD did not inhibit sialyltransferase activity in the assay system. It was suggested that assaying for sialyltransferase activity is valuable not only as a cancer diagnostic method, but also as a barometer in following the progression of the disease during chemotherapeutic and other treatments.

Slater et al. [19] found that 8.5 mg kg^{-1} of cis-PDD administered to rats caused a very significant rise in β-glucuronidase activity in the urine. They examined the effects on kidney and found that cis-PDD bound to proximal tubular membrane suspensions more strongly than the trans isomer, cis-dichlorodicyclopentylamine platinum(II), or even thymine platinum blue. The cis-PDD binding was complete within 5 min at 37°C and was proportional to the concentration of protein in the membrane. The binding was not affected by 5 mM cysteamine, penicillamine, or N-acetyl cysteine. The cis-PDD was also an inhibitor of succinate respiration by kidney mitochondrial preparations. The inhibition of O_2 uptake linked to succinate oxidation was overcome by 1 mM cysteamine or 5 mM histidine, but not by penicillamine.

The platinum compounds were known to be nephrotoxic, as well as ototoxic, which may indicate effects on enzymes as well as on other proteins [20-23]. Hayes et al. [24] investigated the use of mannitol diuresis to ameliorate the renal toxicity of cis-PDD. They studied dose levels of 3-5 mg kg^{-1} in 60 patients. At the 5 mg kg^{-1} level, renal and marrow toxicity, as well as ototoxicity were observed. The renal toxicity was limited to transient increases in serum creatinine levels. The statement was made that

> We believe that the mannitol technique prevents immediate massive platinum binding to renal tubular proteins and despite continuous renal exposure to that portion of the platinum which is excreted over a long half-life of 51-72 hr, severe nephrotoxicity is prevented.

Binet and Volfin [25] have reported on the effect of cis-platinum(II) (3,4-diaminotoluene) dichloride on mitochondrial membrane properties. The platinum compound was found to stimulate State 4 respiration, which is similar to an apparent uncoupling of oxidative phosphorylation, and also may be correlated to changes in membrane permeability. There was also an inhibition of phosphate transport, a decrease in the accumulation of Ca^{2+}, and an increase in passive permeability of Ca^{2+}. These effects are most likely associated with interactions of the platinum compounds with proteins in the membranes.

Renshaw and Thomson [26], using cis-$^{191}Pt(NH_3)_2Cl_4$, found about 70-80% of the platinum was associated with metabolic intermediates and proteins in *E. coli*, *B. cereus*, and *S. aureus* bacteria filamentous cells. In inhibited cells, the Pt was combined only with the cytoplasmic proteins. They also found that rhodium and ruthenium cause long filamentous growth in bacteria.

The effect of platinum compounds on DNA, RNA, and protein synthesis has been investigated by Harder and Rosenberg [27], Howle et al.[28], and in the author's laboratory by Kohl et al. [29]. DNA, RNA, and protein synthesis were inhibited by cis-PDD, cis-diamminotetrachloroplatinum(IV), and ethylenediaminedichloroplatinum(II), but not by tetraamminodichloroplatinum(II) or trans-diamminotetrachloroplatinum(IV). This result has usually been interpreted as an effect on the DNA molecule alone; however, it is not possible to eliminate some interaction with the enzymes that are present. As a matter of fact, Harder and Rosenberg found that the trans-diamminotetrachloro compound produced a rapid initial inhibition of [^{3}H]L-leucine incorporation into protein, which probably indicates an effect on the enzyme systems involved in protein biosynthesis. Kohl et al. [29] found that aquation of the cis-PDD complex enhanced the inhibition of protein biosynthesis in hamster medulloblastoma cells.

2.2.2. *In Vitro Studies*

At about the time of Rosenberg's discovery that cis-PDD is an inhibitor of bacterial cell division, the x-ray crystallographers were studying enzyme structure using various platinum compounds such as $PtCl_4^{2-}$ and $Pt(CN)_4^{2-}$ [30]. These latter studies indicated that the platinum complexes bound to the sulfur of methionine or cysteine and the nitrogen of the histidine side chain. Chymotrypsin, ribonuclease S, carboxypeptidase, subtilisin, and cytochrome c were some of the enzymes studied, and it was proposed that the platinum complex bound preferentially to methionyl sulfurs on the enzymes' surface. It was further suggested that the sulfur of methionine donates an electron pair to platinum to form an octahedral, rather than a square planar, structure and the Pt(II) was probably oxidized to Pt(IV) while acquiring a sixth ligand from the solvent system. Thomson et al. [31] criticized the conclusion that Pt(II) was oxidized to Pt(IV), as this does not happen when $PtCl_4^{2-}$ reacts with simple amino acids. In some instances where a methionyl residue is involved in some vital function in the protein, as a ligand to heme in cytochrome c for example, the methionine does not react with $PtCl_4^{2-}$.

It has been shown that $Pt(CN)_4^{2-}$ can also bind to glutamic acid, lysine, and serine in cytochrome c [32]. In pyruvate kinase, alkaline phosphatase, alcohol dehydrogenase, lysozyme, carboxypeptidase A, trypsin, carbonic anhydrase B and C, serum albumin, hemoglobin, and rhodanese, the $Pt(CN)_4^{2-}$ complex probably binds to arginine sites [33-35]. In horse liver alcohol dehydrogenase, the main binding site appears to be arginine-47, which is located in a hydrophobic region that is responsible for the binding of the diphosphate bridge of the pyridine nucleotide coenzyme. Another possible platinum binding site is arginine-271, located in the "adenine pocket." Binding sites in serum albumin are lysine, arginine, and tryptophan [36,37]. Anion binding to albumin decreases with chemical modification of arginine, but not of lysine.

Some of the earliest studies of the effects of platinum complexes on enzyme activity were those reported by Melius et al. [39] and Friedman et al. [40]. The work of Melius and coworkers was on the effects of $PtBr_4^{2-}$, dibromo(ethylenediamine)platinum(II), and bromo(diethylenetriamine)platinum(II) binding to leucine aminopeptidase, an enzyme which had been reported to be elevated in tumors. LAP, which is an exopeptidase enzyme requiring Mg^{2+} or Mn^{2+} for catalytic activity and also containing Zn^{2+}, was inhibited completely within 1 hr at 37°C by 5×10^{-3} M $PtBr_4^{2-}$. The $Pt(en)Br_2$ complex at the same concentration resulted in 80% inhibition after 50 hr at 37°C, while the $Pt(dien)Br^+$ only inhibited the enzyme 20% after

50 hr at 37°C. Thus, the tetrabromo complex reacts with the enzyme much faster than the other complexes. In studies with malate dehydrogenase (MDH), Friedman et al. [38] showed that complexes of Pt(IV) were more potent than those of Pt(II) as inhibitors of this enzyme. In addition, increased alkylation of the ethylenediamine ligand decreased the ability of the platinum complexes to inhibit MDH, but no consistent effect of alkylation was found with LAP.

Melius et al. [41] studied the reactions of MDH with $PtCl_4^{2-}$ and $PtBr_4^{2-}$ and found that inhibition decreased for both complexes with increasing pH, in both equilibrium and rate experiments. The isoionic pH of MDH is 6.1-6.4, so the increase in pH from 6.5 to 8.3 could have led to a decrease in the binding because of an increased negative charge on the enzyme. It was also found that the $PtBr_4^{2-}$ complex inhibition was 6 to 8 times faster than the $PtCl_4^{2-}$ inhibition, which agrees with the fact that bromide ligands are aquated seven times faster than chloride ligands in substitution reactions of platinum complexes [42]. These results are expected on the basis that the aquated form goes through a deprotonization as the pH is brought above 7.5 as follows:

$$PtX_4^{2-} + H_2O \xrightarrow{X^-} Pt(H_2O)X_3^- \xrightarrow{H^+} Pt(OH)X_3^{2-}$$

The effects of amino acids on the inhibition of MDH by $PtCl_4^{2-}$ has been studied [43]. Both methionine and cysteine at 1-5 mM concentrations decreased the rate of inhibition of MDH by $PtCl_4^{2-}$; however, 0.05 M histidine, lysine, or aspartic acid had no effect. Methionine complexes of platinum were found to be very stable, and addition of methionine to completely inhibited MDH regenerated 38% of the initial enzyme activity.

When Friedman and Teggins [44] observed the inhibition of MDH, rabbit muscle lactate dehydrogenase (LDH) and horse liver alcohol dehydrogenase (ADH) by cis- and trans-PDD, the greatest difference between the two inhibitors was seen in the LDH system. The trans-PDD gave a 20-fold larger binding constant than the cis-PDD when LDH was used. Using yeast ADH, the binding constant for the trans-PDD was four times that of the cis-PDD compound, while the MDH and liver ADH inhibition produced binding constants which were almost the same for both the cis- and trans-PDD compounds. In the case of the LDH, where the trans isomer is a better inhibitor, it was suggested that two bonds were formed between the platinum complex and the enzyme. It was also suggested that the charge on the platinum complex might be the most important factor for the inhibition of the MDH [45]. The steric differences seemed to have only a minor effect, while geometric variation, as in the case of the trans- and cis-PDD, yielded no significant difference in the inhibition of the MDH. This is of interest since Rosenberg [1] found that the neutral complexes were superior as antitumor compounds. No inhibition was observed using the positively charged platinum complexes, and the concentration of cis-PDD required for 50% inhibition was 2000-fold greater than the enzyme concentration [45], yet this platinum complex has proven to be one of the best antitumor compounds.

Friedman et al. [46] studied the protective effect of NADH, NAD^+, L-malate and mixtures of NAD^+ and L-malate on the $PtCl_4^{2-}$ inhibition of MDH at a pH where the enzyme was not active. NADH proved to be very effective in preventing the inhibition of MDH. High concentrations (10-100 mM) of either L-malate or NAD^+ alone did not provide much protection for the enzyme against platinum complex inhibition; however, in combination they proved to be excellent protectors even at lower concentrations (1 mM). Fluorescence and ultraviolet measurements indicated that $PtCl_4^{2-}$ does not react directly with NADH [47,48]. There presumably is a sulfhydryl allosteric site in the mitochondrial MDH [49], but K_2PtCl_4 does not react with it at low temperature or high concentrations of the complex. The authors concluded that the protective effect of NADH is exerted probably by the shielding of the two reactive sites in the enzyme by two different NADH molecules, thereby blocking the platinum complex attack on the active site. The specificity of $PtCl_4^{2-}$ for a sulfur-containing side chain indicated that there probably is a methionine or cysteine at the NADH (or NAD^+) malate binding site on MDH which is the key reactive group for the platinum complex. Only one platinum complex molecule reacts at that site on the MDH molecule [39].

There have been many studies of interactions between platinum compounds and polynucleotides [50-53], mononucleotides, and nitrogen bases [54-56]. Friedman et al. [57] studied mitochondrial MDH, heart LDH, and muscle LDH enzyme inhibition by $PtCl_4^{2-}$, cis-PDD, and trans-PDD in the presence of various polynucleotides and mononucleotides. It was found that ATP, AMP, RNA, and DNA decreased the $PtCl_4^{2-}$ inhibition of mitochondrial MDH, with the effect being concentration-dependent. The mononucleotides proved to be the better protectors. No significant difference in the protection was found between the cis- or trans-PDD. Similar results were obtained using heart LDH. In these studies, the binding constant of the enzyme-platinum complex and the binding constant of the nucleic acid-platinum complex were much greater for $PtCl_4^{2-}$ than for either cis- or trans-PDD. It was also calculated that the platinum complexes were bound to the mononucleotides several times more tightly than to the polynucleotides. In contrast, the muscle LDH system showed an enhanced inhibition by the platinum complexes when in combination with native polynucleotides as well as with partially denatured DNA. It was concluded, in the case of the MDH and heart LDH enzymes, that the reduction of platinum complex inhibition by mono- and polynucleotides resulted from competition of the nucleic acid for the platinum inhibitor. The suggestion was made that either a strong monodentate linkage occurs between platinum and the nitrogen base, or that a bidentate ligand occurs with a second binding site selected from adjacent positions of the bases. Paired nucleic acid chains are not necessary for platinum binding, as the similarity in RNA and DNA binding constants demonstrate. In the case of muscle LDH, a ternary complex (enzyme-Pt complex-nucleotide) may occur, which actually enhances the enzyme inhibition by either effectively blocking the active site or producing a significant conformational change in the protein. In support of this, Thomas et al. [58] while studying the interaction of the bifunctional reagent bis(2-chloroethyl)methylamine (nitrogen mustard) with extracts of L-1210 cells, found that a DNA-protein cross-link was the most plausible explanation for the binding of the reagent to the DNA polymer.

Melius and McAuliffe [59] investigated the effect of N-alkyl-substituted ethylenediaminedihaloplatinum(II) complexes on LAP as a function of temperature. At 25°C, the complexes had no inhibitory effect after 50 hr of incubation, however, at 38°C and 5×10^{-3} M platinum complex, most N-alkylated ethylenediamine complexes produced about 50% inhibition after 24 hr. The half-lives for the inhibition by the N-alkylethylenediamine Pt complexes varied from 60 to 150 hr; whereas, the half-life for the simple ethylenediamine platinum(II) dibromo complex ranged from 15 to 17 hr. The slower inhibition rates by the N-alkylamine complexes should permit those compounds to avoid interaction with the plasma proteins and allow them to enter the cell and bind to cellular proteins and/or nucleic acids in order to exert their antitumor activity. Several of these complexes have been shown to have antileukemia (L-1210) activity Methionine and alanylmethionine were found to be ineffective in preventing or reversing the inhibition of LAP by these platinum complexes.

Another hydrolase enzyme system, α-chymotrypsin, has been studied using a number of platinum compounds [60]. When benzoyl-L-tryosine ethyl ester was used as the substrate for the enzyme, $PtBr_4^{2-}$, cis-PDD, trans-PDD, $PtCl_4^{2-}$, Pt(glycylmethionine), and Pt-(methionine)$(NH_3)_2$Cl were all effective inhibitors. The ethylenediaminedichloroplatinum complexes were found to be weak, slow inhibitors of the enzyme. A surprising observation was that α-chymotrypsin immobilized onto glass beads was not inhibited by any of the platinum compounds. This seems to imply that the surface of the protein molecule is very sensitive to the mode of binding by the platinum complexes, and any change in the conformation prevents that interaction from taking place.

Melius et al. [41] found that LAP and MDH were more rapidly inhibited by the $PtBr_4^{2-}$ complex after hydrolysis of the Pt complex. Teggins et al. [61] had previously determined that approximately 17% of the complex was hydrolyzed within 5 min. The resulting Pt-$(H_2O)Br_3^-$ was suggested as a potential active site reagent for structure-function studies of enzymes. As the rate of aquation appears to be involved in the rate of reaction of platinum compounds with proteins, the use of compounds which aquate very slowly may be a way of reducing the nephrotoxicity of these compounds and also a means of getting them past the blood and tissue proteins to the assumed target site in the nucleus.

Because of the nephrotoxic effects [62] of the platinum complexes, investigations were begun [63] using methionine complexes of platinum as inhibitors of liver ADH and LDH enzymes; the intent being to obtain a more specific effect of the platinum complex on the cell replication mechanism without inhibiting other enzymes necessary for the normal functions of the cell. The trans-[Pt(methionine)$_2$(NH$_3$)$_2$]Cl$_2$ complex inhibited the ADH 50% under the experimental conditions while at the same time only 8% of the LDH was inhibited. The complexes α-[Pt(methionine)(NH$_3$)Cl]Cl and [Pt(methionine-H)$_2$(NH$_3$)]Cl$_2$ gave about 40-50% inhibition of ADH and LDH, whereas complexes such as Pt(methionine)Cl, Pt(methionine)Cl$_2$, β-[Pt(methionine)(NH$_3$)Cl]Cl, and [Pt(methionine)(NH$_3$)$_2$]Cl produced 100% inhibition of ADH and LDH. Some very specific differences have been found in these methionine complexes, and they show promise as inhibitors for specific enzyme systems, while at the same time they can provide information concerning the design of more effective antitumor agents. The dipeptide methionylmethionine was found to completely protect LDH against ethylenediamine platinum(II) inhibition, whereas alanylmethionine and histidine-containing peptides gave less protection. The dipeptides did not protect ADH from β-[Pt(methionine)(NH$_3$)Cl]Cl inhibition.

The interaction of cis- and trans-PDD with thymidylate synthetase has been investigated [64]. The cis isomer did not inhibit the enzyme, but the trans isomer was a good inhibitor at 1 mM concentration. Although not an inhibitor, the cis-PDD apparently reacted with an average of 1.3 of the enzyme's four sulfhydryl groups, while the inhibitory trans isomer reacted with an average of 2.5 sulfhydryl groups. Since this enzyme is known to contain one essential sulfhydryl group, the inhibition by trans-PDD is presumed to indicate that this compound, but not the cis isomer, can block the essential sulfhydryl group of thymidylate synthetase. The presence of 2-mercaptoethanol protected the enzyme from inhibition; however, inhibited enzyme samples were not reactivated by the addition of this thiol or by removal of excess inhibitor by gel filtration. The loss of enzyme activity seen with the trans isomer correlated well with the loss of free sulfhydryl groups, but the possibility that other amino acid side chains also reacted was not investigated.

Recently, the enzymes fructose-1,6-diphosphate aldolase, catalase, glucose-6-phosphate dehydrogenase (GPD), glyceraldehyde-3-phosphate dehydrogenase (GaPD), peroxidase, tyrosinase, and dihydrofolate reductase from *L. casei* were tested for inhibition by K_2PtCl_4, cis-PDD, and trans-PDD [65]. Interestingly, of the seven enzymes tested, only those that contain essential sulfhydryl groups were inhibited. GPD, GaPD, and aldolase were inhibited by K_2PtCl_4, while only GaPD was inhibited by the cis and trans isomers of PDD. Kinetic analyses suggest that the inhibition of GPD by K_2PtCl_4 is first-order with respect to enzyme concentration; however, all of the other inhibition reactions were more complex. For example, the data obtained from the inhibition of GaPD by cis- and trans-PDD were consistent with two first-order processes where the rate of the first (fast) process is about the same with both isomers, but the rate of the second (slow) process is faster with the trans isomer than with the cis isomer. The data obtained with aldolase and K_2PtCl_4 are consistent with a system where the reaction is first-order, but does not completely inactivate the enzyme.

2.3. Effects of Other Metallocompounds on Enzymes

It has been shown that cellular deaminases reduce the utility of the nucleoside analogs arabinosylcytosine (ara-C) and arabinosyladenine (ara-A) as antitumor agents by changing them to inactive compounds. Inhibitors of these deaminases, and especially nonstructural analogs, would therefore increase the antitumor activities of ara-C and ara-A by preventing their inactivation. The rhodium(II) acetate "cage" complex inhibited purified Swiss mouse kidney cytidine deaminase when ara-C was used as substrate [66]. The inhibition was linear with dose level. A Lineweaver-Burk plot established that rhodium(II) acetate gave a mixed-type inhibition. The implication is that rhodium(II) acetate binds close to, but not precisely in, the substrate site of the enzyme. A Hill plot with a slope of 1 was obtained, which indicates that rhodium is bound to a critical site in the vicinity of the catalytic center.

In vivo, rhodium(II) carboxylates were found to be potent inhibitors of DNA synthesis, but they stimulated RNA synthesis [67]. In short-term studies (30 min), the rhodium(II) carboxylates inhibited the in vitro DNA polymerase I activity as well as DNA polymerase activity. These short-term effects are probably due to more rapid axial-type binding rather than exchange for a carboxylate ion. In order to distinguish between enzyme binding as the mechanism for inhibition compared to nucleic acid binding, the specific binding of the rhodium(II) acetate to poly-A was studied. Two systems were set up, one in which the template and substrate would bind to rhodium(II) carboxylates (poly-dA, poly-dT, dATP, dTTP), and one in which binding to template or substrate did not occur (poly-dC, poly-dG, dCTP, dGTP). In the poly-dA:poly-dT system, 83.9% inhibition of the polymerase occurred where the rhodium complex could bind to template, substrate, and enzyme; whereas 55.7% inhibition occurred where the rhodium complex could bind to the enzyme, but not the template or substrate. Opposite results were obtained with the rhodium(II) propionate and butyrate complexes. This was explained as being due to the larger ratio of dATP to rhodium(II) which leaves little complex to bind to the enzyme. The authors concede that more work is necessary to confirm their interpretation.

Gallium nitrate was particularly active against solid tumors transplanted subcutaneously, and suppressed the growth of six out of eight tumors more than 90% [68]. It is taken up by the tumors and concentrated, thus having a good potential. Gallium nitrate, at a concentration of 50 $\mu g\ ml^{-1}$, was found to inhibit DNA polymerases B and D from Walker 256 cells as well as the RNA-dependent DNA polymerase (reverse transcriptase).

Harder et al. [69] attempted to determine if cis- and trans-PDD acted directly on DNA or on the DNA polymerase enzyme during the inhibition of DNA polymerase activity. Binding of trans-PDD to DNA was higher by a factor of 2 than binding of cis-PDD. When the cis-PDD was preincubated with DNA polymerase α for 2 hr, enzyme inhibition occurred only at high platinum concentrations, however, their incubation times were only about 30 min. The trans-PDD was 7-10 times more effective than the cis isomer in inhibiting the polymerase α. In contrast, preincubation of DNA polymerase β with platinated templates enhanced the incorporation rate. Their conclusion was that both inter- and intrastrand DNA cross-links as well as monofunctional reactions contribute to the overall inactivation of DNA.

From the preceding discussion, it becomes clear why platinum complexes may exert toxic effects in animals through their interaction with proteins and enzymes, thereby modifying the chemical structure of proteins. The sulfur-containing amino acids methionine and cysteine are favored sites but histidine, arginine, glutamic and aspartic acid residues are also ligands for platinum, depending on the nature of the platinum compound.

3. INTERACTIONS OF METALLOENZYMES AND METAL-ION-ACTIVATED ENZYMES WITH ANTICANCER COMPOUNDS

3.1. Target Enzymes

3.1.1. Nucleic Acid Polymerases

DNA and RNA polymerases [70-75] from a variety of sources are zinc metalloenzymes. In addition, they also require a divalent metal ion for activity; Mg^{2+} is most commonly used, but can usually be replaced by Mn^{2+} and, in some cases, by metals such as Co^{2+} and Ni^{2+} [76-79]. The Zn^{2+} content of a number of polymerases has been measured by atomic absorption spectroscopy and is 2 g-atoms per mole for yeast RNA polymerase III [80], 2.2 for *E. gracilis* RNA polmerase II [81], 2.4 for yeast RNA polymerase I [82], 0.98 for yeast RNA polymerase II [83], 2.0 for *E. coli* RNA polymerase [84], 1.0 for *E. coli* DNA polymerase I [85,86], 1.0 for T4 phage DNA polymerase [86], 4.0 for sea urchin DNA polymerase [85], 1-2 for avian myeloblastosis virus DNA polymerase [87,88], 1.4 for murine leukemia virus DNA polymerase [88], 1.0 for wooly monkey virus DNA polymerase [88], 2-4 for T7 RNA polymerase [89], and 5.5 for N4 RNA polymerase [90]. 1,10-Phenanthroline generally inhibits DNA and RNA polymerase preparations from a variety of sources, suggesting that zinc is a common constituent of these enzymes. The related enzymes, terminal deoxynucleotidyl transferase and poly-A polymerase are also zinc metalloenzymes [91,92].

All polymerases produce nucleic acids by transferring nucleoside monophosphate units from nucleoside 5'-triphosphates to the 3' end of a growing polynucleotide, leaving pyrophosphate as the other product. The polymerases also require a template, which determines the sequence of the nucleotide units in the new strand by Watson-Crick base pairing. DNA is the template for DNA-dependent DNA and RNA polymerases, and RNA is the template for the DNA polymerases of the RNA tumor viruses (reverse transcriptase). To initiate nucleic acid synthesis DNA polymerases require a polynucleotide (primer), while RNA polymerases can use purine nucleotides [93] or dinucleotides [94,95].

Zinc is apparently involved in the formation of the enzyme-primer (initiator) complex, since 1,10-phenanthroline is a partially competitive inhibitor with respect to the DNA primer with DNA polymerase [85] and it inhibits GTP binding to the initiation site of RNA polymerase [84]. A mechanism has been proposed [96] in which zinc is coordinated to the 3'-OH of the primer [97]. Zinc not only functions in binding, but it also has a proposed catalytic function by assisting the nucleophilic attack on the α-phosphorus of the incoming nucleotide by increasing the nucleophilicity of the 3'-oxygen [82]. Since the nascent nucleic acid chain is not believed to dissociate from the enzyme after the formation of each phosphodiester bond, a translocation step must occur before the addition of the next nucleotide. Zinc plays a role in translocation since it must now coordinate with the new 3'-OH. Zinc may also be involved in maintaining the structure of the polymerases. Some polymerases contain two or more zinc ions bound at nonequivalent sites, suggesting that Zn^{2+} may have more than one function.

The nature and concentration of the divalent metal cation can influence the ability of various DNAs and synthetic polynucleotides to act as templates, depending on the type and source of the polymerase [71,73,90,98]. The role of the metal in this regard is not clear, and may involve interaction with the template rather than the enzyme. However, a specific function of the divalent metal is the formation of an enzyme-metal-nucleotide bridge complex [99-101]. In this complex, Mn^{2+} has been shown to be coordinated to the γ-phosphate of the nucleoside triphosphate rather than to all three phosphates as occurs in binary metal-(d)NTP complexes. Also, the torsion angle of the glycosidic bond of the (d)NTP is altered to conform to that found in DNA [100-103]. In addition to contributing to the conformational changes of the (d)NTP, the metal ion also probably assists pyrophosphate departure that accompanies the nucleophilic displacement [101]. Different metal ions may induce slightly different conformations that may influence base pairing, since the fidelity of replication is affected by the metal ion used. A greater error frequency was seen with Co^{2+} or Mn^{2+} than with Mg^{2+}; and Co^{2+}, Mn^{2+}, and Ni^{2+} increased the error frequency even in the presence of excess Mg^{2+} [76,77,104]. The influence of the metal ion on fidelity has been observed not only with a DNA polymerase lacking exonuclease activity [76], but also with *E. coli* DNA polymerase I [77,104], that has exonuclease activity to which a "proofreading" function has been ascribed [70].

Because of their central role in nucleic acid biosynthesis, nucleic acid polymerases are obvious targets for antineoplastic chemotherapeutic reagents. A group of compounds currently in use is the ansamycins [105-107], which have an aromatic nucleus spanned by a long aliphatic bridge, and include such antibiotics as the rifamycins and streptovaricins.

Although the ansamycins are primarily used as antibacterial agents, some show potent antimitotic activity, and thus are potential anticancer agents. With rifamycin, the bactericidal effect occurs at low concentrations (<1 μg ml^{-1}) and results from inhibition of bacterial DNA-dependent RNA polymerase. Studies with rifamycin-resistant mutants of *E. coli* and *B. subtilis* have shown that the drug binding site is on the β subunit of RNA polymerase [108, 110]. However, other subunits as well as the β subunit, have been labeled by the covalent attachment of radioactive rifamycin derivatives [111,112], suggesting that drug binding occurs near the junction of several subunits. The binding is mainly lipophilic [113]; the β' subunits contain zinc [114,115], and there is no evidence that the metal is involved in drug binding. Rifamycin has been shown to block the initiation of RNA chains by preventing the translocation step that follows the formation of the first phosphodiester bond [116,117]. Low concentrations (<1 μg ml^{-1}) of rifamycins and streptovaricins in general, do not inhibit eucaryotic RNA polymerase; however, maytansine, a related ansamycin, and some of its derivatives are inhibitors of mitosis in eucaryotic cells and show antitumor activity [118,119]. High concentrations of ansamycins have been reported to inhibit reverse transcriptase [120,121] and various mammalian nucleic acid polymerases. However, as Wehrli [105] has pointed out, "It must be stressed very strongly that an effect on a mammalian enzyme at drug concentrations of 5-200 μg/ml cannot be interpreted in the same way as the inhibition of bacterial RNA polymerase at 10^{-2} μg/ml."

Another class of clinically important anticancer drugs binds to DNA and renders it unsuitable as a template for nucleic acid polymerases. Adriamycin and daunorubicin bind to DNA by intercalating between adjacent base pairs. Goodman et al. [122] have suggested that adriamycin might inhibit DNA polymerase by promoting the formation of an intercalated template which binds irreversibly to the enzyme. Therefore, a direct interaction between DNA polymerase and adriamycin in the presence of a template cannot be ruled out. Another anticancer drug, mithramycin, probably does not intercalate DNA, but its binding to DNA requires Mg^{2+} [123]. Its principle effect is on RNA synthesis. (Bleomycin also belongs to this category, but it will not be discussed here since it is covered in detail in Chap. 7.)

A number of compounds that have anticancer and antiviral activity are analogs of the nucleosides found in nucleic acids [124]. These include the arabinosides of adenine (ara-A) and cytosine (ara-C) reviewed by Cohen [125], cordycepin (3'-deoxyadenosine), and the adenosine analogs tubercidin and formycin. Kinases convert these compounds to the corresponding nucleoside triphosphates which inhibit nucleic acid polymerases, and in some cases by their incorporation, produce counterfeit DNA. DNA polymerase is inhibited by ara-ATP and ara-CTP, as well as ara-TTP, which are competitive inhibitors of the corresponding deoxyribonucleoside triphosphates [126, 127]. Cordycepin 5'-triphosphate inhibits RNA polymerases by blocking subsequent elongation after its incorporation in place of ATP [128-130].

Zimmerman et al. [131] recently reported that, in addition to their incorporation into nucleic acids, tubercidin and formycin are also converted to their 3',5'-monophosphates by mammalian cells, but the physiological significance of this has not yet been determined. It is also of interest that adenosine 2',3'-riboepoxide 5'-triphosphate irreversibly inhibits DNA polymerases by alkylating the enzyme. It was suggested that this suicide reagent may have potential as an antineoplastic agent [132].

3.1.2. Ribonucleotide Reductases

The 2'-deoxyribonucleotide precursors of DNA are produced from ribonucleotides by the metalloenzyme ribonucleotide reductase [133-136]. There are two major classes of ribonucleotide reductases; one contains nonheme iron and requires nucleoside diphosphate substrates, while the other uses adenosylcobalamin, a cobalt-containing coenzyme, and acts on nucleoside triphosphate substrates. The cobalamin requiring enzymes have not been found in vertebrates, and are not considered here. Most mammalian ribonucleotide reductases appear similar to the enzyme from *E. coli* [137-141], which has been purified and characterized. Ribonucleotide reductase is composed of two nonidentical subunits (B1 and B2) [137,142-144]. The B1 subunit contains binding sites for both substrates and allosteric effectors [145,146]. The B2 subunit contains two iron atoms, recently reported to be in the ferric state, and an organic free radical [133,147-149]. The active enzyme is formed in the presence of Mg^{2+} by the association of the subunits in a 1:1 ratio [150].

The enzyme catalyzes the reduction of all four ribonucleoside diphosphates to the corresponding 2'-deoxyribonucleotides. The reducing agent is the dithiol-containing protein thioredoxin, although another protein called glutaredoxin can substitute [151]. The details of the reaction mechanism are unknown, but it is known that the catalytic site is formed by both B1 and B2 subunits [152], that sulfhydryl groups on the B1 subunit alternate between oxidized and reduced forms during catalysis [150,153], and that the free radical, which is associated with the aromatic ring of a tyrosyl residue [154] and requires iron [149], is required for enzyme activity [155].

Ribonucleotide reductase is highly regulated by allosteric control [143,146,156-158]. The allosteric sites are of two different types, both of which are on the B1 subunit [146,156]. One type controls activity and only binds ATP and dATP, which activate and inhibit the enzyme, respectively. The other type regulates substrate specificity and binds ATP, dATP, dTTP, dGTP, and dCTP. ATP stimulates the reduction of CDP and UDP, dTTP stimulates the reduction of GDP, and dGTP stimulates the reduction of ADP. This elaborate regulation maintains the proper balance of the four deoxynucleoside triphosphates, and their concentrations are controlled by the ratio of ATP to dATP.

The α(N)-heterocyclic carboxaldehyde thiosemicarbazones constitute a class of compounds that includes several antineoplastic agents [159]. Their antitumor activity is usually attributed to their inhibition of ribonucleotide reductase. Kinetic data indicate that it is unlikely that the inhibitor, 1-formylisoquinoline thiosemicarbazone interacts with the substrate (CDP) binding site or the allosteric site, but rather suggests that it binds near the site occupied by the dithiol reducing agent (dithiothreitol was used instead of thioredoxin) [159,160]. This inhibitor is a good chelator, but since the addition of free iron did not prevent inhibition, an iron complex was proposed as the active form of the inhibitor [159].

A well-known antitumor compound and inhibitor of ribonucleotide reductase is hydroxyurea. This compound is a free-radical scavenger and destroys the radical on the B2 subunit [149,155]. This may be an oversimplified view of its mechanism of action, however, since inhibition is reported to be at least partially reversible [161]. There is also conflicting data as to whether hydroxyurea inhibits the reduction of all four substrates equally [161,162]. Parker et al. [163,164] have suggested that the action of hydroxyurea may be related to its conformation, which in turn is influenced by the polarity of its environment. A ribonucleotide reductase which is apparently less sensitive to inhibition by hydroxyurea and guanazole [165] has recently been reported, and further studies with this enzyme may provide valuable information on the mechanism of hydroxyurea inhibition.

Ribonucleotide reductase is also inhibited by antineoplastic nucleotide analogs such as ara-ATP, which mimics dATP as an allosteric inhibitor [166], and 3'-amino-3'-deoxyadenosine 5'-triphosphate which is both a substrate and an allosteric effector [167,168]. In this regard, it is of interest to note that the 2'-deoxynucleoside triphosphates of tubercidin, toyocamycin, and formycin have been prepared, and that they substitute for dATP in the regulation of the cobalamine-containing ribonucleotide reductase from *Lactobacillus leichmannii* [169].

3.1.3. Glutamine-Requiring Amidotransferases

Azaserine and 6-diazo-5-oxo-2-norleucine (DON) are antineoplastic agents. They are analogs of glutamine and inhibit enzymes that catalyze reactions in which glutamine is an amino group donor. Several of these enzymes require metal ions for activity, including amidophosphoribosyltransferase (EC 2.4.2.14) [170,171], phosphoribosylformylglycinamidine synthetase (EC 6.3.5.3) [170,172], CTP synthetase (EC 6.3.4.2) [170,173-175], and GMP synthetase (EC 6.3.5.2) [170,176,177]. These enzymes catalyze reactions (1)-(4), respectively:

$$\text{Glutamine} + \text{PRPP} \rightarrow \text{glutamate} + \text{PRA} + \text{PP} \qquad (1)$$

$$\text{Glutamine} + \text{FGAR} + \text{ATP} \rightarrow \text{glutamate} + \text{FGAM} + \text{ADP} + \text{P} \qquad (2)$$

$$\text{Glutamine} + \text{UTP} + \text{ATP} \rightarrow \text{glutamate} + \text{CTP} + \text{ADP} + \text{P} \qquad (3)$$

$$\text{Glutamine} + \text{XMP} + \text{ATP} \rightarrow \text{glutamate} + \text{GMP} + \text{AMP} + \text{PP} \qquad (4)$$

The first two are involved in the biosynthesis of purine nucleotides, and the last two provide the cytosine and guanine precursors of DNA and RNA. All four reactions require a divalent metal cation that is usually Mg^{2+}, but which in some cases, can be replaced by Co^{2+}, Ni^{2+}, Mn^{2+}, Zn^{2+}, or Fe^{2+} [170]. In addition, amidophosphoribosyltransferases from several sources, including human, are metalloproteins that contain nonheme iron [178-181] that is required for activity. Ammonia can substitute for glutamine in all of the above reactions. Glutamine is generally considered to be the physiologically important substrate [175], however, ammonia may play a physiologically significant role in the formation of phosphoribosylamine in humans [182,183].

The catalytic mechanisms of all four enzymes have many similarities. The initial step is a nucleophilic attack by a cysteinyl sulfhydryl on the γ-carbonyl carbon of glutamine, releasing ammonia and producing a glutamyl-enzyme complex [170,171,175,184-189]. The ammonia released in the initial step then aminates the substrate. In the case of the synthetases, ATP is hydrolyzed in the amination step, not in the step involving glutamine. Mg^{2+} is also involved in the second step. The antitumor agents DON and azaserine, as analogs of glutamine, alkylate the catalytically important cysteinyl residue, and thus irreversibly inactivate these enzymes [171,185, 188-191].

Several of the enzymes mentioned above are allosterically regulated, and some antineoplastic nucleotide analogs either inhibit the catalyzed reaction directly or act indirectly as a false feedback inhibitors. For example, psicofuranine (9-β-D-psicofuranosyladenine) inhibits GMP synthetase [192], the 5'-triphosphate of 3-deazauridine inhibits CTP synthetase [193], and amidophosphoribosyltransferase is inhibited by the nucleoside monophosphates derived from 6-mercaptopurine [194] and cordycepin [195].

3.1.4. Phosphoribosylpyrophosphate Synthetase

5-Phosphoribosyl-α-1-pyrophosphate is required for the de novo biosynthesis of purine, pyrimidine, and pyridine nucleotides, as well as for the salvage pathway for purines [196]. It is synthesized from ribose-5-phosphate and ATP by the enzyme phosphoribosylpyrophosphate synthetase. This enzyme requires a divalent metal ion, best fulfilled by Mg^{2+} or Mn^{2+}. The metal ion performs two functions: (1) the formation of a metal-ATP complex, which is the true substrate; and (2) the activation of the enzyme. The first function can also be performed by Co^{2+} [197], and the second by Ca^{2+}, provided there is sufficient Mg^{2+} for the formation of MgATP [198, 199]. In addition, the enzyme has an absolute requirement for orthophosphate [198-200].

The mechanism of the enzyme catalyzed reaction has been investigated using Co(III) tetraamino-β,γ-phosphate-ATP [197]. The reaction is viewed as a direct nucleophilic attack on the β phosphorus of ATP by the anomeric oxygen of ribose 5-phosphate resulting in the displacement of AMP and inversion of configuration about the β phosphorus. The metal ion of the ATP complex is presumed to facilitate nucleophilic attack by acting as a Lewis acid and withdrawing electrons by coordination with the oxygen of the β phosphate. The enzyme-bound metal seems to be involved in substrate binding or enzyme conformation [197]. A lysyl residue [201] and a cysteinyl residue [202] are essential for activity. It was suggested that the lysine amino group might serve as a base to increase the nucleophilicity of the anomeric hydroxyl group by abstracting the proton, or it may be involved in the ionic substrate binding [201].

Phosphoribosylpyrophosphate synthetase is inhibited by a number of nucleotides. Though the enzymes from different sources vary in the type and degree of inhibition by the compounds, generally purine nucleoside di- and triphosphates, particularly ADP, are the best inhibitors [198-200,203-205]. Most of the inhibition is attributed to competition with the substrates; however, some data suggests that an allosteric site may be present [198]. In addition, the enzyme can exist in multiple states of aggregation which may be involved in regulation of its activity [206-209]. With the enzyme from human erythrocytes, higher aggregation states are more active and are favored by increased concentrations of ATP, Mg^{2+} and orthophosphate. 2,3-Diphosphoglycerate inhibits by promoting dissociation of the active aggregates, but the purine nucleotide inhibitors act directly on the higher aggregated forms [207-208].

Phosphoribosylpyrophosphate synthetase is inhibited by the 5'-triphosphates of several analogs of adenosine that have antineoplastic activity [210-212]. Some of these are modified in the glycosyl moiety (9-β-D-xylofuranosyladenine and cordycepin) and others contain analogs of adenine (formycin and tubercidin). It is not known whether the nucleoside triphosphates inhibit the enzyme by direct competition with ATP at the catalytic site, by binding to the allosteric site, or by influencing the degree of aggregation.

3.2. Drug Activation (Kinases)

Many of the drugs used in cancer chemotherapy are analogs of the common purines and pyrimidines and their nucleosides. Examples are mercaptopurine, 6-thioguanine, 6-methylthiopurine, 5-fluorouracil, 6-methylthiopurine riboside, arabinosylcytosine, 5-fluorouridine, 3'-amino-3'-deoxyadenosine, xylosyladenine, arabinosyladenine, formycin, cordycepin, tubercidin, 5-azacytidine, 6-azauridine, and 3-deazauridine. As noted above, it is usually the nucleotides of these compounds that inhibit enzymes and are cytotoxic; however, since cell membranes are generally impermeable to charged nucleotides, bases and nucleosides are commonly administered. In such cases the drugs are "activated" by enzyme-catalyzed phosphorylation and/or phosphoribosylation. Base analogs are enzymatically converted to nucleoside monophosphates by phosphoribosyl transfer from phosphoribosylpyrophosphate. Three general types of kinases are involved in nucleoside activation. The nucleoside drugs mimic the physiological substrates of nucleoside kinases and are consequently phosphorylated to their nucleoside monophosphates. The monophosphates are converted step-wise to diphosphates and then to triphosphates by nucleotide kinases and nucleoside diphosphate kinases, respectively. In addition, some ribonucleoside diphosphates of base analogs are converted to their deoxyribonucleoside diphosphates by ribonucleotide reductase.

Certain generalizations can be made concerning the substrate specificity of the kinases. Nucleoside kinases generally show broad specificity for the nucleoside triphosphate donor, but are more restricted for the nucleoside acceptor [213-221] whereas nucleotide kinases usually show a high degree of specificity for both the phosphate donor (usually ATP or dATP) and the nucleoside monophosphate acceptor [213,222-225]. An exception is adenylate kinase, which is rather nonspecific for the phosphate donor [226-228]. The nucleoside diphosphate kinases, however, are relatively nonspecific for both substrates and use almost any combination of nucleoside di- and triphosphates containing either purine or pyrimidine bases [229-231].

All of the kinases require a divalent metal ion for activity. The preferred ion is usually Mg^{2+}, but Mn^{2+} is a satisfactory substitute and sometimes activates equally well. Differences between the various kinases have been demonstrated with respect to their ability to utilize Co^{2+}, Ni^{2+}, Zn^{2+}, Fe^{2+}, Ca^{2+}, Cd^{2+}, and Cu^{2+} as metal ion activators [213,216,217,220-222,224-226,229,231].

All kinases catalyze the transfer of the γ-phosphoryl group from a nucleoside triphosphate to a nucleoside or nucleotide acceptor. The mechanism of most of the nucleoside and nucleotide kinases appears to be sequential [213,226], in which the phosphoryl group is transferred directly to the acceptor. The nucleoside diphosphate kinases generally show "ping-pong" kinetics [229], implying that the phosphoryl group is transferred indirectly in two steps with the formation of a phosphorylated enzyme intermediate. In some cases the phosphorylated enzyme intermediate has been isolated [229-231]. The actual donor substrates for all of the kinases appear to be metal ion complexes of nucleoside triphosphates. The role of the divalent cation is apparently to assist the nucleophilic displacement on the γ-phosphorus by complexing with α- and β-phosphate oxygen atoms of the donor, thus making the nucleoside diphosphate product a better leaving group.

3.3. Drug Inactivation (Xanthine Oxidase)

One of the problems encountered in chemotherapy is that the drug is sometimes inactivated by enzyme-catalyzed processes before it can exert its chemotherapeutic effect. Because of the wide variety of drugs, the general phenomenon of drug inactivation involves a large number of enzyme systems. A classical example of drug inactivation by a metalloenzyme is the inactivation of 6-mercaptopurine by xanthine oxidase. {For a thorough discussion of the properties and mechanism of action of xanthine oxidase, see the article by Bray [232] and references therein.} The normal function of xanthine oxidase is presumably in purine catabolism where it converts hypoxanthine to xanthine, and then to uric acid. It is not very specific, however, and catalyzes the oxidation of a large number of purines and pyrimidines, including the antitumor drug 6-mercaptopurine, which is converted to the inactive product 6-thiouric acid prior to excretion. Xanthine oxidase also can use various oxidizing agents including oxygen, NAD^+, and such nonbiological electron acceptors as dyes and ferricyanide. Enzymes from different sources frequently show a preference for either oxygen (oxidase) or NAD^+ (dehydrogenase) as the electron acceptor. The enzyme(s) from most mammalian sources exists in both oxidase and dehydrogenase forms, with the latter predominating in vivo.

Xanthine oxidase is a dimer and contains molybdenum, flavin, nonheme iron, and sulfur in the ratio of 1:1:4:4 per subunit. The iron and sulfur in xanthine oxidase are present in Fe-S centers. The overall reaction occurs in two steps, the first being reduction of the enzyme by the reducing substrate (e.g., xanthine) and the second being the reoxidation of the reduced enzyme by oxygen or NAD^+. In the catalytic cycle, the Mo, flavin, and Fe-S centers all participate in the oxidation-reduction. In its normal operation, xanthine oxidase accepts six electrons in going from the fully oxidized to fully reduced state, and it can exist in several intermediate oxidation states as well. The principal states of Mo are Mo(IV), Mo(V), and Mo(VI). The reducing substrates react at a site containing Mo and probably a persulfide. The electron acceptors oxygen and NAD^+ react at a site containing the flavin and probably a thiol. The added oxygen in the oxidized product (uric acid) comes from water, not oxygen, which is just an electron acceptor. The oxygen is reduced primarily to H_2O_2 by a two-electron transfer, but superoxide anion is also produced by a one-electron transfer. The proportion of superoxide is increased by high oxygen or low xanthine concentrations.

Allopurinol is a potent inhibitor of xanthine oxidase. When mercaptopurine is used in the treatment of cancer, sometimes allopurinol is also administered to prevent the oxidation of the antitumor compound and the hyperuricemia that would result. Allopurinol inhibits xanthine oxidase by acting as a pseudosubstrate and reducing the enzyme while being oxidized to alloxanthine. The alloxanthine, however, does not dissociate, but forms a tight complex (apparent dissociation constant of 6×10^{-10} M) with the partially reduced enzyme [233,234].

ABBREVIATIONS

ADH	alcohol dehydrogenase
ara-A	arabinosyladenine
ara-C	arabinosylcytosine
B. cereus	*Bacillus cereus*
B. subtilis	*Bacillus subtilis*
dien	diethylenetriamine
DON	6-diazo-5-oxo-2-norleucine
E. coli	*Escherichia coli*
E. gracilis	*Euglena gracilis*
en	ethylenediamine
FGAM	phosphoribosylformylglycinamidine
FGAR	phosphoribosylformylglycinamide
GaPD	glyceraldehyde-3-phosphate dehydrogenase

GPD	glucose-6-phosphate dehydrogenase
LAP	leucine aminopeptidase
L. casei	*Lactobacillus casei*
LDH	lactate dehydrogenase
MDH	malate dehydrogenase
PDD	dichlorodiamminoplatinum(II)
PRA	5-phosphoribosylamine
PRPP	5-phosphoribosyl-1-pyrophosphate
S. aureus	*Staphylococcus aureus*
XMP	xanthine-5'-phosphate

REFERENCES

1. B. Rosenberg, *Nature*, *222*, 385 (1969).
2. P. Melius and M. E. Friedman, *Inorg. Perspect. Biol. Med.*, *1*, 1 (1977).
3. R. E. Dickerson, D. E. Eisenberg, J. Varnum, and M. L. Kopka, *J. Mol. Biol.*, *45*, 44 (1969).
4. C. R. Morris and G. R. Gale, in Platinum Coordination Complexes in Cancer Chemotherapy (T. A. Connors and J. J. Roberts, eds.), Springer-Verlag, New York, 1974, pp. 76-77.
5. A. B. Robins, in Platinum Coordination Complexes in Cancer Chemotherapy (T. A. Connors and J. J. Roberts, eds.), Springer-Verlag, New York, 1974, p. 77.
6. S. J. Bannister, L. A. Sternson, A. J. Repta, and G. W. James, *Clin. Chem.*, *23*, 2258 (1977).
7. C. L. Litterst, T. E. Gram, and R. L. Dedrick, *Cancer Res.*, *36*, 2340 (1975).
8. V. Balice and T. Theophanides, *J. Inorg. Nucl. Chem.*, *32*, 1237 (1970).
9. A. Furst, Chemistry of Chelation in Cancer, Thomas, Springfield, Ill., 1963, p. 61.
10. K. Kaziwara, *Gann.*, *42*, 272 (1951).
11. X. Chahovic, *Glas. Srpske. Acad. Kanka (Beograd)*, *215*, 143 (1955).
12. J. L. Marx, *Science*, *192*, 774 (1976).
13. I. H. Krakoff, *J. Clin. Hematol. Oncol.*, 7, 604 (1977).

14. B. Rosenberg, and L. Van Camp, *Cancer Res.*, *30*, 1799 (1970).

15. B. Rosenberg, *Plat. Met. Rev.*, *15*, 7 (1971).

16. R. J. Kociba, S. D. Sleight, and B. Rosenberg, *Cancer Chemother. Rep.*, *54*, 323 (1970).

17. D. Kessel, M. K. Samson, P. Shah, J. Allen, and L. H. Baker, *Cancer*, *38*, 2132 (1976).

18. H. B. Bosmann and T. C. Hall, *Proc. Nat. Acad. Sci. U.S.*, *71*, 1833 (1974).

19. T. F. Slater, A. Mustag, and S. A. Ibrahim, *J. Clin. Hematol. Oncol.*, *7*, 538 (1977).

20. S. W. Stadniki, R. W. Fleischman, U. Schaeppi, and P. Merriam, *Cancer Chemother. Rep.*, *59*, 467 (1975).

21. R. W. Fleischman, S. W. Stadnicki, and M. F. Ethier, *Toxicol. Appl. Pharmacol.*, *33*, 320 (1975).

22. A. H. Rossof, R. E. Stayton, and C. P. Perlia, *Cancer*, *30*, 1451 (1972).

23. D. J. Higby, H. J. Wallace, and J. F. Holland, *Cancer Chemother. Rep.*, *57*, 459 (1973).

24. D. M. Hayes, E. Cvitekovic, R. B. Golbey, E. Scheiner, L. Helson, and I. H. Krakoff, *Cancer*, *39*, 1372 (1977).

25. A. Binet and P. Volfin, *Biochim. Biophys. Acta*, *461*, 182 (1977).

26. E. Renshaw and A. J. Thomson, *J. Bacteriol.*, *94*, 1915 (1967).

27. H. C. Harder and B. Rosenberg, *Int. J. Cancer*, *6*, 207 (1970).

28. J. A. Howle, G. R. Gale, and A. B. Smith, *Biochem. Pharmacol.*, *21*, 1465 (1972).

29. H. H. Kohl, M. E. Friedman, P. Melius, E. C. Mora, and C. A. McAuliffe, *Chem. Biol. Interact.*, *24*, 209 (1979).

30. R. E. Dickerson, D. E. Eisenberg, J. Varnum, and M. L. Kopka, *J. Mol. Biol.*, *45*, 77 (1969).

31. A. J. Thomson, R. J. P. Williams, and S. Reslova, in Structure and Bonding (J. D. Dunetz, ed.), Springer-Verlag, New York, 1972, pp. 1-46.

32. T. Takano, O. B. Ballai, R. Swenson, and R. W. Dickerson, *J. Biol. Chem.*, *248*, 5234 (1973).

33. T. E. Bull, B. Lindman, R. Einarsson, and M. Zeppezauer, *Biochim. Biophys. Acta*, *377*, 1 (1975).

34. H. Eklund, B. Nordstrom, E. Zeppezauer, G. Soderland, I. Ohlsson, T. Boiwe, and C. E. Branden, *FEBS Lett.*, *44*, 200 (1974).

35. R. Einarsson, H. Eklund, E. Zeppezauer, T. Boiwe, and C. I. Branden, *Eur. J. Biochem.*, *49*, 41 (1974).

36. A. Jonas and G. Weber, *Biochem.*, *10*, 1335 (1971).

37. A. Jonas and G. Weber, *Biochem.*, *10*, 4492 (1971).

38. M. E. Friedman, P. Melius, and C. A. McAuliffe, *Bioinorg. Chem.*, *8*, 255 (1978).

39. P. Melius, R. W. Guthrie, and J. E. Teggins, *J. Med. Chem.*, *14*, 75 (1971).

40. M. E. Friedman, B. Musgrove, K. Lee, and J. E. Teggins, *Biochim. Biophys. Acta*, *250*, 286 (1971).

41. P. Melius, J. E. Teggins, M. E. Friedman, and R. W. Guthrie, *Biochim. Biophys. Acta*, *268*, 194 (1972).

42. T. S. Woods and J. E. Teggins, *Inorg. Chem.*, *7*, 1424 (1968).

43. M. E. Friedman and J. E. Teggins, *Biochim. Biophys. Acta*, *341*, 277 (1974).

44. M. E. Friedman and J. E. Teggins, *Biochim. Biophys. Acta*, *350*, 263 (1974).

45. M. E. Friedman and J. E. Teggins, *Biochim. Biophys. Acta*, *350*, 273 (1974).

46. M. E. Friedman, H. B. Otwell, and J. E. Teggins, *Biochim. Biophys. Acta*, *391*, 1 (1975).

47. G. F. Wright, *Ann. N.Y. Acad. Sci.*, *65*, 436 (1957).

48. F. A. Cotton and J. E. Leto, *J. Amer. Chem. Soc.*, *80*, 4823 (1969).

49. E. Silverstein and G. Sulebele, *Biochem.*, *8*, 2543 (1969).

50. J. P. Macquet and T. Theophanides, *Biopolymers*, *14*, 781 (1975).

51. I. A. G. Roos, A. J. Thomson, and J. Eagles, *Chem. Biol. Interact.*, *8*, 421 (1971).

52. J. J. Roberts and J. M. Pascoe, *Nature*, *235*, 383 (1972).

53. P. Horacek and J. Drobnik, *Biochim. Biophys. Acta*, *254*, 341 (1971).

54. A. B. Robins, *Chem. Biol. Interact.*, 7, 11 (1973).

55. S. Mansy, B. Rosenberg, and A. G. Thompson, *J. Amer. Chem. Soc.*, *96*, 6484 (1974).

56. T. Theophanides and P. C. Kong, *Bioinorg. Chem.*, *5*, 51 (1975).

57. M. E. Friedman, P. Melius, J. E. Teggins, and C. A. McAuliffe, *Chem. Biol. Interact.*, *8*, 341 (1978).

58. C. B. Thomas, K. W. Kohn, and W. M. Bonner, *Biochem.*, *17*, 3954 (1978).

59. P. Melius and C. A. McAuliffe, *J. Med. Chem.*, *18*, 1150 (1975).

60. P. Melius, L. Tebbetts, and Y. Y. Lee, Abstract no. 87, 30th Southeastern Regional Meeting, ACS, 1978.

61. J. E. Teggins, D. R. Gana, M. A. Tucker, and D. S. Martin, *Inorg. Chem.*, *6*, 69 (1967).

62. W. D. Hardaker, R. A. Stone, and R. McColy, *Cancer*, *34*, 1030 (1974).

63. P. Melius, C. A. McAuliffe, I. Photaki, and M. Sakarellon-Diatsioton, *Bioinorg. Chem.*, *7*, 203 (1977).

64. J. L. Aull, A. Rice, and L. Tebbetts, *Biochem.*, *16*, 672 (1977).

65. J. L. Aull, H. H. Daron, A. Bapat, J. Wilson, and R. Allen, *Biochim. Biophys. Acta*, *571*, 352 (1979).

66. S. H. Lee, D. L. Chao, J. L. Bear, and A. P. Kimball, *Cancer Chemother. Rep.*, *59*, 661 (1975).

67. J. L. Bear, H. B. Gray, L. Ramin, I. M. Chang, R. Howard, G. Seno, and A. D. Kimball, *Cancer Chemother. Rep.*, *59*, 611 (1975).

68. R. H. Adamson, G. P. Canellos, and S. M. Swebe, *Cancer Chemother. Rep.*, *59*, 599 (1975).

69. H. C. Harder, R. G. Smith, and A. F. Leroy, *Cancer Res.*, *36*, 3821 (1976).

70. T. Kornberg and A. Kornberg, in The Enzymes, Vol. 10 (P. D. Boyer, ed.), 3rd ed., Academic Press, New York, 1974, pp. 119-144.

71. L. A. Loeb, in The Enzymes, Vol. 10 (P. D. Boyer, ed.), 3rd ed., Academic Press, New York, 1974, pp. 173-209.

72. H. M. Temin and S. Mizutani, in The Enzymes, Vol. 10 (P. D. Boyer, ed.), 3rd ed., Academic Press, New York, 1974, pp. 211-235.

73. P. Chambon, in The Enzymes, Vol. 10 (P. D. Boyer, ed.), 3rd ed., Academic Press, New York, pp. 261-331.

74. M. J. Chamberlin, in The Enzymes, Vol. 10 (P. D. Boyer, ed.), 3rd ed., Academic Press, New York, 1974, pp. 333-374.

75. R. Losick and M. J. Chamberlin, eds., RNA Polymerases, Cold Spring Harbor Laboratories, Cold Spring Harbor, New York, 1976.

76. M. A. Sirover and L. A. Loeb, *J. Biol. Chem.*, *252*, 3605 (1977).

77. M. A. Sirover, D. K. Dube, and L. A. Loeb, *J. Biol. Chem.*, *254*, 107 (1979).

78. M. A. Sirover and L. A. Loeb, *Biochem. Biophys. Res. Commun.*, *70*, 812 (1976).

79. W. R. McClure, C. L. Cech, and D. E. Johnston, *J. Biol. Chem.*, *253*, 8941 (1978).

80. T. M. Wadzilak and R. W. Benson, *Biochem.*, *17*, 426 (1978).

81. K. H. Falchuk, B. Mazus, L. Ulpino, and B. L. Vallee, *Biochem.*, *15*, 4468 (1976).

82. D. S. Auld, I. Atusya, C. Campino, and P. Valenzuela, *Biochem. Biophys. Res. Commun.*, *69*, 548 (1976).

83. H. Lattke and U. Weser, *FEBS Lett.*, *65*, 288 (1976).

84. M. J. Scrutton, C. W. Wu, and D. A. Goldthwait, *Proc. Nat. Acad. Sci. U.S.*, *68*, 2497 (1971).

85. J. P. Slater, A. S. Mildvan, and L. A. Loeb, *Biochem. Biophys. Res. Commun.*, *44*, 37 (1971).

86. C. F. Springate, A. S. Mildvan, R. Abramson, J. L. Engle, and L. A. Loeb, *J. Biol. Chem.*, *248*, 5987 (1973).

87. B. J. Poiesz, N. Battula, and L. A. Loeb, *Biochem. Biophys. Res. Commun.*, *56*, 959 (1974).

88. D. S. Auld, H. Kawaguchi, D. M. Livingston, and B. L. Vallee, *Proc. Nat. Acad. Sci. U.S.*, *71*, 2091 (1974).

89. J. E. Coleman, *Biochem. Biophys. Res. Commun.*, *60*, 641 (1974).

90. C. Casoli, M. Manini, and A. Pesce, *FEMS Microbiol. Lett.*, *4*, 167 (1978).

91. L. M. S. Chang and F. J. Bollum, *Proc. Nat. Acad. Sci., U.S.*, *65*, 1041 (1970).

92. K. M. Rose, M. S. Allen, I. L. Crawford, and S. T. Jacob, *Eur. J. Biochem.*, *88*, 29 (1978).

93. C. W. Wu and D. A. Goldthwait, *Biochem.*, *8*, 4450, 4458 (1969).

94. K. M. Downey and A. G. So, *Biochem.*, *9*, 2520 (1970).

95. K. M. Downey, B. S. Jurmark, and A. G. So, *Biochem.*, *10*, 4970 (1971).

96. J. P. Slater, I. Tarmir, L. A. Loeb, and A. S. Mildvan, *J. Biol. Chem.*, *247*, 6784 (1972).

97. H. Lattke and U. Weser, *FEBS Lett.*, *83*, 297 (1977).

98. Y. Nagamine, D. Mizuno, and S. Natori, *Biochim. Biophys. Acta*, *519*, 440 (1978).

99. R. Koren and A. S. Mildvan, *Biochem.*, *16*, 241 (1977).

100. B. L. Bean, R. Koren, and A. S. Mildvan, *Biochem.*, *16*, 3322 (1977).

101. D. L. Sloan, L. A. Loeb, A. S. Mildvan, and R. J. Feldmann, *J. Biol. Chem.*, *250*, 8913 (1975).

102. P. J. Stein and A. S. Mildvan, *Biochem.*, *17*, 2675 (1978).

103. T. R. Krugh, *Biochem.*, *10*, 2594 (1971).

104. L. A. Weymouth and L. A. Loeb, *Proc. Nat. Acad. Sci. U.S.*, *75*, 1924 (1978).

105. W. Wehrli, in Topics in Current Chemistry, Vol. 72 (F. Boschke, ed.), Springer-Verlag, New York, 1977, pp. 21-49.

106. W. Wehrli and M. Staehelin, *Bacteriol. Rev.*, *35*, 290 (1971).

107. S. Riva and L. G. Silvestri, *Ann. Rev. Microbiol.*, *26*, 199 (1972).

108. S. M. Halling, K. C. Burtis, and R. H. Doi, *J. Biol. Chem.*, *252*, 9024 (1977).

109. D. Rabussay and W. Zillig, *FEBS Lett.*, *5*, 104 (1969).

110. A. Heil and W. Zillig, *FEBS Lett.*, *11*, 165 (1970).

111. L. S. Rice and C. F. Meares, *Biochem. Biophys. Res. Commun.*, *80*, 26 (1978).

112. W. Stender, A. A. Stutz, and K. H. Scheit, *Eur. J. Biochem.*, *56*, 129 (1975).

113. W. Wehrli, *Eur. J. Biochem.*, *80*, 325 (1977).

114. C. W. Wu and F. Y.-H. Wu, and D. C. Speckhard, *Biochem.*, *16*, 5449 (1977).

115. S. M. Halling, F. J. Sanchez-Anzaldo, R. Fukuda, R. H. Doi, and C. F. Meares, *Biochem.*, *16*, 2880 (1977).

116. W. R. McClure and C. L. Cech, *J. Biol. Chem.*, *253*, 8949 (1978).

117. A. Sippel and G. Hartmann, *Biochim. Biophys. Acta*, *157*, 218 (1968).

118. S. M. Kupchan, Y. Komoda, Y. A. Court, G. J. Thomas, R. M. Smith, A. Karim, C. J. Gilmore, R. C. Haltiwanger, and R. F. Bryan, *J. Amer. Chem. Soc.*, *94*, 1354 (1972).

119. S. Remillard and L. I. Rebhun, *Science*, *189*, 1002 (1975).

120. R. C. Gallo, S. S. Yang, and R. C. Ting, *Nature*, *228*, 927 (1970).

121. C. Gurgo, R. K. Ray, L. Thiry, and M. Green, *Nature New Biol.*, *229*, 111 (1971).

122. M. F. Goodman, G. M. Lee, and N. R. Bachur, *J. Biol. Chem.*, *252*, 2670 (1977).

123. M. J. Waring, *J. Mol. Biol.*, *54*, 247 (1970).

124. P. Roy-Burman, Analogues of Nucleic Acid Components, Vol. 25, Recent Results in Cancer Research, Springer-Verlag, New York, 1970.

125. S. S. Cohen, *Cancer*, *40*, 509 (1977).

126. A. Matsukage, K. Ono, A. Ohashi, I. Takahashi, C. Nakayama, and M. Saneyoshi, *Cancer Res.*, *38*, 3076 (1978).

127. R. A. Dicioccio and B. I. S. Srivastava, *Eur. J. Biochem.*, *79*, 411 (1978).

128. H. T. Shigeura and C. N. Gordon, *J. Biol. Chem.*, *240*, 806 (1965).

129. A. Sentenac, A. Ruet, and P. Fromageot, *Eur. J. Biochem.*, *5*, 385 (1965).

130. D. L. Panicali and C. N. Nair, *J. Virol.*, *25*, 124 (1978).

131. T. P. Zimmerman, G. Wolberg, and G. S. Duncan, *J. Biol. Chem.*, *253*, 8792 (1978).

132. M. M. Abbound, W. J. Sim, L. A. Loeb, and A. S. Mildvan, *J. Biol. Chem.*, *253*, 3415 (1978).

133. L. Thelander, *Biochem. Soc. Trans.*, *5*, 606 (1977).

134. L. Thelander, B.-M. Sjöberg, and S. Eriksson, *Meth. Enzymol.*, *51*, 227 (1978).

135. S. Hopper, *Meth. Enzymol.*, *51*, 237 (1978).

136. P. Reichard, *Fed. Proc.*, *37*, 9 (1978).

137. S. Hopper, *J. Biol. Chem.*, *247*, 3336 (1972).

138. E. C. Moore, *Meth. Enzymol.*, *12*, 155 (1967).

139. B. Larsson, *Biochim. Biophys. Acta*, *324*, 447 (1973).

140. W. H. Lewis and J. A. Wright, *Biochem. Biophys. Res. Commun.*, *71*, 128 (1974).

141. J. G. Cory and M. M. Mansell, *Cancer Res.*, *35*, 2327 (1975).

142. N. C. Brown, Z. N. Canellakis, B. Lundin, P. Reichard, and L. Thelander, *Eur. J. Biochem.*, *9*, 561 (1969).

143. E. C. Moore, *Adv. Enzyme Regul.*, *15*, 101 (1977).

144. J. G. Cory, A. E. Fleischer, and J. B. Monro, III, *J. Biol. Chem.*, *253*, 2898 (1978).

145. U. von Döbeln and P. Reichard, *J. Biol. Chem.*, *251*, 3616 (1976).

146. N. C. Brown and P. Reichard, *J. Mol. Biol.*, *46*, 39 (1969).

147. N. C. Brown, R. Eliasson, P. Reichard, and L. Thelander, *Eur. J. Biochem.*, *9*, 512 (1969).

148. O. Berglund, *J. Biol. Chem.*, *250*, 7450 (1975).

149. C. L. Atkin, L. Thelander, P. Reichard, and G. Lang, *J. Biol. Chem.*, *248*, 7464 (1973).

150. L. Thelander, *J. Biol. Chem.*, *248*, 4591 (1973).

151. A. Holmgren, *Proc. Nat. Acad. Sci. U.S.*, *73*, 2275 (1976).

152. L. Thelander, B. Larsson, J. Hobbs, and F. Eckstein, *J. Biol. Chem.*, *251*, 1398 (1976).

153. L. Thelander, *J. Biol. Chem.*, *249*, 4858 (1974).

154. B.-M. Sjöberg, P. Reichard, A. Gräslund, and A. Ehrenberg, *J. Biol. Chem.*, *253*, 6863 (1978).

155. A. Ehrenberg and P. Reichard, *J. Biol. Chem.*, *247*, 3485 (1972).

156. U. von Döbeln, *Biochem.*, *16*, 4368 (1977).

157. J. G. Cory, M. M. Mansell, and T. W. Whitford, Jr., *Adv. Enzyme Regul.*, *14*, 45 (1976).

158. D. Kummer, F. Kraml, W. Heitland, and E. Jacob, *J. Krebsforsch. Klin. Onkol.*, *91*, 23 (1978).

159. A. C. Sartorelli, K. C. Agrawal, A. S. Tsiftsoglou, and E. C. Moore, *Adv. Enzyme Regul.*, *15*, 117 (1977).

160. E. C. Moore, M. S. Zedeck, K. C. Agrawal, and A. C. Sartorelli, *Biochem.*, *9*, 4492 (1970).

161. O. Berglund and B.-M. Sjöberg, *J. Biol. Chem.*, *254*, 253 (1979).

162. Y.-C. Yeh and I. Tessman, *J. Biol. Chem.*, *253*, 1323 (1978).

163. G. R. Parker and E. C. Moore, *J. Pharm. Sci.*, *66*, 1040 (1977).

164. G. R. Parker, T. L. Lemke, and E. C. Moore, *J. Med. Chem.*, *20*, 1221 (1977).

165. W. H. Lewis and J. A. Wright, *J. Cell. Physiol.*, *97*, 87 (1978).

166. E. C. Moore and S. S. Cohen, *J. Biol. Chem.*, *242*, 2116 (1967).

167. R. J. Suhadolnik, S. I. Finkel, and B. M. Chassy, *J. Biol. Chem.*, *243*, 3532 (1968).

168. B. M. Chassy and R. J. Suhadolnik, *J. Biol. Chem.*, *243*, 3538 (1968).

169. S. A. Brinkley, A. Lewis, W. J. Critz, L. L. Witt, L. B. Townsend, and R. L. Blakley, *Biochem.*, *17*, 2350 (1978).

170. J. M. Buchanan, *Adv. Enzymol.*, *39*, 91 (1973).

171. J. M. Lewis and S. C. Hartman, *Meth. Enzymol.*, *51*, 171 (1978).

172. J. M. Buchanan, S. Ohnoki, and B. S. Hong, *Meth. Enzymol.*, *51*, 193 (1978).

173. C. Long and D. E. Koshland, Jr., *Meth. Enzymol.*, *51*, 79 (1978).

174. H. Weinfeld, C. R. Savage, Jr., and R. P. McPartland, *Meth. Enzymol.*, *51*, 84 (1978).

175. D. E. Koshland, Jr. and A. Levitzki, in The Enzymes, Vol. 10 (P. D. Boyer, ed.), 3rd ed., Academic Press, New York, 1974, pp. 539-559.

176. N. Sakamoto, *Meth. Enzymol.*, *51*, 213 (1978).

177. T. Spector, *Meth. Enzymol.*, *51*, 219 (1978).

178. M. Itakura and E. W. Holmes, *J. Biol. Chem.*, *254*, 333 (1979).

179. S. C. Hartman, *J. Biol. Chem.*, *238*, 3024 (1963).

180. P. B. Rowe and J. B. Wyngaarden, *J. Biol. Chem.*, *243*, 6373 (1968).

181. J. Y. Wong, E. Meyer, and R. L. Switzer, *J. Biol. Chem.*, *252*, 7424 (1977).

182. O. Sperling, J. B. Wyngaarden, and C. F. Starmer, *J. Clin. Invest.*, *52*, 2468 (1973).

183. G. L. King, C. G. Bounous, and E. W. Holmes, *J. Biol. Chem.*, *253*, 3933 (1978).

184. A. Levitzki and D. E. Koshland, Jr., *Biochem.*, *10*, 3365 (1971).

185. H. Zalkin and C. D. Truitt, *J. Biol. Chem.*, *252*, 5431 (1977).

186. N. Patel, H. S. Moyed, and J. F. Kane, *Arch. Biochem. Biophys.*, *178*, 652 (1977).

187. K. Mizobuchi and J. M. Buchanan, *J. Biol. Chem.*, *243*, 4853 (1968).

188. D. D. Schroeder, A. J. Allison, and J. M. Buchanan, *J. Biol. Chem.*, *244*, 5856 (1969).

189. S. Ohnoki, B.-S. Hong, and J. M. Buchanan, *Biochem.*, *16*, 1070 (1977).

190. A. Levitzki, W. B. Stallcup, and D. E. Koshland, Jr., *Biochem.*, *10*, 3371 (1971).

191. N. Patel, H. S. Moyed, and J. F. Kane, *J. Biol. Chem.*, *250*, 2609 (1975).

192. T. T. Fukuyama, *J. Biol. Chem.*, *241*, 4745 (1966).

193. R. P. McPartland, M. C. Wang, A. Bloch, and H. Weinfeld, *Cancer Res.*, *34*, 3107 (1974).

194. R. J. McCollister, W. R. Gilbert, Jr., D. M. Ashton, and J. B. Wyngaarden, *J. Biol. Chem.*, *239*, 1560 (1964).

195. F. Rottman and A. M. Guarino, *Biochim. Biophys. Acta*, *89*, 465 (1964).

196. I. H. Fox and W. N. Kelley, *Ann. Intern. Med.*, *74*, 424 (1971).

197. T. M. Li, A. S. Mildvan, and R. L. Switzer, *J. Biol. Chem.*, *253*, 3918 (1978).

198. R. L. Switzer, in The Enzymes, Vol. 10 (P. D. Boyer, ed.), 3rd ed., Academic Press, New York, 1974, pp. 607-629.

199. R. L. Switzer and K. J. Gibson, *Meth. Enzymol.*, *51*, 3 (1978).

200. D. G. Roth, C. White, and T. F. Deuel, *Meth. Enzymol.*, *51*, 12 (1978).

201. M. F. Roberts and R. L. Switzer, *Arch. Biochem. Biophys.*, *185*, 391 (1978).

202. M. F. Roberts, R. L. Switzer, and K. R. Schubert, *J. Biol. Chem.*, *250*, 5364 (1975).

203. G. Planet and I. H. Fox, *J. Biol. Chem.*, *251*, 5839 (1976).

204. W. C. Sadler and R. L. Switzer, *J. Biol. Chem.*, *252*, 8504 (1977).

205. R. C. Garcia, P. Leoni, and A. C. Allison, *Biochem. Biophys. Res. Commun.*, *77*, 1067 (1977).

206. K. R. Schubert, R. L. Switzer, and E. Shelton, *J. Biol. Chem.*, *250*, 7492 (1975).

207. L. J. Meyer and M. A. Becker, *J. Biol. Chem.*, *252*, 3919 (1977).

208. M. A. Becker, L. J. Meyer, W. H. Huisman, C. Lazar, and W. B. Adams, *J. Biol. Chem.*, *252*, 3911 (1977).

209. L. C. Yip, S. Roome, and M. E. Balis, *Biochem.*, *17*, 3286 (1978).

210. K. Overgaard-Hansen, *Biochim. Biophys. Acta*, *80*, 504 (1964).

211. J. F. Henderson, A. R. P. Paterson, I. C. Caldwell, and M. Hori, *Cancer Res.*, *27*, 715 (1967).

212. D. B. Ellis and G. A. LePage, *Mol. Pharmacol.*, *1*, 231 (1965).

213. E. P. Anderson, in The Enzymes, Vol. 9 (P. D. Boyer, ed.), 3rd ed., Academic Press, New York, 1973, pp. 49-96.

214. A. Orengo and S.-H. Kobayashi, *Meth. Enzymol.*, *51*, 299 (1978).

215. P. Valentin-Hansen, *Meth. Enzymol.*, *51*, 308 (1978).

216. E. P. Andersen, *Meth. Enzymol.*, *51*, 314 (1978).

217. D. H. Ives and S.-M. Wang, *Meth. Enzymol.*, *51*, 337 (1978).

218. M. R. Deibel, Jr. and D. H. Ives, *Meth. Enzymol.*, *51*, 346 (1978).

219. M. S. Chen and W. H. Prusoff, *Meth. Enzymol.*, *51*, 354 (1978).

220. E. Bresnick, *Meth. Enzymol.*, *51*, 360 (1978).

221. Y.-C. Cheng, *Meth. Enzymol.*, *51*, 365 (1978).

222. A. Orengo and P. Maness, *Meth. Enzymol.*, *51*, 321 (1978).

223. E. P. Andersen, *Meth. Enzymol.*, *51*, 331 (1978).

224. M. P. Oeschger, *Meth. Enzymol.*, *51*, 473 (1978).

225. K. C. Agrawal, R. P. Miech, and R. E. Parks, Jr., *Meth. Enzymol.*, *51*, 483 (1978).

226. L. Noda, in The Enzymes, Vol. 8 (P. D. Boyer, ed.), 3rd ed., Academic Press, New York, 1973, pp. 279-305.

227. W. E. Criss and T. K. Pradhan, *Meth. Enzymol.*, *51*, 459 (1978).

228. K. K. Tsuboi, *Meth. Enzymol.*, *51*, 467 (1978).

229. R. E. Parks, Jr. and R. P. Agarwal, in The Enzymes, Vol. 8 (P. D. Boyer, ed.), 3rd ed., Academic Press, New York, 1973, pp. 307-333.

230. J. L. Ingraham and C. L. Ginther, *Meth. Enzymol.*, *51*, 371 (1978).

231. R. P. Agarwal, B. Robison, and R. E. Parks, Jr., *Meth. Enzymol.*, *51*, 376 (1978).

232. R. C. Bray, in The Enzymes, Vol. 12 (P. D. Boyer, ed.), 3rd ed., Academic Press, New York, 1975, pp. 299-419.

233. V. Massey, H. Komai, G. Palmer, and G. B. Elion, *J. Biol. Chem.*, *245*, 2837 (1970).

234. T. Spector and D. G. Johns, *J. Biol. Chem.*, *245*, 5079 (1970).

AUTHOR INDEX

Numbers in parentheses are reference numbers and indicate that an author's work is referred to although his name may not be cited in the text. Underlined numbers give the page on which the complete reference is listed.

A

Abbound, M. M., 357(132), 371
Abramson, R., 353(86), 370
Adams, W. B., 361(208), 375
Adamson, R. H., 49(129), 50(129), 51(129,130), 62; 352(68), 369
Agarwal, R. P., 306(1), 331; 363(229,231), 375, 376
Aggarwal, S. K., 114(126,127), 125; 165(73), 193
Agrawal, K. C., 212(31,36,38), 213(40), 227; 358(159,160), 363(225), 373, 375
Agre, K. A., 307(3), 331
Ahrland, S., 28(85), 60; 67(9), 119
Akami, Y., 310(35),332
Akiyama, I., 251(142), 282
Albert, A., 7(37), 58
Albertin, G., 81(44), 120
Alcock, R. M., 87(61), 121
Alexopoulos, C., 132(14), 183 (112), 184(112), 190, 195
Allan, R., 350(65), 369
Allen, A. D., 234(60), 278
Allen, J., 341(17), 367
Allen, M. S., 353(92), 370
Allen, R. J., 236(81,82), 242 (81), 245(133), 259(133), 279, 281
Allison, A. C., 361(205), 375
Allison, A. J., 360(188), 374
Alnaes, E., 264(164), 283
Alvarez, V. E., 245(133), 259 (133), 281
Ames, B. N., 172(96), 174(96), 195
Anderson, E. P., 363(213,216, 223), 375
Anderson, T. A., 171(92), 173(92), 194
Anghileri, L. J., 233(29), 264 (166), 277, 283
Angiuli, R., 326(81), 327(81), 335
Ankel, E., 213(42), 227
Annibale, G., 81(46), 121
Anson, F. C., 236(88), 240(88), 246(88), 279
Antholine, W. E., 213(41), 214 (43,44), 215(44,45), 216(45, 47), 217(45), 220(59,60), 221 (65), 222(60,65), 223(60,68), 224(59,60), 227, 228, 229
Appelton, T. G., 67(7), 73(7), 74(7), 119
Apple, M. A., 295(30), 303
Arcamone, F., 326(79,81), 327 (81,82), 334, 335
Armor, J. N., 234(61), 237(96), 238(97), 245(131), 278, 279, 280, 281
Armstrong, D. R., 73(27), 74(27), 119
Armstrong, R. E., 240(105), 280
Arnold, M., 3(8), 57
Arnold, S. T., 141(24), 142(24), 143(24), 153(24), 191
Asakura, H., 308(10), 331
Asama, K., 319(62), 334
Ashton, D. M., 360(194), 374

Atkin, C. L., 357(149), 358 (149), 372
Atkins, H., 269(170), 270(170), 271(170), 283
Atkins, L. M., 150(43), 151(43), 155(55), 157(55-57), 192, 193
Atusya, I., 353(82), 354(82), 370
Auld, D. S., 353(82,88), 354 (82), 370
Aull, J. L., 349(64), 350(65), 369
Azegami, M., 310(28), 332

B

Bachur, N. R., 327(84), 335; 356(122), 371
Bailar, J. C., Jr., 72(20), 119
Baker, L. H., 341(17), 367
Baker, M. S., 172(96), 174(96), 195
Balch, A. L., 82(54), 121
Balice, V., 340(8), 366
Balis, M. E., 361(209), 375
Ball, T. E., 344(33), 367
Ballai, O. B., 344(32), 367
Ballou, J. E., 232(20), 276
Balo, J., 4(19), 57
Bancroft, G. M., 73(26), 119
Banerjea, D., 23(61), 59
Banfi, E., 43(107), 61; 172(95), 194; 233(35), 268(35), 273 (35), 277
Banga, I., 4(19), 57
Banner, R. J., 41(103), 61
Bannister, S. J., 146(34), 191; 339(6), 366
Bapat, A., 350(65), 369
Barclay, P. J., 236(88), 240(88), 246(88), 279
Bardos, T. J., 3(14), 57
Barker, G., 132(14), 183(112), 184(112), 190, 195
Barr, D. H., 47(121), 62
Bartocci, C., 82(50), 121
Barton, D., 188(124), 196
Barton, J. K., 68(13), 93(67a), 94(67a), 97(82c), 98(67a), 110(117-119), 111(121), 112
[Barton, J. K.], (118,119,122), 113(122), 115(128), 116(132), 117(132), 119, 122, 123, 125
Basolo, F., 23(60-62), 29(60), 59; 67(10), 69(10), 70(10), 74 (10), 75(10), 77(39a), 79(10), 81(10,45a), 83(10), 91(10), 119, 120; 237(94), 238(94), 279
Battula, N., 353(87), 370
Bau, R., 96(77), 112(123), 122, 125
Bauer, W. R., 97(82a,b,83a), 102 (82b), 109(115), 114(125), 115 (128), 116(83a), 117(82b), 123, 124, 125; 166(78), 194
Baxendale, J. H., 236(85), 238 (85), 279
Bean, B. L., 354(100), 370
Bear, J. L., 36(91,92), 37(91-95), 38(92), 39(92,93), 40(96), 60; 198(1), 226; 298(44,45), 299 (47), 304; 351(66,67), 369
Beattie, J. K., 81(42), 120
Beaumont, K. P., 14(52), 58
Beck, D. J., 26(70), 59; 171(91), 172(94), 173(91), 174(91), 194
Becker, M. A., 361(207,208), 375
Beer, M., 257(145), 282
Behnke, W. D., 40(97), 42(97), 61; 232(5), 241(5), 269(5), 276
Behr, W., 329(99), 330(99), 336
Beirne, C., 5(23), 57
Beirne, G., 5(23), 57
Bell, W. B., 3(11), 57
Belluco, U., 65(4a), 77(4a,39a), 118, 120
Belt, R. J., 146(34), 191
Benedict, W. F., 172(96), 174 (96), 195
Benjamin, R. S., 179(106), 195
Bennett, L. E., 240(103,104), 242(104), 259(103,104), 280
Bennett, L. L., Jr., 287(9), 302
Benson, J. E., 68(12), 71(12), 119
Benson, R. W., 353(80), 369
Berg, G. G., 232(13), 276
Berglund, O., 357(148), 358(161), 372, 373
Bergman, J. G., 7(34), 58; 199(5), 226

Bergstrand, A., 260(148), 282
Berlin, E., 141(25), 191
Berlin, Y. A., 328(96,97), 329 (96,97), 335
Bermke, Y., 206(24), 207(24), 227
Bertino, J. R., 179(107), 186 (107), 195; 212(38), 227
Bertoli, G., 35(90), 43(90), 60; 232(6), 241(6), 268(6), 272 (6), 276
Best, S. A., 110(118), 112(118), 125
Bethune, V., 178(104), 195; 212(39), 227
Bhargava, M. M., 97(84b), 123
Bhat, T. N., 102(98), 124
Bhutani, R., 185(117), 196
Bielli, E., 82(55b), 90(55b), 121
Biemann, K., 324(67), 325(67), 334
Bihler, I., 150(42), 192
Binet, A., 342(25), 367
Bitler, S., 260(146), 282
Bittman, L., 327(90), 335
Blakley, R. L., 359(169), 373
Blankenhorn, G., 260(147), 282
Blanz, E. J., Jr., 212(32-35, 37), 227
Blattmann, P., 96(75), 116(75), 122
Blessing, J., 186(118), 196
Bloch, A., 360(193), 374
Bloomfield, V. A., 91(66), 122
Blow, D. M., 104(101), 124
Blum, R. H., 307(3), 331
Blumberg, W. E., 205(20), 216 (49), 222(49), 226, 228
Blundell, T. L., 103(99), 104 (99), 107(99), 116(99), 124
Boiwe, T., 344(34), 367
Bollum, F. J., 353(91), 370
Bolton, J. R., 216(46), 228
Bond, P. J., 97(83a), 116(83a), 123
Bond, W. H., 182(111), 195
Bondareff, W., 264(161), 283
Bonner, W. M., 347(58), 368
Booth, B. A., 45(113), 61; 205 (18), 209(28), 213(40), 226, 227
Bordignon, E., 81(44), 120
Borisova, V. N., 329(100), 330 (100), 336
Bosmann, H. B., 341(18), 367
Bottomley, F., 234(46), 277
Bounous, C. G., 360(183), 374
Bracken, R. B., 187(122), 196
Braddock, P. D., 14(55), 15(55), 20(55), 59; 80(41a), 120; 165(70), 167(70), 168(70), 193
Bradner, W. T., 16(59), 59; 319 (61), 326(80), 334, 335
Brändén, C. E., 344(34,35), 367
Branfman, A., 158(62), 193
Bray, R. C., 364(232), 376
Brazhnikova, M. G., 328(92), 329 (100), 330(100), 335, 336
Brennan, T. F., 234(68), 238(68), 256(68), 278
Bresnick, E., 363(220), 375
Breter, H. J., 309(17), 331
Breuel, W., 245(132), 281
Brinkley, S. A., 359(169), 373
Bromfield, R. J., 11(43), 30(43), 58
Broomhead, J. A., 237(94), 238 (94), 241(110), 242(110), 279, 280
Brown, D. B., 109(108), 124
Brown, G. M., 234(74), 235(74), 237(74), 245(74), 251(74), 257(74), 274(174), 279, 283
Brown, N. C., 357(146,147), 358 (146), 372
Brown, R. S., 96(76), 116(76), 122
Brubaker, R., 26(70), 59; 172 (94), 194
Bruce, R. S., 232(8-10), 261(8-10), 262(8-12), 276
Bruckner, H. W., 132(15), 184 (15), 190
Brundrett, R. B., 293(24), 303
Brunelli, M., 79(40), 120
Bruylants, A., 253(144), 258 (144), 282
Bryan, R. B., 237(93), 239(93), 245(93), 279
Bryan, R. F., 356(118), 371
Brynes, S., 47(117), 61
Buchanan, J. M., 359(170,172), 360(170,187-189), 373, 374

Buchbinder, M., 242(116), 256 (116), 260(146), 280, 282
Buddeckar, I., 245(132), 281
Bugge, G., 108(106), 124
Buldakov, L. A., 232(11), 262 (11), 276
Burbank, R. D., 109(108), 124
Burchenal, J. H., 160(67,68), 161(67), 162(67), 163(67), 180 (109), 193, 195; 212(39), 227
Burckhalter, J. H., 232(1), 273 (1), 276
Burdett, J. K., 67(8), 77(8), 80(8), 85(8), 119
Burger, R. M., 329(100), 330 (100), 336
Burtis, K. C., 355(108), 371
Buschbeck, C., 245(132), 281
Bush, J. A., 326(80), 335
Buskirk, H. H., 200(8,10,11,14), 201(8,10,11), 202(10), 203 (10), 226
Bustad, L. K., 232(15), 276
Butler, K. D., 73(26), 119
Butour, J.-L., 26(71), 59; 97 (80), 115(80), 123; 173(97,98), 195
Bygrave, F. L., 264(162), 283

C

Calabresi, P., 287(12), 291(12), 293(12), 302
Calderella, L. A., 324(68), 334
Caldwell, I. C., 362(211), 375
Calendi, E., 327(87), 335
Calvert, A. H., 185(115), 196
Campino, C., 353(82), 354(82), 370
Canellakis, Z. N., 357(142), 372
Canellos, G. P., 49(129), 50 (129), 51(129), 62; 352(68), 369
Carassiti, V., 82(50), 121
Carmichael, J. W., 102(97), 124
Carmichael, N., 4(17), 57
Carofoli, E., 262(152), 282
Carr, B., 130(7), 177(7), 190
Carr, T. E. F., 232(8-10), 261 (8-10), 262(8-10), 276
Carter, M. G., 81(42), 120
Carter, S. K., 175(100), 195; 307(3), 331
Casoli, C., 353(90), 354(90), 370
Cass, A. E. G., 318(57), 333
Cassinelli, G., 326(79), 334
Cater, D. B., 243(126), 259(126), 281
Catsch, A., 232(12,23), 261(12), 262(12,23), 276, 277
Cattalini, L., 65(4a,b), 74(33), 75(33), 77(4a,b,39a), 79(40), 81(46), 118, 120, 121
Cavalli, F., 148(40), 192
Cech, C. L., 353(79), 355(116), 369, 371
Chahovic, X., 340(11), 366
Chamberlin, M. J., 353(74,75), 369
Chambon, P., 353(73), 354(73), 369
Chan, N., 102(97), 124
Chan-Stier, C. H., 209(29), 211 (29), 227
Chance, B., 260(148), 282
Chang, I. M., 36(91), 37(91,93), 39(93), 60; 198(1), 226; 298 (44,45), 304; 351(67), 369
Chang, L. M. S., 353(91), 370
Chao, D. L., 37(95), 60; 351(66), 369
Chao, Y. Y. H., 103(100), 116 (100), 124
Charlson, A. J., 41(103), 42(104), 48(104), 61
Chassy, B. M., 359(167,168), 373
Chatt, J., 28(85), 60; 67(9), 68(11), 119
Chen, D. M., 221(66), 229
Chen, M. S., 363(219), 375
Cheng, Y.-C., 363(221), 375
Chien, M., 310(36), 311(36), 320 (36), 332
Chisholm, J., 160(67), 161(67), 162(67), 163(67), 193
Chisholm, M. H., 73(24), 119
Chiu, Y. H., 324(69), 325(69), 334
Choi, E., 172(96), 174(96), 195
Chou, M., 240(108), 241(108), 280
Chow, R. P., 115(131), 125

Chow, V., 216(46), 228
Chretien, P. B., 185(114), 196
Christoph, G. G., 299(49), 304
Chu, G. Y. H., 94(69), 95(69,71), 98(69,71), 115(69,71,130), 117 (69,71,130), 122, 125
Clark, D. F., 253(144), 258(144), 282
Clark, H. C., 67(7), 73(7,24), 74(7), 119
Clarke, M. J., 43(106), 61; 234 (63-68), 235(63,67), 237(66), 239(46-68), 242(64,65,115-117), 247(63-67,138,139), 248(66,67, 138,139), 249(66), 250(63,66, 141), 251(63-67), 252(63), 253 (63,65,66,143), 254(63,66,67), 255(63,67,141), 256(64,65,68, 116), 258(65), 260(63,146), 265(117,167), 268(117,167), 269(170), 270(167,170), 271 (170), 278, 280, 281, 282, 283
Cleare, M. J., 2(2), 8(2), 13 (2), 14(2,49-52,56), 15(49), 16(49,56,59), 17(56), 19(56), 20(56), 21(56), 22(56), 23 (49), 24(2,56), 33(51), 34 (51,88), 35(88), 40(51), 41 (99), 42(2,105), 44(51), 48 (128), 56, 58, 59, 60, 61, 62; 146(36), 192; 232(7), 259(7), 276; 296(33), 303
Coffey, K., 250(141), 253(143), 281, 282
Cohen, C. C., 132(15), 184(15), 190
Cohen, D., 109(109), 124
Cohen, G. L., 115(128), 125
Cohen, S. S., 356(125), 359 (166), 371, 373
Coleman, G. M., 236(87), 279
Coleman, J. E., 353(89), 370
Coley, R. F., 84(57), 85(57), 121
Collier, W. A., 4(16), 57
Collins, M. E., 232(10), 261 (10), 262(10), 276
Colvin, M., 293(24), 303
Comor, D., 233(25), 277
Cone, R., 325(75), 334
Connors, T. A., 14(55), 15(55), 20(55), 31(86), 42(105), 48 127,128), 59, 60, 61, 62; 165 (70), 167(70), 168(70), 193; 291(17), 302,
Coon, A. B., 326(80), 335
Coonan, D. A., 141(23), 191
Corbett, T. H., 157(60), 193
Cordes, A. W., 102(97), 124
Cory, J. G., 357(144), 358(157), 372, 373
Cotton, F. A., 65(2), 67(2), 68 (2), 71(2), 118; 346(48), 368
Court, Y. A., 356(118), 371
Craciunescu, D., 48(126), 62
Craft, M. L., 185(117), 196
Craft, R. W., 240(107), 280
Crawford, I. L., 353(92), 370
Creasey, W. A., 25(68), 59; 212 (38), 227
Creutz, C., 234(50,69,71), 239 (50), 240(108), 241(108), 259 (50), 278, 280
Crim, J. A., 200(9-11,14), 201 (9-11), 202(10), 203(10), 204 (17), 205(17), 226
Criss, W. E., 363(227), 375
Critz, W. J., 359(169), 373
Croatto, U., 65(4a), 77(4a), 118
Cronin, R. A., 81(45b), 120
Crooke, S. T., 16(59), 59; 221 (63), 222(64), 228; 311(40), 315(50,52), 316(54), 318(54), 319(40), 320(40,52), 321(52), 322(52,63), 323(63), 333, 334
Cross, C., 47(124), 62
Crothers, D. M., 91(66), 122
Crouzel, C., 233(25), 277
Cullen, W. R., 328(93), 335
Cummins, J. E., 233(30), 277
Cundliffe, E., 325(73), 334
Cvitkovic, E., 131(8), 132(13), 178(8,104), 190, 195; 342(24), 367

D

Dabrowiak, J. C., 221(63,64), 222(63), 228; 309(24,25),

[Dabrowiak, J. C.], 311(25), 312 (24,25,47), 313(24,25), 314 (24,47,49), 315(47,50,52), 316 (24,47,53,54), 317(24,47,53), 318(24,47,53,54), 320(52), 321 (47,52), 322(24,52,53,63), 323 (63), 332, 333, 334
Dainty, R. H., 11(43), 30(43), 58
D'Andrea, A. D., 309(12), 311 (12), 331
Danneman, J., 40(97), 42(97), 61; 232(5), 241(5), 269(5), 276
Daron, H. H., 350(65), 369
Das, M., 41(100), 61
Dasdia, T., 326(77), 334
Datta-Gupta, N., 3(14), 57; 311 (37), 320(37), 332
David, R. D., 141(23), 191
Davidson, J. P., 109(112), 124; 140(19), 165(19), 191
Davies, N. R., 28(85), 60; 67 (9), 119
Davis, R. D., 141(24), 142(24), 143(24), 153(24), 191
DeBolster, M. W. G., 286(1), 302
DeConti, R. C., 25(68), 59; 212 (38), 227
Dedrick, R. L., 25(65), 59; 340 (7), 366
Degetto, G., 79(40), 120
Dehand, J., 169(87), 194
Deibel, M. R., Jr., 363(218), 375
Del Monte, U., 243(119), 259 (119), 280
DeNardo, L., 97(85d), 115(85d), 123
Deposito, F., 6(31), 58
Deppe, G., 132(15), 184(15), 190
Deuel, T. F., 361(200), 374
Devlin, H. B., 5(24), 57
Dew, K., 160(68), 193
DeWys, W., 47(120), 62
Diamond, S. E., 234(48), 244 (48,128), 278, 281
Dicioccio, R. A., 356(127), 371
Dickerson, R. E., 104(103), 124; 339(3), 343(30), 366, 367
Dickerson, R. W., 344(32), 367
Dickey, A., 243(125), 259(125), 281
Dickinson, R. J., 73(27), 74(27), 119
DiMarco, A., 326(77,78), 327(82, 85,87), 334, 335
Doadrio, A., 48(126), 62
DoAmaral, J. R., 212(35), 227
Doi, R. H., 355(108,115), 371
Donohue, J., 131(12), 180(12), 187(123), 190, 196
Dougherty, D., 251(142), 282
Douple, E. B., 159(64), 193
Dowling, M., 234(68), 239(68), 256(68), 269(170), 270(170), 271(170), 278, 283
Downey, K. M., 354(94,95), 370
Drago, R. S., 74(29), 120; 315 (51), 333
Dragulescu, C., 41(101), 61
Drake, G. A., 232(16), 276
Drobnick, J., 11(87), 26(78), 34 (87), 60; 142(27), 147(37,38), 159(63), 191, 192, 193; 232(4), 241(4), 269(4), 276; 347(53), 368
Drucker-Colin, R., 264(165), 283
Dube, D. K., 353(77), 355(77), 369
Dubois, R. J., 47(117), 61
Dumas, P. E., 234(51), 278
Duncan, G. S., 357(131), 372
Duncan, R. E., 94(69), 95(69,71), 98(69,71), 115(69,71,130), 117 (69,71,130), 122, 125
Dunn, R., 3(5), 56
Dunn, W. J., III, 45(116), 46 (116), 61
Duran, L., 264(165), 283
Durant, J. R., 157(59), 193
Durig, J. R., 40(97), 42(97), 61; 232(5), 241(5), 269(5), 276
Dwyer, F. P., 5(20), 6(27), 10 (27,41), 57, 58; 232(2), 273 (2,172), 276, 283

E

Eagles, J., 116(133), 125; 347 (51), 368
Earley, J. E., 263(154), 282
Ebsworth, E. A. H., 65(3), 77(3), 118

Eckstein, F., 357(152), 372
Edmonds, S. D., 43(106), 61; 242(117), 247(139), 248(193), 265(117,167), 268(117,167), 270(167), 280, 281, 283
Edward, J. M., 65(3), 77(3), 118
Edwards, C. F., 3(6), 57
Ehrenberg, A., 358(154,155), 372, 373
Ehrlich, C. E., 184(113), 196
Einarsson, R., 344(33,35), 367
Einhorn, L. H., 131(12), 152 (48), 180(12), 181(48), 182 (111), 184(113), 187(123), 190, 192, 195, 196
Eisenberg, D. E., 104(103), 124; 339(3), 343(30), 366, 367
Eisenberger, P., 116(132), 117 (132), 125
Eklund, H., 344(34,35), 367
Elding, L. I., 68(16), 72(18), 73(18), 74(28), 79(28), 80 (28), 82(52), 84(16,28,58), 85(28), 87(62), 119, 120, 121
Eliades, T., 234(60), 278
Elias, E. G., 185(114), 196
Eliasson, R., 357(147), 372
Elion, G. B., 365(233), 376
Ellis, D. B., 362(212), 375
Elson, C. M., 236(86), 279
Endicott, J. F., 235(76), 238 (76), 240(76), 241(76), 279
Engle, J. L., 353(86), 370
Enomoto, Y., 232(22), 261(22), 277
Epstein, A. J., 68(14), 119
Erck, A., 36(91,92), 37(91,92), 38(92), 39(92), 40(96), 60; 198(1), 226; 298(44), 304
Erickson, L. E., 101(93), 115 (131), 123, 125
Erickson, P. M., 245(135), 281
Eriks, K., 247(139), 248(139), 281
Eriksson, S., 357(134), 372
Ernster, L., 260(148), 282
Etcubanas, E., 212(39), 227
Ethier, M. F., 142(28), 191; 342(21), 367
Evans, I. P., 245(129), 270 (129), 281
Evenor, M., 236(78), 279
Everett, G. W., 245(129), 270 (129), 281
Ewalt, R. X., 240(104), 242(104), 259(104), 280
Ewig, R. A. G., 293(23), 303

F

Faber, P. J., 109(112), 124; 140(19), 165(19), 191
Faggiani, R., 88(64a,c,d), 121 122
Failla, M. L., 47(122), 62
Fair, C. K., 102(97), 124
Falchuk, K. H., 353(81), 369
Fantini, G., 326(79), 334
Farr, H., 185(116), 196
Fealey, T., 263(154), 282
Fedorova, G. B., 329(100), 330 (100), 336
Fee, W. E., 179(107), 186(107), 195
Feldmann, R. J., 354(101), 370
Fenselau, C., 293(24), 303
Ferguson, J. A., 274(174), 283
Ferrer, J. F., 200(15), 226
Ferri, A., 82(50), 121
Finkel, S. I., 359(167), 373
Fiorina, V. J., 47(117), 61
Fisch, J. E., 171(91), 173(91), 174(91), 194
Fischer, H., 234(52), 264(52), 278
Fischer, R. G., Jr., 109(112), 124; 140(19), 165(19), 191
Fishman, M. M., 327(88), 335
Fiskin, A. M., 257(145), 282
Fleischer, A. E., 357(144), 372
Fleischman, R. W., 141(23), 142 (28), 191; 342(20,21), 367
Fletcher, R., 182(111), 195
Flowers, A., 287(8), 302
Ford, P. C., 233(42,43), 234(42, 43), 235(77), 236(81,82,84, 89), 237(84,90), 238(89), 239 (84,100), 242(81), 245(133), 246(43), 277, 279, 280, 281
Foresti, E., 326(81), 327(81), 335

Fortune, R., 73(27), 74(27), 119
Fox, B. W., 291(18), 302
Fox, I. H., 360(196), 361(203), 374
Fradkin, R., 287(8), 302
Francis, D., 7(34), 58; 199(5), 226
Francis, J. N., 274(175), 283
Franco, D., 234(75), 279
Fraval, H. N. A., 152(46), 192
Freelander, B. L., 199(6,7), 226
Frei, E., III, 295(32), 303
French, D. A., 212(35), 227
French, F. A., 45(115), 61; 199 (6,7), 212(32-35,37), 226, 227
Friedman, M. E., 100(91), 101 (91), 102(91), 103(91), 104 (91), 108(91), 123; 153(50,51), 192; 339(2), 340(2), 343(29), 344(40), 345(38,41,43,44), 346 (45,46), 347(57), 349(41), 366, 367, 368
Friedman, O. M., 293(25), 303
Fromageot, P., 356(129), 371
Frye, H. G., 101(94), 123
Fujii, A., 218(50), 220(50), 228; 307(4,6), 308(4), 316(4,55,56), 318(56), 319(4), 323(4,64), 324(55), 331, 333, 334
Fujii, T., 308(7), 331
Fujisawa, K., 319(58,59), 320 (59), 321(59), 333
Fukuda, R., 355(115), 371
Fukuoka, T., 307(6), 331
Fukushima, M., 6(32), 58
Fukuyama, T. T., 360(192), 374
Fuller, W., 327(86), 335
Furchner, J. E., 232(16), 276
Furst, A., 3(3,4,10), 6(10,33), 56, 57, 58; 286(3,4), 296(3), 300(4), 302; 340(9), 366
Fyfe, D. A., 264(163), 283

G

Gabe, E. J., 205(19), 226
Gaetani, M., 326(77,78), 334
Galdes, A , 318(57), 333
Gale, E. F., 325(73), 334
Gale, G., 160(68), 193
Gale, G. R., 14(53), 20(53), 26 (73), 43(108), 44(108), 58, 59, 61; 150(43), 151(43), 152(45, 47), 155(55), 157(55-57), 192, 193; 339(4), 343(28), 366, 367
Gale, R. P., 41(103), 61
Gallo, R. C., 356(120), 371
Gambetta, R. A., 327(85), 335
Gana, D. R., 349(61), 369
Gang, M., 131(10), 154(10), 190; 299(50), 304
Garafolo, J. A., 234(68), 238 (68), 256(68), 278
Garcia, R. C., 361(205), 375
Gaunder, R. G., 233(43), 234(43), 240(107), 246(43), 277, 280
Gause, G. F., 328(94,95), 330 (95), 335
Gautieri, R. F., 6(30), 57
Gazzotti, P., 262(152), 282
Gee, M. V., 327(84), 335
Gerold, F., 185(116), 196
Geschickter, C. F., 4(18), 57
Gesler, J. W., 236(87), 279
Ghanta, V. K., 157(59), 193
Ghirvu, G., 48(126), 62
Gibson, K. J., 361(199), 374
Gilbert, I. G., 232(19), 262(19), 270(19), 276
Gilbert, W. R., Jr., 360(194), 374
Gilboe, D. P., 325(76), 334
Gillard, R. D., 11(43,44), 30 (43,44), 31(44), 34(44), 42 (44), 58; 82(55a,b,c), 89(55a), 90(55b,c), 101(92), 109(107), 121, 123, 124
Gilman, A., 287(5), 302
Gilmore, C. J., 356(118), 371
Ginsberg, E., 232(13), 276
Ginther, C. L., 363(230), 376
Giraldi, T., 34(89), 35(90), 43(90), 60; 232(6), 241(6), 268(6), 272(6), 276
Gleu, K., 245(132), 281
Glickman, J. D., 221(66), 229
Glusker, J. P., 205(19), 226
Goffinet, D. R., 179(107), 186 (107), 195
Golbey, R. B., 131(8), 132(13), 178(8), 188(124), 190, 196;

[Golbey, R. B.], 342(24), 367
Goldberg, I. H., 329(98), 330 (98), 336
Goldberg, N. H., 185(114), 196
Goldenberg, G. J., 150(42), 192
Goldin, A., 295(31), 303
Goldsmith, M., 175(100), 195
Goldthwait, D. A., 353(84), 354 (84,93), 370
Golub, E. S., 166(76), 194
Gonias, S. L., 114(125), 125
Goode, R. L., 179(107), 186 (107), 195
Goodgame, D. M. L., 169(86), 194
Goodman, M. F., 356(122), 371
Goodwin, D. A., 311(44), 333
Gordon, C. N., 356(128), 371
Gordon, S. L., 327(84), 335
Görller-Walrand, C., 77(38), 120
Gosling, R., 70(17), 79(17), 119
Gottlieb, J. A., 295(32), 303
Grabin, M. E., 141(25), 191
Graciela, M., 264(165), 283
Grady, R. W., 198(3), 226
Graham, R. D., 41(98), 61
Grahe, G., 253(144), 258(144), 282
Gram, T. E., 25(65), 59; 340(7), 366
Gräslund, A., 358(154), 372
Graves, B. J., 102(95), 123; 247(140), 249(140), 250(140), 281
Gray, H. B., 23(62), 37(93), 39 (93), 59, 60; 74(30), 75(30), 81(45a), 120; 298(45), 304; 351(67), 369
Greco, A., 182(111), 195
Green, M., 86(60), 87(60), 121; 356(121), 371
Greenaway, F. T., 221(63,64), 222(63), 228; 309(24,25), 311 (25), 312(24,25), 313(24,25), 314(24.49), 315(50,52), 316 (24,53,54), 317(24,53), 318 (24,53,54), 320(52), 321(52), 322(53,63), 323(63), 332, 333, 334
Greenspan, E., 132(15), 184(15), 190
Gregg, C. T., 243(124), 259 (124), 281
Grein, A., 326(79), 334
Gress, M. E., 234(71), 278
Griffith, W. P., 233(37), 277
Grimley, E. B., 9(40), 58; 129(2), 190
Griswold, D. P., Jr., 157(60), 193
Groenweld, W. L., 286(1), 302
Grollman, A. P., 309(13), 310 (29,36), 311(13,36), 320(36), 331, 332
Gröning, O., 74(28), 79(28), 80 (28), 84(28), 85(28), 120
Gruber, J. E., 233(31), 260(31), 277
Grulich, R., 221(64), 228; 315 (52), 316(53), 317(53), 318 (53), 320(52), 321(52), 322 (52,53,63), 323(63), 333, 334
Grunbaum, Z., 269(170), 270 (170), 271(170), 283
Guantieri, V., 97(85d), 115 (85d), 123
Guarino, A. M., 25(65), 59; 141 (24), 142(24), 143(24), 145 (31,33), 148(31,33), 153(24), 191; 360(195), 374
Guengerich, C. P., 245(130), 270 (130), 281
Gullino, P. M., 243(118), 259 (118), 260(118), 280
Gunasekera, S. W., 311(43), 333
Gupta, G., 245(134), 281
Gurgo, C., 356(121), 371
Gusberg, S. B., 132(15), 184(15), 190
Gustafson, G. T., 263(158), 282
Gustafson, L., 72(18), 73(72), 82(52), 119, 121
Guthrie, R. W., 344(39), 345(41), 347(39), 349(41), 368
Gutman, V., 239(101), 280
Gyarfas, E. C., 6(27), 10(27,41), 57, 58; 273(172), 283

H

Haake, P., 81(45b), 120
Hadjiliadis, N., 95(73), 108 (104), 122, 124
Hahn, E., 45(114), 61
Hahne, W. F., 101(93), 123
Haidle, C. W., 309(14-16), 310 (14), 331
Haim, A., 240(105), 280
Hall, A. J., 77(36), 120
Hall, L. M., 14(54), 20(54), 58; 109(116a), 124
Hall, S. W., 179(106), 195
Hall, T. C., 341(18), 367
Hallee, J. G., 311(39), 320(39), 332
Halling, S. M., 355(108,115), 371
Haltiwanger, R. C., 356(118), 371
Halvorson, H. O., 97(84b), 123
Hamilton, L. D., 327(86), 335
Handel, P. B., 187(122), 196
Harbour, J. R., 216(46), 228
Hardaker, W. D., 349(62), 369
Harder, H. C., 25(63,64), 26(72, 76), 59, 60; 141(26), 145(32), 152(44), 167(80), 179(26,108), 191, 192, 194, 195; 343(27), 352(69), 367, 369
Haro, R. T., 3(3), 56
Haroun, L., 172(96), 174(96), 195
Harrigan, T. D., 72(19), 119
Harrington, J. T., 150(41), 179 (41), 192
Harris, A. W., 232(3), 273(3), 276
Harris, J. W., 47(121), 62
Harris, R. O., 234(60), 278
Harrison, R. C., 93(68), 94(68), 122
Harrold, C., 185(116), 196
Hart, M. H., 51(130), 62
Hartley, F. R., 64(1), 65(1), 67(1), 72(1), 74(1), 75(1), 76(1), 77(1), 78(1), 87(61), 91(1), 118, 121
Hartman, G., 329(99), 330(99), 336; 355(117), 371
Hartman, H., 245(132), 281
Hartman, S. C., 359(171), 360 (171,179), 373
Hartwick, J., 11(87), 34(87), 60; 232(4), 241(4), 269(4), 276
Harvey, J. W., 232(14), 276
Hasan, S. K., 325(75), 334
Haseltine, W. A., 309(12), 311 (12), 331
Hatfield, W. E., 205(21), 226
Hatori, M., 319(59), 320(59), 321(59), 333
Hawkins, B. I., 221(66), 229
Hayes, D. M., 131(8), 132(13), 178(8), 190; 342(24), 367
Heaton, B. T., 11(43), 30(43), 58
Hebborn, P., 3(14), 57
Hebrank, G., 298(43), 304
Heffer, J. P., 299(48), 304
Heggen, G. E., 3(6), 57
Heil, A., 355(110), 371
Heitland, W., 358(158), 373
Heller, K., 185(116), 196
Helson, L., 131(8), 178(8), 190; 342(24), 367
Henderson, J. F., 362(24), 375
Herken, R., 234(57), 272(57), 278
Hevesi, L., 253(144), 258(144), 282
Heyman, I. A., 141(23), 191
Higby, D. J., 130(6), 177(6), 179(6), 187(6), 190; 342(23), 367
Hilaris, B. S., 188(124), 196
Hill, B. T., 185(115), 196
Hill, D. L., 287(10), 295(10), 302
Hill, H. O. A., 318(57), 333
Hill, J. M., 14(54), 20(54), 25 (67), 58, 59; 109(116a,b), 124; 130(5), 131(101), 157 (53), 176(5,101), 190, 192, 195
Hill, N. O., 25(67), 59; 109 (116b), 124; 130(5), 131(101), 176(5,101), 190, 195
Himmelstein, K. J., 146(34), 191
Hintze, R. E., 235(77), 279
Hiramoto, R. N., 157(59), 193

Hiroshima, Y., 239(98), 280
Hitchcock, P. B., 73(23), 119
Hnuma, T., 232(21), 261(21), 276
Hobbs, J., 357(152), 372
Hodgson, D. J., 68(15), 93(67b), 94(67b), 98(67b), 102(95), 119, 122, 123; 247(140), 249 (140), 250(140), 281
Hodgson, K., 234(72), 273(72), 279
Hodnett, E. M., 45(115,116), 46 (116), 61
Hoeschele, J. D., 14(49,50), 15 (49,58), 16(49), 23(49), 58, 59; 144(29), 148(29), 191
Hoffman, L. M., 274(175), 283
Hofman, K. A., 108(106), 124
Hogan, M., 311(37), 320(37), 332
Holland, J. F., 132(15), 184(15), 190; 342(23), 367
Holles, O. L., 232(20), 276
Hollinshead, A. C., 166(77), 194
Hollis, L. S., 15(58), 59
Holmes, E. W., 360(178,183), 373, 374
Holmgren, A., 357(151), 372
Holoye, P. Y., 131(11), 179(11), 190
Holtzer, A., 103(100), 116(100), 124
Homesley, H., 186(118), 196
Hong, B. S., 359(172), 360(189), 373, 374
Hong, W. K., 185(117), 196
Honikel, K., 329(99), 330(99), 336
Hopf, F. R., 108(105b), 124; 274(174), 283
Hopfan, S., 188(124), 196
Hopper, S., 357(135,137), 372
Horacek, P., 26(78), 60; 147 (37), 192; 347(53), 368
Hori, M., 308(10), 331; 362 (211), 375
Hori, S., 308(9), 331
Horimai, A., 6(32), 58
Horner, S. M., 205(21), 226
Horwitz, S. B., 219(56-58), 228; 309(21-23), 310(21-23,29,36), 311(21,23,36), 312(21,23), 313 (23), 320(36), 330(101), 332, 336
Hoshino, Y., 319(58), 333
Hosoyamada, Y., 239(98), 280
Howard, J., 185(116), 196
Howard, R., 36(92), 37(92-94), 38(92), 39(92,93), 60; 298(45), 304; 351(67), 369
Howe, K. E., 109(116a), 124
Howe-Grant, M., 97(82b,86), 102 (82b), 117(82b), 123; 166(78), 194
Howell, J. N., 236(159), 283
Howie, J. K., 115(131), 125
Howle, J. A., 26(73), 59; 152 (45,47), 192; 343(28), 367
Howse, R., 26(77), 60
Hudson, C. W., 234(49), 278
Huheey, J. E., 233(41), 277
Huisman, W. H., 361(208), 375
Hunt, H. R., 299(46), 304
Hunter, L. D., 68(12), 71(12), 119
Husen, M. V., 221(63), 222(63), 228
Hush, N. S., 65(6), 66(6), 119
Hutchinson, F., 309(11), 331
Hydes, P. C., 14(56), 16(56,59), 17(56), 19(56), 20(56), 21 (56), 22(56), 24(56), 59; 146(36), 192

I

Ibers, J. A., 234(73), 279
Ibrahim, S. A., 342(19), 367
Ichikawa, R., 232(22), 261(22), 277
Ichikawa, T., 308(9), 310(32), 331, 332
Iitaka, Y., 316(55,56), 318(56), 324(55), 333
Ilievski, V., 141(23), 191
Ilse, D., 150(42), 192
Imanishi, H., 319(60), 333
Ingraham, J. L., 363(230), 376
Inlow, R. O., 296(39), 298(41), 303
Isaacs, N. W., 326(81), 327(81), 335
Ishida, R., 219(55), 228
Ishidate, M., 300(51), 304
Ishii, S., 310(28), 332

Ishikawa, Y., 6(32), 58
Ishizuka, M., 219(52,53), 224 (52,53), 228; 308(9), 310 (33,34), 331, 332
Isied, S. S., 236(83), 237(91, 92), 238(91), 239(83), 245 (136), 279, 281
Israels, L. G., 150(42), 192
Itakura, M., 360(178), 373
Itoh, H., 323(64), 334
Itzkovitch, I. J., 236(86), 279
Ives, D. H., 363(217,218), 375
Iwanaga, J., 219(53), 224(53), 228; 310(34), 332
Iwashima, K., 232(21), 261(21), 276

J

Jack, A., 96(76), 116(76), 122
Jacob, E., 358(158), 373
Jacob, S. T., 353(92), 370
Jacobs, C., 179(107), 186(107), 195
Jacobsen, B., 73(23), 119
Jacobson, R. A., 68(12), 71(12), 119
James, D. W., 82(55b), 90(55b), 121
James, G. W., 339(6), 366
Jardine, I., 293(24), 303
Jeanne, C., 311(42), 333
Jeeves, I., 169(86), 194
Jennette, K. W., 97(82a), 123; 233(31), 260(31), 277
Jesson, J. P., 65(5a,b), 77(5a, b), 118
Jocelyn, P. C., 207(26), 227
Joesten, M. D., 286(2), 296(39, 40), 298(41,43), 302, 303, 304; 315(51), 333
Johns, D. G., 365(234), 376
Johnson, A. G., 200(14), 226
Johnson, A. K., 87(63), 121
Johnson, D. A., 102(97), 124
Johnson, D. E., 187(122), 196
Johnson, E. E., 131(11), 179 (11), 190
Johnson, L. N., 103(99), 104(99), 107(99), 116(99), 124
Johnson, R. C., 72(19), 119
Johnson, S. A., 299(46), 304
Johnston, D. E., 353(79), 369
Jonas, A., 344(36,37), 368
Jones, M., 14(55), 15(55), 20 (55), 59; 165(70), 167(70), 168(70), 193
Jones, M. T., 157(59), 193
Jordanov, J., 169(87), 194
Juckett, D. A., 165(75), 193
Jurmark, B. S., 354(95), 370

K

Kabakow, B., 132(15), 184(15), 190
Kaihara, T., 310(32), 332
Kajiwara, K., 5(25,26), 57
Kalaher, K., 160(67,68), 161(67), 162(67), 163(67), 193
Kan, M. N., 293(24), 303
Kane, J. F., 360(186,191), 374
Kane-Maguire, L. A. P., 82(55a), 89(55a), 121; 237(95), 238(95), 279
Kanisawa, M., 3(15), 57
Kann, H. E., Jr., 293(22), 303
Karayannis, N. M., 298(42), 303
Karim, A., 356(118), 371
Kashi, H., 311(38), 320(38), 332
Katakis, D., 108(104), 124
Kausar, A. R., 171(89), 194
Kawaguchi, H., 319(58-60,62), 320(59), 321(59), 333, 334; 353(88), 370
Kawin, B., 232(23), 262(23), 277
Kaziwara, K., 340(10), 366
Kelley, W. N., 360(196), 374
Kelman, A. D., 26(81,83), 43(106), 60, 61; 97(84a,b), 123; 169 (84), 194; 242(116,117), 256 (116), 260(146), 265(117,167), 268(117,167), 270(167), 280, 282, 283
Kennard, O., 326(81), 327(81), 335
Kennedy, B. P., 70(17), 79(17), 119
Kensler, C. J., 158(61), 193
Kersten, H., 291(19), 295(19), 302

Kessel, D., 341(17), 367
Khan, A., 25(67), 59; 109(116b), 124; 130(5), 131(101), 176(5, 101), 190, 195
Khan, B. T., 273(171), 283
Khokhar, A. R., 14(55), 15(55), 20(55), 59; 165(70), 167(70), 168(70), 193
Kijima, K., 112(123), 125
Kikuchi, T., 309(27), 313(27), 314(27), 332
Kim, J. J., 96(75), 116(75), 122
Kim, S. H., 96(75), 116(75), 122
Kim, S. K., 114(125), 125
Kimball, A. P., 36(91,92), 37 (91-95), 38(92), 39(92,93), 40(96), 60; 198(1), 226; 298(44,45), 304; 351(66,67), 369
Kimura, K., 219(53), 224(53), 228; 310(34), 332
King, G. L., 360(183), 374
King, J. J., 25(67), 59; 109 (116b), 124
Kirchner, H., 47(125), 62
Kirkland, J. J., 116(134), 125
Kirschner, S., 7(34), 41(101), 58, 61; 199(5), 226
Kirvan, G., 250(141), 281
Kiseleva, O. A., 328(97), 329 (97), 335
Kitchens, J., 299(47), 304
Klassen, D. M., 234(49), 278
Kleihues, P., 171(93), 194
Kline, I., 131(10), 154(10,52), 190, 192; 299(50), 304
Klose, W., 234(55,57), 261(55), 272(55,57), 278
Klug, A., 96(76), 116(76), 122
Knight, J. M., 213(41), 214(43, 44), 215(44), 227
Kobaru, S., 319(59), 320(59), 321(59), 333
Kobayashi, S.-H., 363(214), 375
Koch, J. H., 10(41), 58; 273 (172), 283
Kociba, R. J., 141(22), 165(71), 191, 193; 341(16), 367
Kogler, J., 131(101), 176(101), 195
Koh, Y. B., 299(49), 304
Kohl, H. H., 343(29), 367
Kohn, K. W., 167(81), 171(92), 173(92), 194; 293(23), 303; 347(58), 368
Kohnlein, W., 309(11), 331
Kojima, M., 309(18), 315(18), 331
Kolosov, M. N., 328(96,97), 329 (96,97), 335
Komai, H., 365(233), 376
Komai, T., 308(9), 331
Komoda, Y., 356(118), 371
Kong, P. C., 347(56), 368
Konishi, M., 319(59,62), 320(59), 321(59), 333, 334
Kono, A., 309(18), 315(18), 331
Konstantinova, N. V., 328(92), 335
Kopka, M. L., 104(103), 124; 339(3), 343(30), 366, 367
Koren, R., 354(99,100), 370
Kornberg, A., 353(70), 355(70), 369
Kornberg, T., 353(70), 355(70), 369
Korolkovas, A., 232(1), 273(1), 276
Koshiyama, H., 319(59), 320(59), 321(59), 333
Koshland, D. E., Jr., 359(173, 175), 360(175,184,190), 373
Kovsharova, I. N., 328(92), 335
Krakoff, I. H., 131(8), 177(102), 178(8,103), 190, 195; 212(39), 227; 341(13), 342(24), 366, 367
Kraml, F., 358(158), 373
Krauss, F., 4(16), 57
Krekulova, A., 147(38), 192
Kremer, W. B., 324(70), 334
Krentzien, H., 247(138), 248 (138), 281
Krigas, T., 9(39), 58; 129(1), 190; 296(34), 303
Krishnamurti, C., 213(42), 214 (42), 215(45), 216(45), 217 (45), 227, 228
Krishnan, N., 264(163), 283
Krugh, T. R., 354(103), 370
Kruglyak, E. B., 328(92), 329 (100), 330(100), 335, 336
Kruse, S. L., 311(44), 333
Kubelkova, A., 147(38), 192

Kuehn, C. G., 234(45), 235(45), 237(45), 238(45), 239(45), 240(45), 246(45), 260(45), 261(45), 262(45), 277
Kuempel, J. R., 236(87,89), 238 (89), 279
Kukushkin, Y. N., 84(59), 121
Kummer, D., 358(158), 373
Kunishima, M., 308(7,8), 331
Kuo, M. T., 309(15,16), 331
Kupchan, S. M., 356(118), 371
Kuylenstierna, B., 260(148), 282

L

Ladner, J. E., 96(76), 116(76), 122
LaGasse, L., 186(118), 196
Lang, G., 357(149), 358(149), 372
Lange, R. C., 25(63,64,68), 59; 145(32), 191
Langford, C. H., 74(30), 75(30), 120
Lantos, P. L., 171(93), 194
Lanza, S., 81(43a,b), 83(43a), 120
Lapis, S., 157(53), 192
Larson, S., 269(170), 270(170), 271(170), 283
Larsson, B., 357(139,152), 372
Laster, W. R., Jr., 157(60), 193
Laszio, J., 324(70), 334
Lattke, H., 353(83), 354(97), 370
Laurent, M., 113(124), 125
Lavender, J. P., 311(43), 333
Lazar, C., 361(208), 375
LeBreton, P. R., 251(142), 282
Lecointe, P., 26(71), 59; 173 (97,98), 195
Lee, G. M., 356(122), 371
Lee, K., 153(50), 192; 344(40), 368
Lee, K. W., 81(47a), 83(47a), 121
Lee, S. H., 37(95), 60; 351(66), 369
Lee, Y. Y., 347(60), 369
Lehninger, A. L., 260(149), 282
Lemke, T. L., 358(164), 373
Leoni, P., 361(205), 375
Leopold, W. R., 174(99), 195
LePage, G. A., 362(212), 375
Lerner, E. I., 109(110,115), 110 (110), 113(110), 124
Leroy, A. F., 25(65), 59; 145 (33), 148(33), 191; 352(69), 369
Leto, J. E., 346(48), 368
Levinson, W., 327(83), 335
Levitzki, A., 359(175), 360(175, 184,190), 373, 374
Lewis, A., 359(169), 373
Lewis, A. E., 212(37), 227
Lewis, J. M., 359(171), 360(171), 373
Lewis, W. H., 357(140), 359(165), 372, 373
Li, T. M., 361(197), 374
Lieberman, D. Z., 109(116a), 124
Lim, H. S., 236(88), 240(88), 246(88), 279
Lim, M.-C., 83(56), 87(56), 88 (56), 94(70), 95(70), 110(56), 121, 122
Lin, J., 251(142), 282
Lin, M. S., 311(44), 333
Lindman, B., 344(33), 367
Lippard, S. J., 68(13), 93(67a), 94(67a), 96(78), 97(82a-c,83a, 86), 98(67a), 102(82b), 109 (115), 110(117-119), 111(121), 112(118,119,122), 113(122), 114(125), 115(128), 116(78,83a, 132), 117(82b,132), 119, 122, 123, 124, 125; 166(78), 194
Lippert, B., 15(57), 59; 88(64a-d,65), 109(113), 112(113), 121, 122, 124
Lippman, A. J., 177(102), 195
Lipscomb, W. N., 324(69), 325 (69), 334
Litterst, C. L., 25(65), 59; 141 (24), 142(24), 143(24), 145 (31,33), 148(31,33), 153(24), 191; 340(7), 366
Little, W. F., 68(15), 119
Livingston, D. M., 353(88), 370
Livingstone, S. E., 41(100,102), 48(102), 61
Lock, C. J. L., 88(64a-d,65), 121, 122
Loeb, E., 25(67), 59; 109(116b),

[Loeb, E.], 124; 129(5), 131 (101), 176(5,101), 190, 195
Loeb, L. A., 353(71,76-78,85-87), 354(71,85,96,101), 355(76,77, 104), 357(132), 369, 370, 372
Logan, M. E., 159(64), 193
Lokys, L., 160(68), 193
Long, C., 359(173), 373
Longhi, R., 315(51) 333
Longo, W., 221(63), 222(63), 228; 315(50), 316(54), 318(54), 333
Loo, T. L., 179(106), 195
Losick, R., 353(75), 369
Louie, S., 96(77), 122
Louw, W. J., 82(53a), 121
Lown, J. W., 309(20,26), 312(20), 313(26), 324(71,72), 325(75), 332, 334
Lu, K., 179(106), 195
Ludlum, D. B., 287(6), 291(6, 16), 302
Luft, J. H., 263(155-157), 282
Luhrs, W., 3(13), 57
Lundin, B., 357(142), 372
Lunec, J., 220(62), 224(62), 228
Lyons, J. R., 82(55c), 90(55c), 121

M

Mabel, J., 158(62), 193
Macfarlane, R. D., 109(114), 124
Machida, S., 6(32), 58
MacLellan, A., 25(67), 59; 109 (116b), 124; 129(5), 131(101), 176(5,101), 190, 195
Macquet, J. P., 26(71,84), 59, 60; 97(79a-c,80,85c), 98 (79a-c), 115(79a-c,80,85c), 117(79a-c), 122, 123; 169(85), 173(97,98), 194, 195; 347(50), 368
Madias, N. E., 150(41), 179(41), 192
Maeda, K., 310(35), 323(64), 324(66), 332, 334
Maeda, T., 309(18), 315(18), 331
Magee, P. N., 171(93), 194
Mailer, K., 215(45), 216(45), 217(45), 228
Malerbi, B. W., 14(56), 16(56), 17(56), 19(56), 20(56), 21(56), 22(56), 24(56), 59; 146(36), 192
Manaka, R. C., 145(30), 148(30), 149(30), 191
Maness, P., 363(222), 375
Manini, M., 353(90), 354(90), 370
Mann, D. E., Jr., 6(30), 57
Manojlović-Muir, L. J., 73(22), 119
Mansell, M. M., 358(157), 373
Mansour, V. H., 2(1), 11(1), 14(1), 56; 109(111), 124; 130(4), 138(4), 190; 296(35), 303
Mansy, S., 26(79), 60; 94(69), 95(69,71), 98(69,71), 109(112), 115(69,71), 117(69,71), 122, 124; 140(19), 165(19), 167(82), 191, 194; 242(113), 280; 347 (55), 368
Mantel, N., 295(31), 303
Manzer, L. E., 67(7), 73(7,24), 74(7), 119
Maragini, G., 79(40), 120
Marchant, J. A., 236(84), 237(84), 239(84), 279
Maresca, L., 65(4a,b), 77(4a,b), 81(46), 118, 121
Markopoulos, J., 108(104), 124
Martell, A. E., 222(67), 229
Martin, D. S., Jr., 68(12), 71 (12), 81(47a,b), 82(48), 84 (47a,b,57), 85(57), 119, 121; 146(35), 191; 200(13), 226; 349(61), 369
Martin, J., 178(104), 195
Martin, R. B., 83(56), 87(56), 88(56), 94(70), 95(70), 110 (56), 121, 122
Martini, N., 188(124), 196
Maruyama, Y., 45(114), 61
Marx, J. L., 340(12), 366
Marzilli, L. G., 93(67c), 94 (67c), 98(67c), 122
Mason, W. R., 82(51), 115(129), 121, 125
Massey, V., 365(233), 376
Mastin, S. H., 103(100), 116 (100), 124

Matsubara, T., 234(50), 236(84), 237(84), 239(50,84,100), 245 (133), 259(50,133), 278, 279, 280, 281
Matsuda, A., 223(70), 229; 310 (30,32), 317(30), 332
Matsukage, A., 356(126), 371
Matsunage, Y., 264(160), 283
Matsushima, Y., 309(18), 315 (18), 331
Mattern, J. A., 72(20), 119
Matthews, C. R., 245(135), 265 (168), 268(168), 281, 283
Maurer, A., 41(101), 61
May, C. J., 82(49), 121
Mayer, K., 212(39), 227
Mayhew, E., 5(20,21), 57; 232(2), 273(2), 276
Mazus, B., 353(81), 369
McAfee, J. G., 233(24), 262(24), 277
McAllister, P. K., 114(127), 125; 165(73), 193
McArdle, N. T., 41(103), 61
McAuliffe, C. A., 14(52), 34(88), 35(88), 48(128), 58, 60, 62; 93(68), 100(90), 101(90), 102 (90), 122, 123; 343(29), 345 (38), 347(57), 348(59), 349 (63), 367, 368, 369
McClelland, C. E., 318(57), 333
McClure, W. R., 353(79), 355 (116), 369, 371
McCollister, R. J., 360(194), 374
McColy, R., 349(62), 369
McDonald, J. W., 115(131), 125
McGarvey, B. R., 312(45), 333
McGlynn, S. P., 251(142), 282
McKee, R. W., 243(125), 259 (125), 281
McKenney, J., 236(86), 279
McNitt, S., 327(90), 335
McPartland, R. P., 359(174), 360(193), 373, 374
McPherson, A., 96(75), 116(75), 122
Mead, J. A. R., 212(38), 227
Meakin, P., 65(5a,b), 77(5a,b), 118
Meares, C. F., 355(111,115), 371
Meek, D. W., 315(51), 333
Meester, M. A. M., 73(25), 119
Mehmood, A., 273(171), 283
Meischen, S. J., 14(53), 20(53), 58; 155(55), 157(55-57), 192, 193
Melius, P., 100(91), 101(91), 102(91), 103(91), 104(91), 108(91), 123; 339(2), 340(2), 343(29), 344(39), 345(38.41), 347(39,57), 348(59,60), 349 (41,63), 366, 367, 368, 369
Mercer, E. E., 40(97), 42(97), 61; 232(5), 234(51), 241(5), 269(5), 276, 278
Mercer-Smith, J. A., 108(105a), 124
Merker, P. C., 158(62), 193
Merriam, P., 342(20), 367
Merrick, M. V., 311(43), 333
Merrifield, R. K., 26(77), 60
Merrin, C., 131(9), 178(9), 181(110), 182(110), 187(120), 190, 195, 196
Merritt, K., 200(14), 226
Mestroni, G., 34(89), 35(90), 43(90), 60; 232(6), 241(6,112), 268(6), 272(6), 276, 280
Metzler, D. E., 260(150), 282
Meyer, E., 360(181), 374
Meyer, L. J., 361(207,208), 375
Meyer, T. J., 233(38), 240(109), 241(110), 242(110), 274(174), 277, 280, 283
Meyers, T. D., 157(53), 192
Michelon, G., 81(46), 121
Midelens, P., 327(83), 335
Miech, R. P., 363(225), 375
Miehl, R., 265(168), 268(168), 283
Mihich, E., 200(12,15), 201(12), 202(12), 226
Mihkelson, A. E., 41(102), 48 (102), 61
Mijos, K. J., 324(68), 334
Mikulski, L. M., 298(42), 303
Mildvan, A. S., 353(85,86), 354 (85,96,99-102), 357(132), 361 (197), 370, 372, 374
Milgrom, H., 5(22), 57
Millard, M. M., 26(84), 60;

[Millard, M. M.], 97(79b), 98 (79b), 115(79b), 117(79b), 122
Miller, D. S., 141(24), 142(24), 143(24), 153(24), 191
Miller, E. C., 174(99), 195
Miller, J. A., 174(99), 195
Miller, J. D., 87(63), 121
Miller, J. S., 68(14), 119
Miller, T. J., 141(24), 142(24), 143(24), 153(24), 191
Mingos, D. M. P., 68(11), 119
Minkel, D. T., 207(27), 209(27, 29,30), 210(30), 211(27,29, 30), 227
Minkin, J. A., 205(19), 226
Minniti, D., 81(43b), 82(53b), 120, 121
Miralles, A. J., 240(105), 280
Miyake, A., 328(91), 335
Miyaki, T., 319(59), 320(59), 321(59), 333
Miyamoto, K., 310(32), 332
Mizobuchi, K., 360(187), 374
Mizukawa, K., 233(26), 263(26), 277
Mizuno, D., 354(98), 370
Mizuno, K., 328(91), 335
Mizuno, N. S., 325(76), 334
Mizutani, S., 353(72), 369
Mochida, I., 72(20), 119
Monro, J. P., III, 357(144), 372
Montgomery, J. A., 287(9), 293 (27), 302, 303
Monti-Bragadin, C., 43(107), 61; 172(95), 194; 233(35), 241 (112), 268(35), 273(35), 277, 280
Moore, C. H., 45(115), 61
Moore, C. L., 262(153), 264 (153), 282
Moore, E. C., 212(36), 213(40), 227; 357(138,143), 358(143, 159,160,163,164), 359(166), 372, 373
Mora, E. C., 343(29), 367
Morehouse, S. M., 299(48), 304
Morgan, A. R., 325(75), 334
Morgan, J. L., 184(113), 196
Morris, C. R., 150(43), 151(43), 192; 339(4), 366
Morton, J. F., 140(20), 191
Moseley, J. E., 326(80), 335
Motherwell, W. D. S., 326(81), 327(81), 334
Moudrianakis, I., 257(145), 282
Moyed, H. S., 360(186,191), 373
Moyer, B. A., 233(38), 277
Muir, K. W., 73(22), 119
Müller, W. E. G., 223(69), 229; 309(17), 331
Munchausen, L., 26(82), 60; 97(85a,b), 115(85a,b), 123
Munn, R., 47(124), 62
Murakami, S., 3(7), 57
Muraoka, Y., 218(50), 220(50), 228; 307(4), 308(4), 316(4, 55,56), 318(56), 319(4), 323 (4,64), 324(55,66), 331, 333, 334
Mureinik, R. J., 75(35), 82(35), 120
Murray, S. G., 100(90), 101(90), 102(90), 123
Musgrove, B., 153(50), 192; 344(40), 368
Mustafa, M., 47(124), 62
Mustag, A., 342(19), 367
Myles, A., 293(25), 303
Myllymaki, R. W., 326(80), 335

N

Nagai, K., 220(61), 224(61), 228; 309(19), 310(19,31), 324(19), 331, 332
Nagai, T., 232(21), 261(21), 276
Nagamine, Y., 354(98), 370
Naganawa, H., 218(50), 220(50), 228; 307(4), 308(4), 311(38), 316(4), 319(4), 320(38), 331, 332
Nagy, J. B., 253(144), 258(144), 282
Nagy, O. B., 253(144), 258(144), 282
Nair, C. N., 356(130), 372
Naito, T., 319(62), 334
Nakamura, H., 316(56), 318(56), 333
Nakatani, T., 218(50), 220(50), 228; 307(4), 308(4),

[Nakatani, T.], 316(4,55,56), 318(56), 319(4), 323(4), 324(55), 331, 333
Nakayama, C., 356(126), 371
Nakayama, Y., 308(7,8), 331
Nance, L. E., 101(94), 123
Natile, G., 65(4a,b), 77(4a,b), 81(44,46), 118, 120, 121
Natori, S., 354(98), 370
Navon, G., 236(78), 238(97), 279, 280
Neinken, G., 269(170), 270(170), 271(170), 283
Nelson, N. S., 232(15), 276
Nettleton, D. E., Jr., 326(80), 335
Neumann, H. M., 299(46), 304
Newman, A. D., 14(54), 20(54), 58; 109(116a), 124
Nias, A. H. W., 159(66), 193
Nichol, C. A., 200(12), 201(12), 202(12), 226
Nipper, E., 234(53,55,58), 261 (55), 271(53), 272(55,58), 278
Nishyama, Y., 319(60), 333
Noda, L., 363(226), 375
Nordén, B., 97(83b), 123
Nordquest, K. W., 68(15), 119
Nordstrom, B., 344(34), 367
Novel, J. P., 311(42), 333
Numata, K., 319(58,59,62), 320 (59), 321(59), 333, 334
Nunn, A. D., 220(62), 224(62), 228; 311(41,43), 332

O

O'Connor, T., 95(74), 122
Oda, T., 264(160), 283
Oeschger, M. P., 363(224), 375
O'Hara, G. P., 6(30), 57
Ohashi, A., 356(126), 371
Ohbayashi, M., 319(60), 333
O'Herron, F. A., 326(80), 335
Ohkubo, K., 239(99), 280
Ohlsson, I., 344(34), 367
Ohno, A., 260(147), 282
Ohnoki, S., 359(172), 360(189), 373, 374
Ohtsuama, H., 240(106), 280
Ohtsubo, E., 309(13), 311(13), 331
Ohtsubo, H., 309(13), 311(13), 331
Ohyoshi, A., 239(98,99), 240(106), 280
Oka, S., 260(147), 282
Okamoto, T., 260(147), 282
Oldham, R., 182(111), 195
Oleson, J. J., 324(68), 334
Oliverio, V. T., 293(29), 303
Olson, K. B., 3(6), 57
Omoto, S., 308(8), 331
Ono, K., 356(126), 371
Orengo, A., 363(214,222), 375
Orezzi, P., 326(77,79), 334
Orio, A. A., 81(44), 120
Ostfeld, D., 274(175), 283
O'Toole, T., 160(67,68), 161(67), 162(67), 163(67), 193
Ottensmeyer, F. P., 98(88), 123
Ottlinger, M. E., 311(39), 320 (39), 332
Otwell, H. B., 346(46), 368
Overgaard-Hansen, K., 362(210), 375

P

Padva, A., 251(142), 282
Page, J. A., 236(86), 279
Palmer, G., 365(233), 376
Panicali, D. L., 356(130), 372
Paoletti, C., 26(71), 59; 173 (97), 195
Parish, R. V., 73(27), 74(27), 119
Parker, G. R., 358(163,164), 373
Parks, J. E., 82(54), 121
Parks, M. E., 243(125), 259(125), 281
Parks, R. E., Jr., 287(12), 291 (12), 293(12), 302; 363(225, 229,231), 375, 376
Parsons, D. F., 260(148), 282
Pascoe, J. M., 26(74,75), 59; 97(81), 123; 167(79,83), 194; 347(52), 368
Patel, N., 360(186,191), 374
Paterson, A. R. P., 362(211), 375

Patton, T. F., 146(34), 191
Pearson, R. G., 23(60-62), 29 (60), 59; 67(10), 69(10), 70 (10), 74(10), 75(10), 77(39a, b,c), 79(10), 81(10,45a), 83 (10), 91(10), 119, 120; 237 (94), 238(94), 279
Peisach, J., 205(20), 216(49), 219(56-58), 222(49), 226, 228; 309(21-23), 310(21-23), 311 (21,23), 312(21,23), 313(23), 330(101), 332, 336
Pell, S. D., 234(61), 245(131), 278, 281
Peloso, A., 74(34), 75(34), 78 (34), 79(34), 80(34), 82(34), 83(34), 91(34), 120
Peng, S., 251(142), 282
Pera, M. F., Jr., 141(26), 179 (26,108), 191, 195
Peresie, H. J., 26(83), 43(106), 60, 61; 109(112), 124; 140 (19),165(19), 191; 234(47), 242(117), 265(117), 268(117), 277, 280
Perkins, P. G., 73(27), 74(27), 119
Perlia, C. P., 342(22), 367
Perrin, D. D., 306(1), 331
Pesce, A., 353(90), 354(90), 370
Pesek, J., 115(129), 125
Petering, D. H., 45(112), 61; 200(16), 204(17), 205(17), 206(22-25), 207(24,27), 208 (23), 209(27,29,30), 210(30), 211(27,29,30), 213(41,42), 214(42-44), 215(44,45), 216 (45,47,48), 217(45,48), 220 (48,59,60), 221(65), 222(60, 65), 223(60,68), 224(59,60), 226, 227, 228, 229
Petering, H. G., 200(8-11,14,16), 201(8-11), 202(10), 203(10), 204(17), 205(17), 226
Petersen, J. D., 235(77), 279
Petersheim, M., 245(135), 281
Petsko, G. A., 104(102), 124
Phelan, J. T., 5(22), 57
Phelan, R., 141(23), 191
Phelps, D. W., 68(15), 119
Phillips, A. F., 243(126), 259 (126), 281
Phillips, F. L., 169(86), 194
Phillips, F. S., 287(5), 302
Phillips, J. L., 47(123), 62
Photaki, I., 349(63), 369
Pidcock, A., 73(21,23), 119
Pietsch, P., 219(54), 228
Pigram, W. J., 327(86), 335
Pihl, E., 263(158), 282
Pitha, J., 269(169), 283
Pladziewicz, J. R., 241(110), 242(110), 280
Planet, G., 361(203), 374
Pneumatikakis, G., 108(104), 124
Poiesz, B. J., 353(87), 370
Pol, C., 326(79), 334
Pollock, R. J., 34(88), 35(88), 60; 170(88), 171(88), 194
Pories, W. J., 47(120). 62
Povirk, L. F., 309(11), 311(37), 320(37), 331, 332
Powell, A. R., 299(48), 304
Powell, J., 82(49), 121
Powers, E. L., 159(65), 193
Pradhan, T. K., 363(227), 375
Prestayko, A. W., 16(59), 59
Price, C. C., 291(15), 302
Price, L. A., 185(115), 196
Pritchard, J. B., 141(24), 142 (24), 143(24), 153(24), 191
Proshlyakova, V. V., 328(92), 335
Prusoff, W. H., 363(219), 375
Prysak, M. F., 286(2), 302
Pusch, W. M., 232(17), 276
Pytlewski, L. L., 298(42), 303

Q

Quicksall, C. O., 234(71), 278
Quigley, G. J., 96(75), 116(75), 122

R

Rabinowitz, H. N., 110(117,119), 112(119), 125
Rabussay, D., 355(109), 371

Rahamimoff, R., 264(164), 283
Rahn, R. O., 26(82), 60; 97(85a, b), 115(85a,b), 123
Raichart, D. W., 234(62), 278
Rainen, L., 36(91), 37(91,93, 94), 39(93), 60; 198(1), 226; 298(44,45), 304
Ramani, L., 241(112), 280
Ramin, L., 351(67), 369
Randolph, V., 185(116), 196
Rao, E. A., 220(60), 221(65), 222(60,65), 223(60), 224(60), 228
Rao, K. V., 324(67), 325(67), 328(93), 334, 335
Rasey, J., 269(170), 270(170), 271(170), 283
Ray, R. K., 356(121), 371
Reaoch, R. S., 242(113), 280
Rebhun, L. I., 356(119), 371
Reed, E. J. S., 65(3), 77(3), 118
Reed, J., 116(132), 117(132), 125
Reed, K. C., 264(162), 283
Reedijk, J., 102(95,96), 123
Reggiani, M., 327(87), 335
Reich, E., 329(98), 330(98), 336
Reichard, P., 357(136,142,145-147,149),358(146,149,154,155), 372, 373
Reid, E. E., 4(18), 57
Reincke, A., 3(13), 57
Reinsalu, P., 234(60), 278
Reishus, J. W., 146(35), 191
Reith, A. R., 324(68), 334
Remillard, S., 356(119), 371
Renault, H., 311(42), 333
Rennert, D., 260(146), 282
Renshaw, E., 11(42,87), 34(87), 58, 60; 232(4), 241(4), 269(4), 276; 343(26), 367
Repta, A. J., 146(34), 191; 339(6), 366
Reslova, S., 11(45,46), 29(46), 58; 100(89), 101(89), 102(89), 104(89), 123; 129(3), 190; 241(111), 253(111), 259(111), 280; 344(31), 367
Reynolds, P. E., 325(73), 334
Rezny, Z., 142(27), 191
Rhodes, D., 96(76), 116(76), 122
Rice, E. W., 349(64), 369
Rice, L. S., 355(111), 371
Rich, A., 96(75), 116(75), 122
Richards, P., 269(170), 270(170), 271(170), 283
Richards, R. E., 73(21), 119
Richardson, D., 234(72), 273(72, 173), 279, 283
Richmond, C. R., 232(16), 276
Richmond, M. H., 325(73), 334
Richmond, R. C., 159(64,65), 193
Ridgway, H., 157(53), 192
Ridgway, H. J., 14(54), 20(54), 58; 109(116a), 124
Rieder, K., 234(52), 264(52), 278
Riesselmann, B., 234(56), 272 (56), 278
Riva, S., 355(107), 371
Riva DiSansenerino, L., 326(81), 327(81), 335
Robbins, A. B., 26(77,80), 60
Robert, J., 311(42), 333
Roberts, J. J., 25(69), 26(74, 75), 27(69), 59; 152(46,49), 167(79,83), 192, 194; 347(52), 368
Roberts, M. F., 361(201,202), 374
Robin, M. B., 109(108), 124
Robins, A. B., 95(72), 122; 339 (5), 347(54), 366, 368
Robison, B., 363(231), 376
Rocco, J., 232(15), 276
Roe, E. M. F., 5(20,21), 57; 232(2), 273(2), 276
Rogers, D. E., 87(61), 121
Rogers, R. A. J., 236(85), 238 (85), 279
Rogers, W. P., 10(41), 58; 273 (172), 283
Romeo, R., 80(41a,b), 81(43a,b), 82(53b), 83(43a), 120, 121
Roome, S., 361(209), 375
Roos, I. A. G., 97(87), 116(133), 123, 125; 347(51), 368
Rosa, D., 232(15), 276
Rose, K. M., 353(92), 370
Rosenberg, B., 2(1), 9(38-40), 10(38), 11(1,47,87), 14(1,48), 26(72,79), 27(47), 34(87), 47 (118), 56, 58, 59, 60, 61; 88 (64a-d,65), 109(111,112), 114 (126,127), 121, 122, 124, 125;

[Rosenberg, B.], 129(1,2), 130 (4), 138(4), 140(19,20), 152 (44), 157(54), 165(19,69,71-75), 167(82), 170(88), 171(88-90), 190, 191, 192, 193, 194; 199(4), 226; 232(4), 233(36), 241(4), 242(113), 269(4,170), 271(170), 276, 277, 280, 283; 296(34-36), 303; 339(1), 341 (14-16), 343(27), 346(1), 347 (55), 366, 367, 368
Rosenfeld, I., 3(5), 56
Rosenkrantz, H., 141(23), 191
Ross, W. C. J., 14(55), 15(55), 20(55), 59; 165(70), 167(70), 168(70), 193; 287(7,14), 302
Rossof, A. H., 342(22), 367
Roth, D. G., 361(200), 374
Rottman, F., 360(195), 374
Rowe, P. B., 360(180), 373
Roy-Burman, P., 356(124), 371
Rudd, D. F. P., 233(43), 234(43), 246(43), 277
Ruet, A., 356(129), 372
Ruhl, H., 47(125), 62
Rusconi, A., 327(89), 335
Russell, J. L., 166(76), 194

S

Sadler, W. C., 361(204), 375
Saito, K., 319(59,62), 320(59), 321(59), 333, 334
Sakaki, S., 240(106), 280
Sakamoto, H., 239(99), 280
Sakamoto, K., 310(32), 332
Sakamoto, N., 359(176), 373
Sakarellon-Diatsioton, M., 349 (63), 369
Sakurai, Y., 300(51), 304
Salem, P., 179(106), 195
Samer, L., 241(112), 280
Samson, M. K., 341(17), 367
Samuels, M. L., 131(11), 179(11), 190
Sanchez-Anzaldo, F. J., 355(115), 371
Saneyoshi, M., 356(126), 371
Santillo, F. S., 309(24,25), 311(25), 312(24,25,47),
[Santillo, F. S.], 313(24,25), 314(24,47), 315(47), 316(24,47), 317(24,47), 318(24,47), 321(47), 322(24), 332, 333
Sargeson, A. M., 245(129), 270 (129), 281
Sartorelli, A. C., 45(113), 61; 205(18), 209(28), 212(31,36,38), 213(40), 226, 227; 358(159,160), 373
Saryan, L. A., 209(30), 210(30), 211(30), 213(42), 214(42), 215 (45), 216(45), 217(45), 220 (59), 224(59), 227, 228
Sasahira, T., 319(58), 333
Sasse, D., 3(8), 57
Sastry, B. V., 232(18), 276
Sastry, K. S. R., 311(39), 320 (39), 332
Satchell, D. P. N., 77(36), 120
Sato, T., 6(32), 58
Sausville, E. A., 219(56-58), 228; 309(21-23), 310(21-23), 311(21,23), 312(21,23), 313 (23), 332
Sava, G., 35(90), 43(90), 60
Savage, C. R., Jr., 359(174), 373
Save, G., 232(6), 241(6), 268(6), 272(6), 276
Savitskii, I. V., 44(109), 61
Sawa, T., 308(9), 331
Sawatari, K., 300(51), 304
Sawitsky, A., 6(31), 58
Scandola, F., 82(50), 121
Scarpinato, B., 326(77,78), 327 (87), 334, 335
Schabel, F. M., Jr., 157(60), 193
Schaeppi, U., 141(23), 142(28), 191; 342(20), 367
Scheinberg, H., 198(2), 226
Scheiner, E., 131(8), 178(8), 190; 342(24), 367
Scheit, K. H., 355(112), 371
Schepartz, S. A., 154(52), 192
Schmidt, L. H., 287(8), 302
Schneifels, J. A., 15(58), 59
Schreiber, R. H., 326(80), 335
Schroeder, D. D., 360(188), 374
Schroeder, H. A., 3(15), 57
Schubert, K. R., 361(202,206), 374, 375
Schug, K., 245(130), 270(130), 281

Schwartz, I., 327(88), 335
Schwartz, P., 157(56,57), 193
Schwarz, S., 45(114), 61
Schweer, K., 232(12), 261(12), 262(12), 276
Scovell, W. M., 95(74), 122; 242 (113), 280
Scrutton, M. J., 353(84), 354 (84), 370
Sealy, R. C., 220(60), 222(60), 224(60), 228
Sedov, K. A., 328(96), 329(96), 335
Seidel, D., 232(12), 261(12), 262(12), 276
Seno, G., 351(67), 369
Sentenac, A., 356(129), 372
Serio, G., 37(93), 39(93), 60; 298(45), 304
Shaddix, E. L., 234(49), 278
Shah, J., 185(116), 196
Shah, P., 341(17), 367
Shapot, V. S., 243(121), 259 (121), 281
Shapshay, S., 185(117), 196
Sharpless, T. W., 253(144), 258 (144), 282
Sheena, A. H., 212(37), 227
Shelton, E., 361(206), 375
Shemyakin, M. M., 328(97), 329 (97), 335
Shepherd, R. E., 234(70), 236 (70), 242(70), 278
Sheridan, P. J., 47(123), 62
Sherwood, E., 36(92), 37(92), 38(92), 39(92), 40(96), 60
Shigeura, H. T., 356(128), 371
Shingleton, H., 186(118), 196
Shinohara, T., 239(98), 280
Shirakawa, I., 310(28), 332
Shirley, F. A., 236(87), 279
Shooter, K. V., 26(77), 60
Shore, G. C., 260(148), 282
Shulman, A., 5(20,21), 6(27), 10(27), 57; 232(2,3), 273(2, 3), 276
Sieber, S. M., 49(129), 50(129), 51(129), 62
Sierocki, J. S., 188(124), 196
Sigler, P. B., 104(101), 124
Sillén, L. G., 222(67), 229
Silverstein, E., 346(49), 368
Silverton, J. V., 263(154), 282
Silvertrini, R., 326(77), 334
Silvestri, L. G., 355(107), 371
Sim, S., 309(20), 310(20), 324 (71,72), 332, 334
Sim, W. J., 357(132), 372
Sinex, F. M., 26(81), 60; 97 (84a,b), 123; 169(84), 194
Singer, M., 264(163), 283
Sippel, A., 355(117), 371
Sirica, A. E., 131(10), 154(10), 190; 299(50), 304
Sirover, M. A., 353(76-78), 355(76,77), 369
Sjöberg, B.-M., 357(134), 358 (154,161), 372
Skapski, A. C., 169(86), 194
Slater, J. P., 353(85), 354(85, 96), 370
Slater, T. F., 342(19), 367
Sleigh, M. J., 232(65), 334
Sleight, S. D., 141(22), 165(71), 191, 193; 341(16), 367
Sloan, D. L., 354(101), 370
Slyudkin, O. P., 101(92), 123
Smith, A. B., 43(108), 44(108), 61; 150(43), 151(43), 155(55), 157(55), 192; 343(28), 367
Smith, C. F., 51(130), 62
Smith, P. H. S., 25(66), 59
Smith, P. M., 263(154), 282
Smith, R. G., 352(69), 369
Smith, R. M., 356(118), 371
Sneden, D., 96(75), 116(75), 122
Snow, M. N., 185(117), 196
So, A. G., 354(94,95), 370
Sobel, H., 77(39c), 120
Sobin, B. O., 328(93), 335
Soderland, G., 344(34), 367
Soifer, V. S., 328(97), 329(97), 335
Solaiman, D., 220(59,60), 221 (65), 222(60,65), 223(60,68), 224(59,60), 228, 229
Soldati, M., 326(77), 334
Soloway, M. S., 157(58), 187 (121), 193, 196
Som, P., 269(170), 270(170), 271 (170), 283
Songstad, J., 77(39c), 120

Sonntag, R. W., 148(40), 192
Sorokina, I. V., 328(96), 329 (96), 335
Sottocasa, G. L., 260(148), 282
Spalla, C., 326(79), 334
Spaulding, J., 178(104), 195
Speckhard, D. C., 355(114), 371
Spector, T., 359(177), 365(234), 373, 376
Speer, R. J., 14(54), 20(54), 58; 109(116a), 124; 130(5), 157(53), 176(5), 190, 192
Spencer, R. P., 25(63,64), 59; 145(32), 191
Sperling, O., 360(182), 374
Spiro, R., 185(116), 196
Springate, C. F., 353(86), 370
Srivastava, B. I. S., 356(127), 371
Srivastava, S., 269(170), 270 (170), 271(170), 283
Stadlbauer, E. A., 234(53), 271 (53), 278
Stadnicki, S. W., 142(28), 191; 342(20,21), 367
Staehelin, M., 355(106), 371
Stallcup, W. B., 360(190), 374
Stanko, J. A., 15(58), 59; 111 (120), 125; 234(47), 277
Stara, J. F., 232(15), 276
Starmer, C. F., 360(182), 374
Stayton, R. E., 342(22), 367
Stein, P. J., 354(102), 370
Stein, R. W., 219(58), 228; 309 (23), 310(23), 311(23), 312 (23), 313(23), 332
Stender, W., 355(112), 371
Stephenson, T. A., 299(48), 304
Sternberg, J. J., 187(122), 196
Sternson, L. A., 146(34), 191; 339(6), 366
Stewart, D. P., 14(54), 20(54), 58; 109(116a), 124
Stock, J. A., 287(11), 302
Stone, A. E., 43(108), 44(108), 61; 152(47), 192
Stone, P. J., 26(81,83), 60; 97(84a,b), 123; 169(84), 194
Stone, R. A., 349(62), 369
Stothers, J. B., 73(24), 119
Stritar, J. A., 235(76), 238(76),
[Stritar, J. A.], 240(76), 241 (76), 279
Strong, E., 185(116), 196
Strong, J. E., 311(40), 319(40), 320(40), 333
Strong, S., 185(117), 196
Strothkamp, K. G., 96(78), 116 (78), 122
Struck, R. F., 293(27), 303
Stufkens, D. J., 73(25), 119
Stutz, A. A., 355(112), 371
Stynes, H., 234(73), 279
Subramanian, G., 233(24), 262 (24), 277
Subramarian, S., 132(14), 183 (112), 184(112), 190, 195
Subrumanian, N., 234(58), 272 (58), 278
Suddath, F. L., 96(75), 116(75), 122
Sugiura, Y., 309(27), 312(46), 313(27), 314(27,46,48), 315 (46,48), 316(46), 317(46), 332, 333
Suhadolnik, R. J., 359(167,168), 373
Sulebele, G., 346(49), 368
Sullivan, R., 287(8), 302
Sundberg, R. J., 237(93), 239 (93), 245(93,134), 279, 281
Sunderman, F. W., 233(33), 277
Sutin, N., 240(108), 241(108), 280
Sutton, C., 237(90), 279
Sutton, J. E., 234(72,74), 235 (74), 237(74), 245(74), 251 (74), 257(74), 273(72), 279
Suzuki, G., 220(61), 224(61), 228
Suzuki, H., 309(19), 310(19,31), 324(19), 331, 332
Svec, F., 3(9), 57
Swanwick, M. G., 86(60), 87(60), 121
Swarc, M., 236(79), 279
Swebe, S. M., 352(68), 369
Swenson, R., 344(32), 367
Swiniarsky, J., 158(61), 193
Switzer, R. L., 360(181), 361 (197-199,201,202,204,206), 374, 375
Szalda, D. J., 110(117,119),

[Szalda, D. J.], 112(119), 125
Szekerke, M., 293(28), 303
Szumiel, I. I., 159(66), 193

T

Takahashi, I., 356(126), 371
Takahashi, K., 223(70), 229; 310 (30), 317(30), 332
Takahashi, T., 219(55), 228
Takamatsu, T., 236(80), 271(80), 278
Takamiya, K., 44(110,111), 61
Takano, T., 344(32), 367
Takayama, H., 219(52), 224(52), 228; 310(33), 332
Takeshita, M., 309(13), 310(29), 311(13), 331, 332
Takeuchi, T., 219(52,53), 224 (52,53), 228; 308(9), 310(33-35), 331, 332
Takita, T., 218(50), 220(50), 228; 307(4,6), 308(4,7,8), 311 (38), 316(4,5,56), 318(56), 319(4), 320(38), 323(4,64), 324(55,66), 331, 332, 333, 334
Tamagata, N., 232(21), 261(21), 276
Tamai, T., 233(26), 263(26), 277
Tamaro, M., 43(107), 61; 172 (95), 194; 233(35), 268(35), 273(35), 277
Tamburro, A. M., 97(85d), 115 (85d), 123
Tan, C., 212(39), 227
Tanabe, M., 233(26-28), 263(26-28), 277
Tanaka, N., 220(61), 224(61), 228; 309(19), 310(19,31), 320(19), 331, 332
Tapia, R., 264(165), 283
Tarchiani, G., 3(12), 57
Tarmir, I., 354(96), 370
Tata, J. R., 260(148), 282
Tatsuoka, S., 328(91), 335
Taube, H., 233(39,40,43), 234 (39,43,45,48,52,62,65,66,70, 72,74,75), 235(45,74,76), 236 (70,83,89), 237(45,66,74,91-93,96), 238(39,45,76,89,91),
[Taube, H.], 239(45,65,66,83,93), 240(45,76,102), 241(76,110), 242(39,65,70,110), 244(48), 245(74,93,128), 246(43,45,137), 247(65,66,138), 248(66,138), 249(66), 250(66), 251(65,66, 74), 253(65,66), 254(66), 256 (65), 257(74), 258(65), 259 (40,102), 260(39,45), 261(45), 262(45), 264(52), 273(72), 277, 278, 279, 280, 281
Taylor, A., 4(17), 57
Taylor, A. J., 234(54,58), 271 (54), 272(58), 278
Taylor, D. M., 25(66), 59
Taylor, I. F., 237(93), 239(93), 245(93), 279
Taylor, M. R., 205(19), 226
Taylor, M. W., 47(122), 62
Tebbetts, L., 347(60), 349(64), 369
Teggins, J. E., 153(50,51), 192; 344(39,40), 345(41-44), 346 (45,46), 347(39,57), 349(41, 61), 368, 369
Temin, H. M., 353(72), 369
Teo, B. K., 112(123), 116(132), 117(132), 125
Tessman, I., 358(162), 373
Thakur, M. L., 311(43), 333
Thelander, L., 357(133,134,142, 147,149,150,152,153), 358(149), 372
Theophanides, T., 26(84), 60; 95(73), 97(79a-c,85c), 98(79a-c), 108(104), 115(79a-c,85c), 117(79a-c), 122, 123, 124; 169(85), 194; 340(8), 347(50, 56), 366, 368
Thews, G., 243(120,123), 259(120, 123), 281
Thie, R. S., 324(68), 334
Thigpen, T., 186(118), 196
Thiry, L., 356(121), 371
Thomas, C. B., 347(58), 368
Thomas, G., 237(95), 238(95), 279
Thomas, G. J., 356(118), 371
Thompson, A. G., 347(55), 368
Thompson, H. S., 152(47), 192
Thompson, R. C., 232(20), 276
Thomson, A. J., 9(40), 11(42,46),

[Thomson, A. J.], 25(69), 26(79), 27(69), 29(46), 58, 59, 60; 100(89), 101(89), 102(89), 104(89), 116(133), 123, 125; 129(2), 152(49), 167(82), 190, 192, 194; 241(111), 242(113), 253(111), 259(111), 280; 343 (26), 344(31), 347(51), 367, 368
Thomson, J. M., 45(114), 61
Ting, R. C., 356(120), 371
Tinoco, I., Jr., 91(66), 122
Tiozzo, R., 262(152), 282
Tobe, M. L., 14(55), 15(55), 20 (55), 48(127), 59, 62; 70(17), 74(32), 75(32), 79(17), 80 (41a,b), 81(43a,b), 83(43a), 119, 120; 165(70), 167(70), 168(70), 193
Tobias, C. A., 3(5), 56
Tobias, R. S., 94(69), 95(69, 71), 98(69,71), 115(69,71,130), 117(69,71,130), 122, 125
Toftness, B. R., 25(68), 59; 212(38), 227
Tom, G. M., 234(52), 245(128), 264(52), 278, 281
Tomcheck, R., 212(38), 227
Tomita, K., 319(58,59), 320(59), 321(59), 333
Toplin, I., 324(68), 334
Torgerson, D. F., 109(114), 124
Torres, I. J., 145(31), 148(31), 191
Toshinkuni, K., 240(106), 280
Toth-Allen, J., 141(21), 148 (39), 191, 192
Townsend, L. B., 359(169), 373
Trader, M. W., 157(60), 193
Trainor, K. E., 41(103), 61
Trainovitch, T. A., 5(23), 57
Trosko, J. E., 2(1), 11(1), 14 (1), 56; 109(111), 124; 130(4), 138(4), 190; 296(35), 303
Trozzi, M., 82(53b), 121
Truitt, C. D., 360(185), 374
Tschopp, L., 148(40), 192
Tsiftsoglou, A. S., 358(159), 373
Tsuboi, K. K., 363(228), 375
Tsubosaki, M., 310(32), 332
Tsukiura, H., 319(58,59,62), 320 (59), 321(59), 333, 334
Tsuno, T., 319(62), 334
Tsutsui, M., 274(175), 283
Tucker, M. A., 349(61), 369
Turo, A., Jr., 77(39a), 120

U

Ucmakli, A., 185(117), 196
Uenoyama, Y., 319(58), 333
Uguagliati, P., 65(4b), 77(4b), 118
Ulpino, L., 353(81), 369
Umezawa, H., 218(50), 219(51-53), 220(50,61), 223(70), 224(52, 53,61,70), 228, 229; 370(2, 4-6), 308(2,4,5,7-10), 309(2, 19), 310(19,28,30-35), 311(38), 316(2,4,55,56), 317(30), 318 (56), 319(4), 320(38), 322(2), 323(4,5,64), 324(19,55,66), 331, 332, 333, 334
Umezawa, Y., 218(50), 220(50), 228; 307(4), 308(4), 316(4), 319(4), 323(4), 331
Underwood, G. E., 200(8), 201(8), 226
Urbanek, M. A., 141(24), 142(24), 143(24), 153(24), 191
Utsumi, K., 264(160), 283

V

Valentin-Hansen, P., 363(215), 375
Valentini, L., 326(77), 327(87), 334, 335
Valenzuela, P., 353(82), 354(82), 370
Vallee, B. L., 47(119), 61; 353 (81,88), 369, 370
Vallejo, A., 185(116), 196
Van Camp, L., 2(1), 9(39,40), 11 (1,87), 14(1,48), 34(87), 56, 58, 60; 109(111,112), 124; 129(1,2), 130(4), 138(4), 140 (19), 144(29), 148(29),

[Van Camp, L.], 157(54), 165(19, 69), 190, 191, 192, 193; 232 (4), 241(4), 269(4), 276; 296 (34-36), 303; 341(14), 367
Vandegrift, G. F., III, 81(47b), 84(47b), 121
van den Berg, H. W., 152(46), 192
van Giessen, G. J., 204(17), 205(17), 226
Van Husen, M., 315(50,52), 316 (54), 318(54), 320(52), 321 (52), 322(52), 333
Van Kralingen, C. G., 102(95, 96), 123
Vanquickenbirne, L. G., 77(37, 38), 120
Vanstone, C. L., 150(42), 192
Van Vliet, D. L., 245(135), 281
Varnum, J., 104(103), 124; 339(3), 343(30), 366, 367
Vasina, I. V., 328(97), 329 (97), 335
Vasington, F. D., 262(152), 282
Vassiliadis, G. A., 97(82a), 123
Vaughan, C., 185(117), 196
Vaupel, P., 243(120,122,123), 259(120,122,123), 281
Venanzi, L. M., 73(21), 119
Venditti, J. M., 131(10), 154 (10,52), 158(61,62), 190, 192, 193; 295(31), 299(50), 303, 304
Vijayan, M., 102(98), 124
Vitale, S., 3(12), 57
Volfin, P., 342(25), 367
von Döbeln, U., 357(145), 358 (156), 372, 373
Vranckx, J., 77(37,38), 120
Vrieze, K., 73(25), 119
Vulcano, A. L., 326(80), 335

W

Wacker, W. E. C., 47(119), 61
Wadzilak, T. M., 353(80), 369
Wagner, R. W., 114(126,127), 125; 165(73), 193
Walker, E. M., Jr., 43(108), 44 (108), 61; 155(55), 157(55), 192
Walker, J., 234(72), 273(72), 279
Wallace, H. J., 130(6), 177(6), 179(6), 187(6), 190; 342(23), 367
Wallach, R. C., 132(15), 184(15), 190
Walton, E. C., 41(103), 61
Walton, R. A., 110(118), 112(118), 125
Wampler, D. L., 326(81), 327(81), 335
Wang, M. C., 360(193), 374
Wang, S. M., 363(217), 375
Ward, D. C., 329(98), 330(98), 336
Ward, J. E. H., 73(24), 119
Ward, J. M., 141(25), 191
Ward, M. D., 236(85), 238(85), 279
Waring, M. J., 325(73), 334; 356 (123), 371
Warren, L. E., 205(21), 226
Waszczak, J. V., 110(119), 112 (119), 125
Watari, K., 232(21,22), 261(21, 22), 276, 277
Watkins, D. M., 14(56), 16(56), 17(56), 19(56), 20(56), 21(56), 22(56), 24(56), 59; 146(36), 192
Watt, G. D., 233(44), 234(44), 260(44), 277
Waysbort, D., 236(78), 238(97), 279, 280
Webber, C. E., 232(14), 276
Weber, G., 344(36,37), 368
Weeks, M. H., 232(20), 276
Wehrli, W., 355(105,106,113), 356(105), 370, 371
Wei, Y. K., 7(34), 58; 199(5), 226
Weinfeld, H., 359(174), 360(193), 373, 374
Weinhaus, S., 243(124), 259(124), 281
Weinzierl, J. J., 96(75), 116 (75), 122
Weiss, K. K., 309(15), 331
Wenner, C. E., 243(124), 259 (124), 281
Wenzel, M., 234(53-59), 261(55), 271(53,54), 272(55-59), 278

Weser, U., 353(83), 354(97), 370
Westhead, E. W., 311(39), 320 (39), 332
Weymouth, L. A., 355(104), 370
Wheeler, G. P., 287(13), 291(20), 293(13,20), 302, 303
Whelan, H., 214(43), 227
Whileyman, J., 36(91), 37(91), 60; 198(1), 226; 298(44), 304
White, C., 296(40), 303; 361 (200), 374
White, D. O., 232(3), 273(3), 276
White, H. L., 325(74), 334
White, J. R., 325(74), 334
Whitelock, J. D., 65(3), 77(3), 118
Whitford, T. W., Jr., 358(157), 373
Whiting, R. F., 98(88), 123
Whitmore, W. F., 178(104), 195
Whitten, D. G., 108(105a,b), 124; 274(174), 283
Wicart, L., 311(42), 333
Wilkins, R. G., 74(31), 75(31), 120
Wilkinson, B. W., 148(39), 192
Wilkinson, G., 65(2), 67(2), 68 (2), 71(2), 109(107), 118, 124; 299(48), 304
Williams, E. D., 311(43), 333
Williams, D. R., 41(98), 61; 233(32), 277
Williams, G. R., 260(148), 282
Williams, P. A., 82(55a), 89 (55a), 121
Williams, R. J. P., 11(46), 29 (46), 58; 100(89), 101(89), 102(89), 104(89), 123; 241 (111), 253(111), 259(111), 280; 344(31), 367
Williams, S. D., 182(111), 187 (123), 195, 196
Wilson, J., 350(65), 369
Wiltshaw, E., 130(7), 132(14), 177(7), 183(112), 184(112), 190, 195
Winkelmann, D. A., 206(24), 207 (24), 227
Witt, L. L., 359(169), 373
Wittes, R., 185(116), 188(124), 196
Wodinsky, I., 158(61,62), 193; 293(25), 303
Wolberg, G., 357(131), 372
Wolf, W., 145(30), 148(30), 149 (30), 191
Wolfe, R., 3(5), 56
Wolfson-Davidson, E., 253(144), 258(144), 282
Wolpert-DeFilippes, M. K., 139 (18), 140(18), 191
Wong, J. Y., 360(181), 374
Woodard, D. A., 157(59), 193
Woodman, R. J., 131(10), 154(10, 52), 190, 192; 299(50), 304
Woods, T. S., 345(42), 368
Woodward, R. B., 324(67), 325 (67), 334
Woster, A. W., 47(122), 62
Wright, G. F., 346(47), 368
Wright, J. A., 357(140), 359 (165), 372, 373
Wright, R. D., 6(27), 10(27), 57
Wu, C. W., 353(84), 354(84,93), 355(114), 370, 371
Wu, F. Y.-H., 355(114), 371
Wu, K. C., 97(82b,83a), 102(82b), 114(125), 116(83a), 117(82b), 123, 125; 166(78), 194
Wübker, W., 309(11), 331
Wyngaarden, J. B., 360(180,182, 194), 373, 374

Y

Yagoda, A., 187(119), 196
Yamada, T., 239(98), 280
Yamaki, H., 220(61), 224(61), 228; 309(19), 310(19,31), 320 (19), 331, 332
Yamamoto, G., 233(26,28), 263(26, 28), 277
Yamashita, T., 240(106), 280
Yamato, M., 233(26), 263(26), 277
Yamazaki, Z., 309(17), 331
Yancey, S. T., 51(130), 62
Yang, S. S., 356(120), 371
Yartseva, I. V., 328(97), 335
Yasbin, R., 265(168), 268(168), 283
Yatskoff, R. W., 233(30), 277
Yeh, Y.-C., 358(162), 373

Yesair, D. W., 327(90), 335
Yip, L. C., 361(209), 375
Yoe, J. H., 116(134), 125
Yolles, S., 140(20), 191
Yoshioka, O., 223(70), 224(70), 229; 310(30), 317(30), 332
Young, D. M., 141(25), 191
Yu, C., 251(142), 282
Yui, K., 6(28,29), 57

Z

Zaccara, A., 327(85), 335
Zahn, R. K., 223(69), 229; 309 (17), 331
Zak, M., 142(27), 159(63), 191, 193
Zalkin, H., 360(185), 374
Zanella, A., 246(137), 281
Zassinovich, G., 34(89), 35(90), 43(90), 60; 232(6), 241(6, 112), 268(6), 272(6), 276, 280
Zedeck, M. S., 358(160), 373
Zentmyer, G. A., 7(35), 58
Zeppezauer, E., 344(34,35), 367
Zeppezauer, M., 344(33), 367
Zillig, W., 355(109,110), 371
Zimmerman, A., 148(40), 192
Zimmerman, T. P., 357(131), 372
Zipp, A. P., 296(38), 303
Zipp, S. G., 296(38), 303
Zoltewicz, J. A., 253(144), 258 (144), 282
Zook, B. C., 141(26), 179(26), 191
Zubrod, C. G., 134(16), 190
Zumdahl, S. S., 74(29), 120
Zunino, F., 327(82,85), 335
Zvagulis, M., 88(64b,65), 121, 122
Zwelling, L. A., 167(81), 171 (92), 173(92), 194
Zwickel, A. M., 234(69), 278

SUBJECT INDEX

A

Absorption bands and spectra (and spectrophotometry, *see also* UV absorption), 24, 90, 112, 113, 116-118, 209, 222, 247, 250, 251, 255, 256, 312, 315, 316, 318, 321, 327
Acetamide, 108
  trimethyl-, 109
Acetate (or acetic acid), 13, 36-38, 49, 51, 52, 78, 83, 266, 298
  buffer, 82, 106, 107
  ^{14}C, 40
  methoxy-, 36-38
  uranyl-, 13
Acetonitrile, as ligand, 108, 109, 237
Acetylacetonate, 5, 35, 43
Acidity constants (*see also* Equilibrium constants), 80, 87, 90, 93, 99, 147, 221, 222, 245, 247, 252, 253, 257, 312
Actinomycin, 180-182, 341
Activation
  energy, 210
  enthalpy, 147
  volume, 76
Adenine (and residues), 26, 92, 98, 171, 255-258, 273
  arabinosyl-, 351, 356, 362
  9-β-D-psicofuranosyl-, 360
  9-β-D-xyloguranosyl-, 362
Adenosinate complexes, 250
Adenosine, 93-95, 249, 250, 252, 254, 255, 257, 267, 362
  3'-amino-3'-deoxy-, 362
  3'-deoxy-, *see* Cordycepin
  1-methyl-, 94, 253
Adenosine 5'-diphosphate, *see* ADP
Adenosine 5'-monophosphate, *see* AMP
Adenosine triphosphate, *see* ATP
Adenosyl cobalamin, 357
Adenyl kinase, 106, 107, 363
ADP, 37, 358, 359, 361
Adrenals, 148, 149, 272
Adriamycin, 132, 133, 156, 157, 180-184, 187, 188, 326-328, 356
Aglycon, 328
α-Alanine (and residues), 100, 101, 106, 107, 317
Alanylmethionine, 348, 349
Albumins, 344
Alcohol dehydrogenase, 344
  liver, 107, 344-346, 349
  yeast, 345
Alcoholic groups, *see* Hydroxyl groups and Phenolates
Alcohols (*see also* individual names), 67, 299
Aldehydes, *see* individual names
Aldolases
  fructose-1,6-diphosphate, 350
Alkali ions, *see* individual names
Alkaline earth ions, *see* individual names
Alkaline phosphatases, 344
Alkylating agents or drugs (*see also* individual names), 133, 171, 177, 273
  metal complexes of, 285-301
Allopurinol, 365
Alloxanthine, 365
Alloxazines, iso-, 234, 256
Aluminum(III), 49, 51, 301, 327, 330
Ames assay (or test), 172-174, 266-268, 273
Amide group, 109, 307, 308
Amide nitrogen, 99-102, 110, 220, 221

[Amide nitrogen]
coordination of, 98, 101, 316, 317, 322
Amides, 245
Amidotransferases, 359, 360
Amines (*see also* individual names), 2, 4, 14, 20, 32-35, 41-44, 48, 75, 80, 81, 104, 155, 161-163, 168, 234, 244, 245, 307, 316, 342
alicyclic, 15-19, 23
alkyl, 16-19, 21, 22
heterocyclic, 15-19, 30
poly-, 47
Amino acids (*see also* individual names), 115, 339, 340, 344
Pt(II)-binding, 100-102, 340
Ru complexes, 244-246, 274
Amino groups, 92, 98-102, 106, 116, 220-222, 307, 308, 314, 317, 327, 339, 359-361
p-Aminohippuric acid transport, 153
2-Amino-6-oxopurine, *see* Guanine
Aminopeptidases, *see* Peptidases
6-Aminopurine, *see* Adenine
Ammonia, 360
as ligand, 5, 6, 9-13, 15-17, 19, 23, 32-35, 69, 70, 72-74, 78-80, 83-88, 90, 94-97, 103-105, 110, 111, 161, 162, 231-275, 341-350
buffer, 82, 104-107
Ammonium ion (*see also* Ammonia), 105-107
AMP, 37, 95, 347, 359
Anhydrase, carbonic, 106, 344
Aniline, 78, 92
p-nitrosodimethyl-, 116
Animals, *see* individual species and names
Ansamycins, 355, 356
Anthracyclines, 326, 327
Antibiotics (*see also* individual names), 10, 133, 218, 274, 355, 356
interaction with DNA, 325, 327-330
metal complexes of, 305-330
Antifolates, 133
Antimony, 10
[Antimony]
compounds as ligands, 78, 79
Arabinosyladenine, 351, 356, 362
Arabinosyl-ATP, 359
Arabinosyl-CTP, 356
Arabinosylcytosine, 36, 37, 39, 154, 156, 157, 351, 356, 362
Arabinosyl-dTTP, 356
Arginine (and residues), 99, 107, 339, 344, 352
Arsenate, 3, 340
Arsenic, 10
compounds as ligands, 78, 79
organo-, 3
Ascorbate (or ascorbic acid), 4, 5
Asparaginase, 41
Asparagine (and residues), 99
Aspartic acid (and residues), 99, 102, 107, 260, 339, 345, 352
poly-, 103, 339
Association constants, *see* Stability constants
Atomic absorption spectroscopy, 117, 146, 340, 353
flameless, 115
5'-ATP, 37, 347, 356, 358-363
3'-amino-3'-deoxy-, 359
arabinosyl-, 359
d-, 351, 352, 358, 359, 363
2',3'-riboepoxide-, 357
5-Azacytidine, 154, 156, 362
8-Azaguanine, 6
Azaserine, 359, 360
6-Azauridine, 362
Azide, 78
Aziridine derivatives, 286, 287, 289, 291, 292, 295-298
Aziridinyl phosphine oxide(s)
complexes of, 296-298
tris-(1--), 286, 289
Aziridinyl phosphine sulfide(s), 289, 291, 296
complexes of, 297, 298

B

Bacillus
cereus, 343

[*Bacillus*]
  megaterium, 159
  subtilis, 355
π-Back bonding, *see* π-Bonding
Bacteria (*see also* individual names), 9-13, 26, 30-34, 40-42, 53, 129, 171-174, 273, 310, 319, 343, 355
Bacteriophage, 11, 26
Barium ion, 297
Bathocuproine disulfonate, 209
BCNU, *see* Nitrosoureas
Bile, 271
Binding constants, *see* Stability constants
Binuclear complexes, *see* Hydroxo complexes
2,2'-Bipyridyl, 10, 30, 34, 35, 81, 89, 90, 97, 116, 267
P,P-Bis-(1-aziridinyl)-N-phenyl-phosphinic amide, 296, 297
Bis(2-chloroethyl)methylamine, *see* Mechlorethamine
Bismuth, 8, 340
Bladder
  human, 148
  rat, 148, 149
  tumor, 25, 157, 187
Blastomas (*see also* Tumors)
  ependymo-, 136
  neuro-, 188, 291
  retino-, 291
Blenoxan, *see* Bleomycin
Bleomycin, 131, 133, 156, 179-182, 185, 274, 307-322, 325, 327, 330, 356
  A_2, 219, 307-309, 311
  B_2, 219
  cobalt complexes, 314, 315, 317
  complexes of, 309-318
  copper complexes, 218-225, 307, 310, 316, 317
  desamido-, 308
  DNA interaction, 310, 311, 323, 324
  epi-, 308
  hydrolase, 308
  iron complexes, 309, 311-315, 317
  iso-, 308
[Bleomycin]
  mechanism of action, 309, 310
  Ni(II) complexes, 315
  Zn(II) complexes, 318
Blood, 142, 146, 243, 261, 262, 270
  circulation, 132
  mice, 142, 146
  plasma, 83
  rat, 142, 146, 149
  tissue proteins, 349
  urea, 141
Bohr magneton, *see* Magnetic susceptibility measurements
π-Bonding (or π interactions), 73, 77, 80, 233-237, 239, 240, 245, 246, 248, 255, 257, 258
Bone, 262, 271
  marrow, 142, 149, 176, 179
  rats, 148, 149, 264
  tissue, 49
"Borderline" species, *see* "Hard" and "Soft" species
Brain
  irradiation, 188
  rat, 148, 149
  tumor, 159, 189, 291
Breast, tumor, *see* Carcinomas and Tumors
Bromide, as ligand, 15, 23, 68-70, 72, 74, 76, 79-81, 84-86, 95, 106, 236, 238, 266, 344, 345, 348, 349
Bronchia, carcinomas, 291
Buffer, 82, 83, 313
  acetate, 82, 106, 107
  ammonia, 82, 104-107
  carbonate, 82
  citrate, 82, 263
  oxalate, 82
  phosphate, 31, 82, 98, 104-106, 313, 314
  tris, 106, 107
Busulfan, 290, 291, 293
Butyrate, 37-39, 351

C

Cadmium(II), 5, 204, 205, 340, 363
Caffeine, 248
Calcium ion, 297, 298, 330, 361, 363
  transport, 264, 342
Calf
  fetal serum, 220
  thymus DNA, 97, 315
Camptothecin, 154, 156
Cancer, *see* Tumor
Carbazones, *see* Semicarbazones and Thiosemicarbazones
Carbon
  ^{13}C, 115, 221, 315, 316, 318, 321, 322
  ^{14}C, 40, 114
Carbonate, 6, 88
  buffer, 82
Carbonic anhydrase, 106, 344
Carbonmonoxide, 73, 74, 79, 234, 238, 267
Carbonyl group (or carbonyl oxygen), 239, 248, 257, 270, 273, 316, 328, 360
Carboxylate groups (*see also* Carboxylic acids), 15, 20, 24, 98-101, 103, 116, 238, 241, 244, 264
  rhodium complexes, 36-40, 52, 198, 297, 298, 351
Carboxylic acids (*see also* individual names and Carboxylate groups), 271, 272
Carboxypeptidases, 339, 343
  A, 106
Carcinogenesis
  of Pt drugs, 172-175
  of Ru complexes, 269, 274
Carcinomas (*see also* Tumors), 307
  -755, 8
  -1025, 201
  B-16 melano-, 136, 156, 270
  bronchiogenic, 291
  C3H mammary, 201
  chorio-, 134, 177
  DMBA-induced mammary, 136, 138
  Guerin, 5
[Carcinomas]
  Lewis lung, 22, 45, 136, 138
  M5076 ovarian, 137
  mammary, 4, 291
  mammary adeno- (MTG-B), 159
  ovary, 291
  spontaneous mammary, 202
  testis, 291, 307
Cardiac muscle, *see* Heart
Catalases, 350
CCNU, *see* Nitrosoureas
CD, *see* Circular dichroism
CDP, 358
Cells, mammalian, 171
Central nervous system, 25
Cerium(IV), 113
Cervix tumor, 25, 186, 187
Cesium ion, 4
Chelating
  drugs (*see also* individual names), 7, 306
  effect, 71, 81
  ligands (*see also* individual names), 6-8, 15, 42, 234, 262, 263
Chemotherapeutic agents, *see* Drugs and individual names
Chicken, 136, 138
Chironomus hemoglobin, 105
Chlorambucil, 132, 183, 184, 288, 291
Chloride, 147, 148, 261-263
  ^{36}Cl, 75
  as ligand, 4-6, 8-15, 18-20, 23, 24, 32-35, 40, 41, 43, 44, 48, 51, 68-75, 77-87, 89, 90, 95, 96, 98, 101-106, 108, 109, 113, 135, 146, 150, 151, 159, 161, 162, 168, 172, 174, 175, 189, 235, 237, 238, 240, 241, 249, 259, 266, 267, 272, 273, 286, 296, 301, 339, 341-350
Cholanthrene, methyl-, 6
Choriocarcinoma, 134, 177
Chromatography, 255, 256
  high pressure liquid, 170
  thin-layer, 146
Chromium, 4, 6
Chromium(III), 5, 28, 29
Chromomycin, 328-330

[Chromomycin]
  -one, 328
α-Chymotrypsin, 104, 106, 339, 343, 348
Circular dichroism, 315
Cis effect (*see also* Trans effect), 73-75, 79, 80
Cisplatin [cis-Pt$(NH_3)_2Cl_2$] (*see also* Platinum), 2, 10, 11, 14-16, 26-28, 35, 40, 43, 52, 70, 71, 83, 86, 95, 97, 98, 102, 108, 109, 115, 128-190, 198, 199, 242, 243, 259, 265, 269, 270, 273, 286, 299, 300, 339-350
  anticancer activity in humans, 175-189
  biochemical reactions, 152, 153
  carcinogenesis, 172-175
  cervix carcinoma, 186, 187
  clinical studies, 24, 25
  combination drug therapy in animals, 153-160
  combination with radiation therapy, 158-160
  drug fate in animals, 143-153
  drug resistance, 160-163
  excretion of, 25, 139, 145, 176
  history, 9, 10, 128-132
  human bladder cancer, 187
  human head and neck cancers, 185, 186
  human lung cancer, 188
  human ovarian cancer, 183, 184
  human testicular cancers, 179-183
  in animals, 134-172
  kidney toxicity, 130, 131, 140, 148, 150, 153, 176-178, 186, 189
  mutagenesis, 172-174
  selective action, 163-172
  toxic side effects, 25, 141-143, 148, 176, 184, 342, 352
  transport, 150-152
Citrate (or citric acid)
  buffer, 82, 263
  ^{67}Ga, 272
Class a and b character, *see* "Hard" and "Soft" species
CMP, 94-96
Cobalamins
  adenosyl-, 357
  cyano-, 6
Cobalt, 4, 28, 340
  ^{60}Co, 158
Cobalt(II), 4, 6, 13, 44-46, 48, 204, 297, 309, 325, 327, 330, 353, 355, 359, 361, 363
  bleomycin, 314, 315, 317
Cobalt(III), 5, 6, 13, 28, 29, 245, 301, 361
  bleomycin, 314, 315
Coenzymes (*see also* individual names), 344, 357
  Ru complexes, 255, 256, 274
Colon tumor, 22, 137, 189, 341
Colorimetry, *see* Absorption bands and spectra
Complexes, rates of formation, *see* Rate constants and Rates of reactions
Concanavalin A, 105
Conductivity measurements, 24
Copper, 4, 28, 42, 48, 340
  bleomycin, 218-225, 307, 310, 316, 317
  carcinostatic complexes, 197-225
  ^{63}Cu, 205
  -deficient diet, 202
  thiosemicarbazones, 45, 198-218, 301
Copper(I), 44, 45, 207, 209-211, 224, 310
  bleomycin, 316, 317
Copper(II), 4, 5, 29, 30, 44, 45, 48, 197-225, 297, 300, 301, 307, 309, 310, 325, 327, 328, 330, 363
  bleomycin, 316, 317
  phleomycin, 324
  tallysomycin, 321, 322
Cordycepin, 356, 360, 362
  5'-triphosphate, 356
Creatinine, 142
  clearance, 141
Croton oil, 175

Crystal field theory, 65, 66
Crystal structures, *see* X-ray studies
CTP, 359
  arabinosyl-, 356
  d-, 351, 358
  synthetase, 359, 360
Curare, 10
Cyanide, 23, 69-74, 78, 79, 89, 103, 104, 107, 114, 234, 238, 343
  ferri-, 313, 314, 364
Cyanocobalamin, 6
1,1-Cyclobutanedicarboxylate, 20
Cyclohexylamine, 81, 168
1,5-Cyclooctadiene, as ligand, 34, 35, 43
Cyclopentylamine, 159, 168, 174, 175, 342
Cyclophosphamide (*see also* Cytoxan), 133, 154, 156, 157, 180
Cyclopropylamine, 168
Cystamine, 342
Cysteine (and residues), 6, 99, 100, 102, 104-106, 223, 245, 339, 343, 345, 346, 352, 360, 361
  N-acetyl-, 342
  β,β-dimethyl-, *see* Penicillamine
Cystine (and residues), 339
Cytidinate complexes, 250, 267
Cytidine, 249, 250, 252
  5-aza-, 154, 156, 362
  deaminases, 37, 351
  1-methyl-, 249, 267
  3-methyl-, 249, 250, 252
Cytidine 5'-diphosphate, *see* CDP
Cytidine 5'-monophosphate, *see* CMP
Cytidine 5'-triphosphate, *see* CTP
Cytochrome(s), 259, 260
  c, 339, 343, 344
  c_{550}, 105
  c, horse ferri-, 105
  c, tuna ferro-, 105, 107
Cytosine (and moiety), 26, 92, 95, 98, 169, 171, 256-258, 309
[Cytosine]
  arabinosyl-, 36, 37, 39, 154, 156, 157, 351, 356, 362
Cytosine arabinoside, *see* Arabinosylcytosine
Cytotoxicity, 173, 174, 198, 199, 202, 204, 207, 208, 210-214, 218, 220, 224, 295, 317, 326, 362
Cytosol, 260, 262
Cytoxan (*see also* Cyclophosphamide), 155, 157, 184, 187, 188, 287, 288, 291, 293, 295, 300, 341
  complexes of, 298, 299
  derivatives, 294
  metabolism, 294

D

Daunomycin, 326-328
Daunorubicin, 356
Deaminases, 37, 351
  cytidine, 37, 351
Dehydrogenases, 108, 259, 364
  alcohol, 107, 344-346, 349
  glucose-6-phosphate, 350
  glyceraldehyde-3-phosphate, 350
  lactate, 105, 345, 347, 349
  malate, 260, 344-347, 349
  succinate, 260
Deoxyribonucleic acid, *see* DNA
2'-Deoxyribose, 91
Dermatomycosis, 10
Dermatosis, 10
Dialysis, 37, 117, 339
1,2-Diaminocycloheptane, 163
1,2-Diaminocyclohexane, 16, 19-21, 155, 161-163, 189, 266
1,2-Diaminocyclopentane, 163
1,2-Diaminoethane, *see* Ethylenediamine
2,6-Diaminohexanoate, *see* Lysine
Dien, *see* Diethylenetriamine
1,2:3,4-Diepoxybutane, 290
Diethylenetriamine, 23, 32, 76, 81, 87, 94, 96, 344
Diffusion-controlled reactions, 241

Dihydronicotinamide adenine dinucleotide, *see* NADH
Dimeric complexes, *see* Hydroxo complexes
β,β-Dimethylcysteine, *see* Penicillamine
1,3-Dimethyl-2,6-dioxopurine, *see* Theophylline
Dimethylglyoxime, 42, 44, 48
Dimethylsulfoxide, 106, 215
  as ligand, 43, 74, 75, 79-81, 83-86, 98, 101, 172, 175, 241, 268, 272, 273, 299
Dimethylthioether, 70, 235
5-(3,3-Dimethyl-1-triazeno)imidazole-4-carboxamide, 290, 291, 293
Dimethylxanthine
  1,3-, 242, 248, 252-254, 267
  1,9-, 252
  3,9-, 252
Dinitrogen, as ligand, 234, 235, 237, 239, 240, 260
Diol moieties, *see* Hydroxyl groups and Sugars
2,4-Dioxopyrimidine, *see* Uracil
Dioxygen, 207, 243, 259, 365
  complexes, 241
  redox reactions, 209-211, 218, 223, 241, 242, 309, 313, 314, 342, 364
Dipeptides (*see also* Peptides and individual names), 101
Diphosphate, 111, 353, 354, 357, 359
2,3-Diphosphoglycerate, 361
Dipine, 289
2,2'-Dipyridyl, *see* 2,2'-Bipyridyl
Dissociation constant, *see* Stability constants
Disulfides (and groups, *see also* individual names), 206, 210, 211, 224
  bond (*see also* Cystine), 104, 106
  bridges, 100
Dithiothreitol, 206, 207, 209, 210, 358
Diuresis
  mannitol, 178-180, 342
DNA (*see also* Nucleic acids), 42, 43, 52, 91-93, 97, 98, 109, 114-117, 147, 153, 167, 169, 173, 189, 199, 223, 224, 243, 270, 273, 291-293, 295, 299, 300, 307-309, 325, 338, 339, 347, 348
  bleomycin interaction, 310, 311, 323, 324
  calf thymus, 97, 315
  double helix, 91, 97, 167, 274, 354
  E. coli, 97
  interaction with antibiotics, 325, 327-330
  phleomycin interaction, 323, 324
  polymerases, 36, 165, 166, 351-357
  repair, 265-268
  Ru complexes, 256-259, 269
  synthesis, 7, 26, 27, 35-37, 39, 41, 44, 53, 152, 210-214, 218, 265-269, 272, 300, 310, 343, 351
  tallysomycin interaction, 320, 323, 324
Dogs, 145, 178
  toxicologic studies, 141, 142, 322
Double helix of DNA, 91, 97, 167, 274, 354
DPN, *see* NAD
DPNH, *see* NADH
Drugs (*see also* individual names), 7, 9-27, 128-190, 219, 286, 294-296, 300, 306, 308
  activation by kinases, 362, 363
  fate in animals, 143-153
  inactivation by xanthine oxidase, 364
  interactions with enzymes, 337-365
  metabolism, 338
  pro-, 241-243, 260
  resistance, 160-163
Dunning leukemia, 136, 139, 165, 341

E

EDTA, *see* Ethylenediamine-N,N,
N',N'-tetraacetate
Egg, tumor-bearing, 4
Ehrlich ascites tumor, 4-6, 35-
40, 43, 44, 49, 50, 136,
150-152, 207, 209, 210, 213,
216, 217, 219, 220, 263,
272, 298
Electrochemical reactions and
studies, 255, 315, 316, 321
Electronic spectra, *see* Absorption spectra
Electron microscopy, 165, 257,
263
Electron paramagnetic resonance,
see EPR
Electron spectroscopy chemical
analysis, *see* ESCA
Electron spin resonance, *see* EPR
Electron transfer (or transport),
207, 218, 240, 241, 259,
260, 262, 364, 365
Electrophoresis, 117, 165, 339
gel, 113, 166
Electrostatic interaction, 117,
258, 327
Emetine, 154, 156
Enzymes (*see also* individual
names), 24, 47, 53, 104, 255
co-, *see* Coenzymes
inhibition, 153
interaction with anticancer
drugs, 337-365
metallo-, 338, 353-365
redox, 260
Epilepsy, 264
Epoxides, 287, 290, 357
EPR, 112, 113, 205, 207, 209,
214-217, 221, 222, 309,
311-315, 321, 322
Equilibrium constants (*see also*
Acidity and Stability constants), 73, 84, 86, 87,
235, 236
ESCA studies, 26
Escherichia coli, 9, 11, 12, 34,
35, 40-44, 97, 128, 172,
173, 241, 265, 343, 353,
355, 357
ESR, *see* EPR
Ethanol, 2-mercapto-, 223, 350
Ethers (*see also* individual
names), 67
Ether sulfur, *see* Thioether
3-Ethoxy-2-oxobutyraldehyde
bis(thiosemicarbazone)
[= KTS], *see* Semicarbazones
Ethylene (as ligand), 70, 73, 74,
80, 84, 86, 87, 101
Ethylenediamine, 11-13, 16, 30-
32, 40, 71, 77, 81, 83-88,
96, 97, 103, 105, 108, 135,
161-163, 206, 214, 215, 266,
300, 339, 343-345, 348, 349
N,N-dimethyl-, 31
N,N'-dimethyl-, 31
N,N,N',N'-tetramethyl-, 31
Ethylenediamine-N,N,N',N'-tetra-
cetate, 222, 223, 239, 310,
315
N-Ethylmaleimide, 209
Euglena gracilis, 353
Ewing tumor, 134
EXAFS, *see* X-ray
Exchange (kinetics or) rate, *see*
Ligand exchange processes
Extinction coefficients (*see*
also Absorption bands), 312

F

FAD, 255, 256
Ferricyanide, 313, 314, 364
Ferricytochrome c, horse, 105
Ferrocytochrome c, tuna, 105, 107
Flavin adeninedinucleotide, *see*
FAD
Flavins, 256, 364, 365
Flavoproteins, 259, 260
Fluorescence, 96, 117, 346
emission, 221
quenching, 320
spectra, 222, 310
X-ray, 339
Fluoride, as ligand, 23, 70, 78
5-Fluorouracil, 133, 154-157,
184, 187, 362
5-Fluorouridine, 362
Folates

[Folates]
anti-, 133
dihydro-, reductase, 350
Formate (or formic acid), 266
Formation constants, *see* Stability constants
Formycin, 356, 357, 359, 362
Fowler's solution, 3, 340
Frogs, skeletal muscle, 263
Fructose-1,6-diphosphate aldolase, 350
Furosemide, 179
Fungi (*see also* individual names), 307, 319, 326

G

Gallium(III), 30, 49-52, 301, 352
^{67}Ga, 49, 51
^{67}Ga-citrate, 272
Gastrointestinal tract, 149, 153, 271
GDP, 358
Gel electrophoresis, 113, 166
Gel filtration, 350
Germanium, 3
Glucose, 243
Glucose-6-phosphate dehydrogenase, 350
β-Glucuronidase, 342
Glutamic acid (and residues), 99, 107, 339, 344, 352, 359
poly-, 103, 339
Glutaminase, 41
Glutamine (and residues), 41, 42, 99, 107, 359, 360
γ-L-Glutamyl-L-cysteinylglycine, *see* Glutathione
Glutaredoxin, 357
Glutathione, 206, 215-218, 223
Glyceraldehyde 3-phosphate dehydrogenase, 350
Glycine (and residues), 40, 100, 101, 244
methylester, 244
Glycolysis, aerobic, 243
Glycopeptides, 218, 307, 309, 324, 327
Glycoproteins, 264, 265
Glycosidic cleavage, 328
Glycylmethionine, 348
Glyoxal bis(thiosemicarbazone), 207, 210
Glyoxime, dimethyl-, 42, 44, 48
GMP
5'-, 94, 95, 359
synthetase, 359, 360
Gold(I), 29, 48
Gold(III), 29, 48, 77, 204, 301
GTP, 354
d-, 351, 358
Guanazole, 359
Guanine (and moiety), 26, 92, 97, 98, 169-172, 248, 256-258, 273, 291, 309, 311
8-aza-, 6
2,2-dimethyl-, 247
6-thio-, 8, 41, 93, 156, 362
Guanosine, 93-96, 247, 253, 257, 267
deoxy-, 253
7-methyl-, 95
Guanosine 5'-diphosphate, *see* GDP
Guanosine monophosphate, *see* GMP
Guanosine 5'-triphosphate, *see* GTP
Guerin carcinoma, 5

H

Hair loss, 142
Halides (*see also* individual names), 4, 81, 102, 108, 236, 238, 248
Hamster, 343
Chinese, 159, 220
"Hard" and "soft" species, 28, 67, 233, 234, 238, 246, 254, 257
Head tumor, 25, 177, 185, 186, 263, 340
Heart, 347
rat, 149
Hemoglobins, 344
chironomus, 105
Hepatomas (*see also* Tumors)
Novikoff, 264
Hexacyanoferrate(III), 313, 314, 364

Hexamethylmelamine, *see* Melamine
Hexamethylphosphoramide, 286, 296
High-potential iron-sulfur proteins, 106
Hill plot, 351
Hippuric acid, p-amino-, 153
Histidine (and residues), 6, 99, 102, 104-107, 172, 174, 245, 267, 273, 339, 342, 343, 345, 353
  peptides, 349
Histones, 339
Hodgkin disease, 134, 176, 291
Horse
  ferricytochrome c, 105
  liver, 344, 345
Hydrazine, 41, 78
Hydride, 73, 74, 79
Hydrogen
  ^{3}H, 27, 35, 39, 152, 266-268, 339, 343
Hydrogen bonds, 88, 110, 111, 171, 247, 250, 252, 257, 258, 263, 273
Hydrogen peroxide, 241, 365
Hydrogen sulfide (*see also* Sulfide), 67, 73, 235, 307
Hydrolase, bleomycin, 308
Hydrolysis
  of coordinated ligands, 89, 90, 244
  of Pt complexes, 82-90, 349
  of sugar-purine bonds, 253
Hydrophobic interactions, 40
Hydroxide (*see also* Hydroxo complexes), 82, 88, 146
Hydroxo complexes, 20, 22, 23, 67-70, 72-74, 76, 79, 88, 94, 147, 159, 238, 240, 268
  bridges, 15, 23, 68, 88, 147
Hydroxylamine, 78
Hydroxyl groups, 67, 91, 99, 100, 263, 361
Hydroxyl radical, 218, 313, 325
Hydroxyurea, 156, 157, 358, 359
Hypoxanthine (and moiety), 92, 113, 247, 248, 251, 254, 267, 364
  1-methyl-, 252
  7-methyl-, 248, 252

I

Imidazole (and moieties, *see also* Histidine), 78, 99, 100, 102, 220-222, 235, 237, 245, 255, 258, 290, 291, 293, 299, 312, 314, 315, 317, 318, 321, 324
  carbon-bound to metals, 237-239, 245
  N-methyl-, 102
Imines, 234, 244, 245, 270
Indium(III), 49, 51
  ^{114m}In, 51
Infrared (spectroscopy) absorbance, 73, 88, 340
Inosine, 247, 252, 267
Intercalation, 97, 102, 116, 117, 166, 274, 295, 311, 327, 356
Intestinal mucosa, 152, 179
Intestine, 261
  human, 25
  mouse, 141
  rat, 141, 149
Iodide
  as ligand, 23, 68-74, 78, 79, 102, 107, 267
  in redox reactions, 257
Ionization constants, *see* Equilibrium constants
IR, *see* Infrared
Iridium(I), 29
Iridium(III), 12, 29, 43, 44, 301
Iridium(IV), 12, 43, 44
Iron (oxidation state not defined), 4, 5, 47, 198, 212-214, 219, 340, 358
  ^{57}Fe, 313
  high-potential protein, 106
  nonheme, 357, 360, 364
Iron(II), 4, 113, 204, 241, 309, 325, 327, 359, 363
  bleomycin, 309, 311-315, 317
Iron(III), 297, 309, 313, 314, 327, 330, 364
  bleomycin, 309, 311-314
Iron-sulfur proteins (*see also* Nonheme iron), 364
  high-potential, 106

Isoalloxazines, 234, 256
Isoleucine (and residues), 107
Isonicotinamide, 234, 237, 238, 259, 260
Isomerase, triosephosphate, 104
Isophosphamide, 154, 156, 288, 291, 300
Isopropylamine, 159, 161, 163
Isotopes, *see* individual names

K

Kethoxal bis(thiosemicarbazone) [= KTS], *see* Semicarbazones
Keto group, *see* Carbonyl group
Kidneys, 30, 141, 148, 149, 262-264, 271, 272, 342, 351
  cancer, 189
  human, 25
  rabbit, 265-268
  toxicity, 130, 131, 140, 148, 150, 153, 176-178, 184, 186, 189, 198, 271, 322, 340, 342, 349
  toxicity of cisplatin, 130, 131, 140, 148, 150, 153, 176-178, 186, 189
Kinases, 356
  adenyl-, 106, 107, 363
  drug activation, 362, 363
  pyruvate, 344
Kinetics of reactions, *see* Rates of reactions
Krebs cycle, 259, 260
KTS, *see* Semicarbazones

L

Lactate, 243, 260
  dehydrogenase, 105, 345, 347, 349
Lactobacillus
  casei, 350
  leichmannii, 359
Landschutz ascites tumor, 5, 41
Lead, 4
  arsenate, 3, 340
[Lead]
  colloidal, 3, 340
  phosphate, 3, 340
Leucine (and residues)
  amino peptidase, 104, 344, 345, 348, 349
  6-diazo-5-oxo-2-nor-, 359, 360
  ^{3}H, 35, 343
  iso-, 107
  valyl-, 101
Leucylvaline, 101
Leukemias, 3, 4, 133, 134, 200, 269, 291, 353
  acute lymphocytic, 134
  AK, 136
  anti-drugs, 154
  BW 5147, 47
  Dunning, 136, 139, 165, 341
  K 1964, 49, 50
  L 1210, 14, 16, 18, 19, 21, 22, 34-36, 41-44, 47-50, 54, 136, 138-140, 154-157, 160-163, 266-268, 270, 272, 297, 298, 348
  lymphatic, 166
  P 388, 42, 47, 49, 50, 54, 136, 138, 158, 160, 266-268, 270, 297, 298
  spontaneous, 166
Lewis lung carcinoma, 22, 45, 136, 138
Ligand exchange processes (*see also* Exchange rates), 67, 74-79, 81-90, 205, 206, 236-239, 345
Light microscope, 263
Lineweaver-Burk plots, 150, 151, 351
Liver, 25, 40, 172, 262, 263, 271-273, 293, 294
  alcohol dehydrogenase, 107, 344-346, 349
  rat, 148, 149, 259, 262
  toxicity, 176
Lungs, 263, 264
  Lewis carcinoma, 22, 45, 136, 138
  rat, 148, 149, 272
  tumor, 25, 137, 166, 175, 188, 263, 291

Lymph
  circulation, 132
  system, 146
Lymphocytes, 152, 166
Lymphomas (*see also* Leukemias and Tumors), 49, 138, 291, 307, 340
  Burkitt's, 134
  histiocytic, 134
Lysine (and residues), 99, 100, 102, 103, 107, 339, 344, 345, 361
  β-, 320-322
  poly-, 103
  L-sarco-, *see* Melphalan
L-Lyxohexose
  2,3,6-trideoxy-3-amino-, 326

M

Magnesium ion, 44, 297, 298, 327, 329, 330, 344, 353, 355-357, 359-361, 363
Magnetic susceptibility measurements, 111
Malate, 5, 259, 260, 346
  dehydrogenase, 260, 344-347, 349
Malonate (or malonic acid), 15, 16, 21, 161, 189, 260
  derivatives, 16, 19, 20
Mammary carcinomas, *see* Carcinomas and Tumors
Manganese, 4
  dioxide, 323
Manganese(II), 4, 5, 44, 327, 330, 344, 353-355, 359, 360, 363
Mannitol, 25, 186
  diuresis, 178-180, 342
D-Mannose, 308
Mass spectrometry, 116, 117
Maytansine, 355
Mechlorethamine, 287, 288, 291, 295, 347
Melamine
  hexamethyl-, 184
  triethylene-, 289, 291
Melanomas (*see also* Tumors), 49, 52, 291
Melphalan, 288, 291, 295
Membrane
  permeability, 342
  transport, 210, 299, 301
  transport of cisplatin, 150, 151
2-Mercaptoethanol, 223, 350
Mercapto groups and ligands (*see also* individual names, Sulfur ligands, and Thiols), 262
6-Mercaptopurine(s), 7, 8, 41, 199, 360, 362, 364, 365
Mercury, 10, 178, 340
Mercury(II), 5, 44, 204, 327
Metabolism
  of cisplatin, 139
  of cytoxan, 294
  of drugs, 338
  of Rh(II) acetate, 40
  oxidative, 4
Metal complexes (*see also* individual ligands and metal ions)
  of alkylating agents, 285-301
  of antibiotics, 305-330
  of aziridinyl derivatives, 296-298
  toxicity of, 2, 5, 141-143, 148, 176, 184, 198, 269, 270, 296, 342, 352
  with amide nitrogen, 98, 101, 316, 317, 322
Metal ions, *see* individual names
Metalloenzymes, *see* Enzymes
Methanol, 77, 78
Methionine (and residues), 99, 101, 102, 104-106, 245, 339, 343, 345, 346, 348, 349, 352
  alanyl-, 348, 349
  glycyl-, 348
  methionyl-, 349
Methionylmethionine, 349
Methrotrexate, 133, 154, 156, 157, 185, 186
Methoxyacetate, 36-38
Methylamine
  bis(2-chloroethyl)-, *see* Mechlorethamine
Methylbis(2-chloroethyl)amine, *see* Mechlorethamine

Methylcholanthrene, 6
Methylcytidine
  1-, 249, 267
  3-, 249, 250, 252
1-Methylhypoxanthine, 252
7-Methylhypoxanthine, 248, 252
6-Methylthiopurine riboside, 362
5-Methyluracil, *see* Thymine
Mice, 35, 37, 39, 129, 142, 144-146, 148
  A/Jax, 175
  AKD_2F_1, 137
  AKR, 166
  AKR/Lw, 136, 137
  BALB/$\bar{C}$, 15, 32, 33, 43, 48, 136
  BDF_1, 16, 18, 19, 21, 22, 36, 44, 136-138, 154, 155, 157
  C^-, 168
  C+, 136
  C57BL/6, 136
  C57BLxDBA/$2F_1$, 160
  CD-1, 175
  $CD8F_1$, 137
  CDF_1, 22, 49, 137, 266-268
  CD_2F, 266-268
  DBA, 4
  DBA_2, 200
  DMB_2, 202
  ICR, 11, 138, 141, 157, 164
  leukemia, 14, 166
  nude, 137
  Swiss-white, 15, 32, 33, 36, 40, 136, 351
  tumor, 3-6, 42, 47, 48, 54, 134, 138, 152, 154, 155, 157-159, 164, 165, 200-202, 205, 210, 213, 270, 272, 297, 298, 341, 353
Michaelis-Menten kinetics (*see also* Steady-state kinetics), 150, 151
Microorganisms, *see* individual names
Microscopy
  electron-, 165, 257, 263
  light-, 263
Microsomes, 260
Mithramycin, 328-330, 356
Mitochondria, 259, 260, 262, 264, 342, 346, 247
[Mitochondria]
  Ehrlich, 211, 218
Mitomycin C, 289, 293, 295
Mixed ligand complexes (*see also* Ternary complexes), 322
Molybdenum, 29, 364, 365
Monkeys, 353
  Rhesus, 143
  toxicologic studies, 141, 142
Morphine, 306
Morpholine, 35
Mössbauer spectra (or studies), 73, 313
Mouse, *see* Mice
Mucopolysaccharides, 263, 264
Mucoproteins, 264
Mucosa, 152, 179, 261, 273
Muscle, 271, 347
  cell, 147
  rabbit, 345
  rat, 149
  skeletal, 263
Mustards (*see also* individual names), 7, 291
  nitrogen, 10, 132, 150, 200, 201, 203, 286-288, 291, 292, 348
  L-phenylalanine (= Melphalan), 288, 291, 295
  phosphoramide, 154, 156, 293, 294
  sulfur, 41, 288, 300, 340
  uracil, 288, 291
Mutations, 171, 233, 273, 274
  in bacteria, 172-174
  in mammalian cells, 172-174
Mycin
  adria-, *see* Adriamycin
  ansa-, *see* Ansamycin
  bleo-, *see* Bleomycin
  chromo-, *see* Chromomycin
  dauno-, *see* Daunomycin
  for-, *see* Formycin
  mithra-, *see* Mithramycin
  mito- C, *see* Mitomycin
  olivo-, *see* Olivomycin
  phleo-, *see* Phleomycin
  porfiro-, *see* Porfiromycin
  rifa-, *see* Rifamycin
  tallyso-, *see* Tallysomycin
  toyoca-, *see* Toyocamycin

Myeloma (*see also* Tumors), 157, 291
Myleran, *see* Busulfan

N

NAD, 255, 346, 364, 365
NADH, 243, 255, 259, 260, 346
NADP, 255
Neck tumor, 25, 177, 185, 186, 263, 340
Neocarcinostatin, 330
Neoplatin, *see* Cisplatin
Nerves, 25, 264
Neuroblastomas, 188, 291
Neurotoxicity, 184, 264, 273
Neurotransmitters, 264
Neutron activation analysis, 148
Nickel(II), 4, 13, 29, 44, 48, 204, 327, 330, 353, 359, 363
  bleomycin, 315
  tallysomycin, 321
Nicotinamide, 246
  iso-, 234, 237, 238, 259, 260
Nicotinamide adenine dinucleotide, *see* NAD
Nicotinamide adenine dinucleotide (reduced), *see* NADH
Nicotinamide adenine dinucleotide phosphate, *see* NADP
Nitrate, 4, 15, 113
  as leaving group, 83, 91
  as ligand, 20, 21, 23, 24, 33, 79, 88
Nitriles, 245
Nitrite, as ligand, 12, 15, 23, 74, 78, 79, 105, 106, 267
Nitrogen, *see* Dinitrogen
Nitrogen mustards, *see* Mustards
Nitrogen oxide, as ligand, 12, 43, 79, 234, 245, 261-263, 267, 268
Nitroprusside, 313
Nitrosoureas, 287, 290, 291, 293
  bischloroethyl-, 290, 291, 293, 295
  chloroethylcyclohexyl-, 290, 291, 293
NMR (*see also* Paramagnetic resonance), 73, 89, 90, 104, 115, 117
  ^{13}C-, 115, 221, 315, 316, 318, 321, 322
  ^{1}H-, 94, 95, 115, 221, 310, 311, 318
  ^{195}Pt-, 115
Nonheme iron, 357, 360, 364
  proteins, 106, 364
Norleucine
  6-diazo-5-oxo-2-, 359, 360
Nuclear magnetic resonance, *see* NMR
Nucleic acids (*see also* DNA, RNA, and Polynucleotides), 7, 11, 45, 47, 90-95, 96, 117, 152, 165, 166, 233, 242, 247, 311, 348
  polymerases, 36, 165, 166, 351-357
  Ru complexes, 256-295, 269, 274
  synthesis, 43
Nucleoproteins, 166, 167, 269
Nucleosides (*see also* individual names), 92, 93, 95, 98, 115, 116, 133, 351, 356, 357
  diphosphates, 357, 358
  Ru complexes, 246-256
  triphosphates, 357, 358
Nucleotides (*see also* individual names, DNA, RNA, and Polynucleotides), 93, 95, 116, 242, 320, 344, 347
  Ru complexes, 246-256, 274

O

Oligonucleotides (*see also* Polynucleotides), 117
Olivomycin, 328-330
Optical spectra, *see* Absorption spectra
Orthophosphate, *see* Phosphate
Osmium, 4, 10, 13, 29, 48
Ovary, 272
  Chinese hamster cells, 159
  tumor, 25, 130-132, 137, 177, 183, 184, 291, 340

Oxalate (or oxalic acid), 5, 13, 15, 32, 82, 245, 266, 267
Oxidases
  per-, 350
  xanthine, 364, 365
Oxidation-reduction potentials, *see* Redox potentials
Oximes, 44
  dimethylgly-, 42, 44, 48
2-Oxo-4-aminopyrimidine, *see* Cytosine
6-Oxopurine, *see* Hypoxanthine
Oxygen, *see* Dioxygen

P

Palladium(II), 7, 8, 13, 29, 40-42, 48, 65, 297, 301, 340
Palladium(IV), 41
Pancreas, 176
  rat, 149
Paralysis, flaccid, 264
Paramagnetic resonance, *see* EPR and NMR
Penicillamine, 342
Penicillin, 10
Pentanoate, 40
Pentose (*see also* individual names), 92
Peptidases
  carboxy-, 106, 339, 343
  leucine amino-, 104, 344, 345, 348, 349
Peptide
  group, *see* Amide group
  nitrogen, *see* Amide nitrogen
Peptides (*see also* individual names and Polypeptides), 349
  glyco-, 218, 307, 309, 324, 327
  histidine, 349
  linkage, 98
Perchlorate, 83
Peritoneal cavity, 134, 135, 139, 141, 146
Peroxidases, 350
Peroxide(s), 257
  hydrogen-, 241, 365
Persulfide, 365
Phage, 34
  bacterio-, 11, 26
1,10-Phenanthroline, 10, 30, 34, 90, 116, 236, 267, 271, 273, 353, 354
  3,4,7,8-tetramethyl-, 5
Phenolates (or phenols, and phenolic groups), 99
  2,4-dichloro-, 260
Phenylalanine
  mustard, *see* Melphalan
  tRNA, 96
Phenylenediamine
  o-, 16, 161, 163
  p-, 5
α-Phenyl-N-tert-butyl nitrone, 313, 314
Phleomycin, 322-324
Phosphamide
  cyclo-, *see* Cyclophosphamide and Cytoxan
  iso-, *see* Isophosphamide
Phosphatases, *see* Alkaline phosphatases
Phosphate (including hydrogen- and dihydrogenphosphate), 3, 359, 361
  as ligand, 20
  back bone, 258
  buffer, 31, 82, 98, 104-106, 313, 314
  O,O'-dimethylthio-, 41
  lead, 3, 340
  transport, 342
Phosphate back bone, 258
Phosphates (and groups, *see also* individual names), 91, 166, 241, 250, 285, 286, 297, 327, 357, 360, 361, 363
Phosphine oxide
  1-aziridinobis(N,N'-dimethylamino)-, 296, 297
  bis-(1-aziridino)-N,N'-dimethylamino-, 296, 297
  tris-(1-aziridinyl)-, 286, 289, 296, 297
  tris[1-(2-methyl)aziridino]-, 296-298
Phosphines, 68, 73, 74, 78, 79, 300
  oxides, 286, 289, 296-298

[Phosphines]
sulfides, 289, 291, 297, 298
triphenyl-, 299
Phosphine sulfides
bis(1-aziridino)phenyl-, 296-298
bis[1-(2-methyl)aziridino]-phenyl-, 296-298
tris(1-aziridino)-, 289, 291, 297, 298
Phosphoramide
hexamethyl-, 286, 296
mustard, 154, 156, 293, 294
octamethylpyro-, 286
Phosphoribosylamine, 360
Phosphoribosylpyrophosphate synthetase, 360-362
Phosphorylation, oxidative, 342
Photoelectron spectroscopy, x-ray, 98, 112, 117
Photochemical reactions, 9
Phthalate
butyl-, 4, 8
n-butyl-, 48
carboxy-, 161, 189
complexes, 4
sec-butyl-, 48
Piperazine, 48
Piperidine, 34, 41, 78
Plasma (*see also* Serum), 83, 339
proteins, 262, 270, 339, 348
Platinol TM, *see* Cisplatin
Platinum, 2, 6, 48, 71, 72, 116, 296, 298, 338, 341-350
^{191}Pt, 343
^{195}Pt, 112, 115, 116
^{195m}Pt, 25, 115, 143, 144, 148
anticancer drugs (*see also* Cisplatin), 9-27, 128-190, 296, 300
hydrolysis of complexes, 82-90, 349
Platinum(II), 7-27, 29, 31, 34, 35, 40, 48, 63-118, 242, 257, 265, 297, 301, 339, 340
amino acid-binding, 100-102, 340
cis-$Pt(NH_3)_2Cl_2$, *see* Cisplatin
protein and polypeptide binding, 103-108, 116, 152, 153, 340-350
[Platinum(II)]
redox chemistry, 71-73
stability of complexes, 68-71
trans-$Pt(NH_3)_2Cl_2$, 10, 26, 95-98, 102, 108, 144, 146, 167-169, 173, 174, 342, 345-350, 352
Zeise's anion, 86, 87, 101
Platinum(III), 111, 113
Platinum(IV), 7, 10, 11, 20, 29, 71, 72, 82, 104, 109, 113, 116, 129, 135, 159, 343-345
Platinum blues, 68, 108-114
amide, 109
α-pyridone, 110-113
pyrimidine, 109, 165
redox properties, 112, 113
thymine, 342
uracil, 109
Platinum metals, *see* individual names
PMR, *see* NMR
Poisoning of metal ions, *see* Toxicity
Polarography, 206, 312, 314, 315, 317, 318
Polyaspartate, 103, 339
Polycythemia vera, 291
Polyglutamic acid, 103, 339
Polyhydroxy ligands, *see* Hydroxyl groups and Sugars
Polylysine, 103
Polymerases, 152
DNA, 36, 165, 166, 351-357
poly-A, 353
RNA, 36, 353-357
Polynucleotides (*see also* Oligonucleotides), 91, 96-98, 116, 254, 258, 347, 353, 354
poly-A, 351, 353
poly-dA, 351
poly-dC, 351
poly-G, 95
poly-dG, 351
poly-dT, 351
Polyols, *see* Hydroxyl groups and Sugars
Polypeptides (*see also* Peptides and Proteins), 98
Pt(II)-binding, 103-108

Polysaccharides, muco-, 263, 264
Porfiromycin, 289
Porphyrins (*see also* individual names), 274
  proto-, 6
  Pt(II), 108
Potassium ion, 264
  ATPases, 153
Potentials, *see* Redox potentials
Potentiometric measurements, 94, 96, 113, 206, 221, 312, 316, 318
Prednisone, 133, 181, 182
Proline (and residues), 101
Propionate, 36-39, 266, 297, 298, 351
Propylamine
  cyclo-, 168
  iso-, 159, 161, 163
Prostate tumor, 25, 188
Proteinase, 167
Proteins (*see also* individual names), 43, 47, 52, 90, 98-100, 163, 262, 274, 291-293, 310, 339, 340, 357, 364
  cytoplasmic, 11
  flavo-, 259, 260
  glyco-, 264, 265
  high potential iron-sulfur, 106
  iron, 106
  iron-sulfur, 364
  muco-, 264
  nucleo-, 166, 167, 269
  plasma/serum, 262, 270, 339, 348, 349
  synthesis, 26, 35, 36, 265, 272, 343
  Pt(II)-binding, 103-108, 116, 152, 153, 341-350
Protonation constants, *see* Acidity constants
Proton magnetic resonance, *see* NMR
Protoporphyrin, cobalt, 6
Purine(s), 42, 92, 99, 234, 247, 248, 250-255, 257, 258, 269, 354, 360, 362, 364
  derivatives, *see* individual names
  1,3-dimethyl-2,6-dioxo-, *see* Theophylline
[Purine(s)]
  6-mercapto-, 7, 8, 41, 199, 360, 362, 364, 365
  6-methylthio-, 362
Pyrazine, 234, 246
Pyridine, 12, 23, 30-32, 75, 77-79, 81, 89, 90, 97, 150, 151, 235-240, 299, 339
  2--aldehyde, 245
  4--aldehyde, 246
  derivatives, 31, 78, 212, 214, 234, 324, 344, 360
  -like nitrogen, 92, 111
  2--methylamine, 245
  Ru complexes, 246
α-Pyridone, 110-113
Pyrimidine(s), 92, 109, 199, 220-222, 234, 246, 250-257, 269, 311, 312, 314, 315, 317, 318, 321, 324, 360 362-364
  4-amino-, 312, 314, 315
  platinum blues, 109, 165
  tetrahydro-, 312
Pyrophosphate, *see* Diphosphate
  5-phosphoribosyl-α-1-, 360
Pyrrolidine, 174, 175
Pyruvate kinase, 344

Q

Quinine, 306
Quinoline(s)
  iso-, 31, 212
Quinones, 324

R

Rabbit
  kidneys, 265-268
  muscle, 345
Radiation (*see also* Radiotherapy), 182, 185, 186, 188
  ultraviolet, 11
  x-ray, 11, 158-160
Radical, 309
  free, 313, 325, 327, 358
  hydroxyl, 218, 313, 325
  superoxide, 218, 313, 314, 365
  trap, 216, 313, 314

Radioactive isotopes (*see also* individual names), 115, 117, 144, 146, 148, 152
Radioactive tracer experiments, 261-263, 355
Radiodiagnostic agents, *see* Tumor imaging nuclides
Radiopharmaceuticals, *see* Tumor imaging nuclides
Radiotherapy (*see also* Radiation), 45, 132, 158-160, 186
Raman spectroscopy or spectra, 93, 94, 117
Rat, 32, 33, 54, 138, 141, 142, 145, 146, 148, 149, 179, 271, 272
  Alderly Park, 136
  Fisher 344, 136, 175
  liver, 148, 149, 259, 262
  Sprague-Dawley, 49, 136
  tail, 148
  tumor, 4, 5, 45, 46, 51, 148, 149, 159, 165, 200, 201, 224, 264, 341, 342
Rate constants (*see also* Rates of reactions), 75, 77-79, 85, 86, 95, 113, 207-209, 215, 216, 223, 235, 237, 238, 241, 253
Rates of reactions (*see also* Rate constants and Ligand exchange processes), 97, 147, 206, 210, 241, 310, 345, 350
Reaction
  mechanism, 76, 86, 206, 236-241, 292, 358
  rates, *see* Rates of reactions and Rate constants
Redox potentials, 29, 30, 71-73, 206-209, 214, 234, 237, 239-241, 249, 260, 270, 312, 317
Redoxin
  gluta-, 357
  thio-, 357, 358
Redox reactions, 71-73, 112, 113, 205, 206, 209-211, 216, 218, 223, 224, 240-242, 257, 260, 309, 310, 313, 314, 330, 342, 357, 364
Reductases
  dihydrofolate, 350
  ribonucleotide, 212, 357-359, 362
Reduction potentials, *see* Redox potentials
Renal disease, *see* Kidneys
Retinoblastoma, 291
Rhenium(IV), 29
Rhesus monkeys, 143
Rhodium, 4, 343
  carboxylates, 36-40, 52, 198, 297, 298, 351
Rhodium(I), 29-31, 34, 35, 43
Rhodium(II), 36, 39, 40, 297, 298, 351, 352
Rhodium(III), 11, 12, 29-40, 301
Ribonuclease, 104, 105, 107, 245, 339, 343
Ribonucleic acid, *see* RNA
Ribonucleotide reductase, 212, 357-359, 362
D-Ribose (and ribosyl group), 91, 96
  2'-deoxy-, 91
  derivatives (*see also* individual names), 357-360
  5'-phosphate, 360, 361
Ribosomes, 165
Rifamycins, 355
RNA (*see also* Nucleic acids), 52, 92, 93, 97, 109, 116, 152, 166, 347
  polymerases, 36, 353-357
  Ru complexes, 256-259, 269
  synthesis, 26, 35, 36, 272, 328, 343, 351
  $tRNA^{Phe}$, 96
Rous sarcoma, 136, 138
Ruthenium, 4, 10, 11, 340, 343
  amino acid complexes, 244-246, 274
  coenzyme complexes, 255, 256, 274
  complexes, antitumor, and mutagenic activity, 266-270, 274
  complexes in living organisms, 259-275
  excretion of, 261, 262, 271

[Ruthenium]
nitrosyl complexes, 261-263, 267, 268
nucleic acid complexes, 256-259, 269, 274
nucleoside and nucleotide complexes, 246-256, 274
organo-, complexes, 271-273
pyridine complexes, 246
radio-, 232, 261-263
red, 263-265, 274
^{97}Ru, 264, 274
^{103}Ru, 264, 272
tissue distribution, 261-263
Ruthenium(II), 5, 29, 42, 43, 172, 231-275, 301
Ruthenium(III), 12, 29, 42, 43, 231-275
complexes as oncostatic pro-drugs, 241-243, 260
trichloride, 261-263, 270
Ruthenium(IV), 233
Ruthenocene(s), 271, 272, 274

S

Saccharides
di-, 328
poly-, 263, 264
tri-, 328
Salmonella typhimurium, 172, 174
L-Sarcolysin, *see* Melphalan
Sarcomas (*see also* Tumors), 175, 291, 340
-180, 8, 11, 14-16, 23, 32-35, 40-42, 44, 47, 136, 138, 141, 156, 157, 164, 200, 202, 205, 209, 213, 219, 341
-180A, 45, 136
adenoma 755, 201
A-RCS, 50
fibro-, M-89, 50
human osteogenic, 188
Jensen, 201
lympho-, 176
lympho-, P-1798, 50
osteo-, 124F, 50
Mecca lympho-, 201
Murphy-Sturm lympho-, 201
reticulum cell, 136
rhabdomyo-, 134, 291
[Sarcoma]
Ridgeway osteogenic, 137, 201
Rous, 136, 138
Walker-256, 32-34, 41, 45, 46, 48-51, 136, 138, 139, 149, 165, 200-204, 209, 341, 352
Yoshida, 5, 340
Scintillation
camera, 263
counter, 144
Schiff bases, 45, 46
Screening tests, 53-55
Sedimentation studies, 117
Selenium, 340
compounds as ligands, 78
Semicarbazones
3-ethoxy-2-oxobutyraldehyde bis(thio-), 44, 45, 199-211, 224
1-formylisoquinoline thio-, 358
D-glucosone bis(thio-), 42, 48
glyoxal bis(thio-), 207, 210
α-N-heterocyclic carbaldehyde thio-, 212-218
thio-, 42, 44, 45, 199-219, 223, 301, 358
Serine (and residues), 99, 102, 107, 344
aza-, 359, 360
Serum (*see also* Plasma)
fetal calf, 220
Shark, 145
Silver(I), 10, 108, 199, 297, 327
Skin, 134
infections, 10
rat, 148, 149
tumor, 5, 175, 185
Sodium ion, 327
ATPases, 153
"Soft" species, *see* "Hard" and "soft" species
Spectrophotometry, *see* Absorption bands and spectra
Spleen, 262, 273
rat, 149, 264
Stability constants (*see also* Equilibrium constants), 68-70, 95, 205, 208, 214, 222, 249, 316, 320, 327, 330, 345-347, 365

Stacking, 258
Standard potentials, *see* Redox potentials
Staphylococcus aureus, 343
Steady-state kinetics (*see also* Michaelis-Menten kinetics), 215, 216
Stereoselectivity, 169
Stereospecificity (*see also* Stereoselectivity), 236, 237, 239, 250
Steric interactions, 207
Steroids, 133, 272
Stomach, 147
  cancer, 189
  rat, 149
Streptoalloteichus hindustanus, 319-322
Streptomyces
  flocculus, 324
  verticillis, 218, 307
Streptonigrin, 324-326, 330
Streptovaricins, 355
Substitution reactions, *see* Ligand exchange processes
Subtilisin, 339, 343
  BPN', 106
  novo, 106
Succinate (or acid), 259, 342
Succinate dehydrogenase, 260
Sugars (and moieties, *see also* individual names), 42, 91, 166, 221, 247, 263, 308, 317, 318, 326, 328
  -purine bonds, 253
Sulfate, 49, 51, 83
  as ligand, 20, 21, 23, 24, 161, 189, 257
  ion pair formation, 263
Sulfhydryl groups, *see* Thiols
Sulfide (*see also* Hydrogen sulfide and Sulfur ligands), 67, 235
  per-, 365
Sulfite, 78, 237, 238, 245, 257
Sulfonamides, 10
Sulfonates, 287, 290, 291
Sulfur ligands (*see also* individual names, Thiols, and Mercapto-), 7, 41, 79, 90, 97, 99, 100, 108, 205, 210, 238, 245, 289, 291, 297, 298, 364
Sulfur mustards, *see* Mustards
Superoxide, 218, 313, 314, 365
  complex, 315
Superoxide dismutase, 309
Surgery, 5, 132, 181, 182, 185, 186, 200
Susceptibility, *see* Magnetic susceptibility measurements
Synthetases, 360
  CTP, 359, 360
  GMP, 359, 360
  phosphoribosylformylglycine-amidine-, 359
  phosphoribosylpyrophosphate-, 360-362
  thymidylate, 349-350

T

Tallysomycin, 315, 319-322, 327
  DNA interaction, 320
  mechanism of action, 320
  metal complexes, 321, 322
L-Talose, 319, 322
Tendon, rat, 149
Ternary complexes (*see also* Mixed ligand complexes), 347
2,2',2"-Terpyridine, 81, 96, 97, 102
Testis tumor, 24, 130-132, 134, 177, 179-183, 291, 307, 340, 341
Thalassemia, 198
Thallium(III), 49, 51
Theophylline, 248
Therapeutic
  agents, *see* Drugs and individual names
  index, 15, 16, 35, 38, 55, 168, 270, 286, 295, 300, 301
  ratio or dose, 18, 19, 21, 27, 32, 45, 46, 50, 51, 139-141, 162, 168, 176, 178, 180, 181, 184-186, 201, 202, 204, 266-268, 272, 297
Thermolysin, 106
Thiamine, 108
Thiazole moiety, 308, 310, 319, 320, 323, 324
Thin-layer chromatography, 146
Thiocyanate, 15, 23, 70, 76, 78, 79, 107

Thioethers (*see also* Sulfur ligands), 67, 78, 99, 235, 286, 299, 340
6-Thioguanine, 8, 41, 93, 156, 362
Thiols (*see also* individual names, Mercapto groups, and Sulfur ligands), 67, 68, 78, 99, 206, 207, 209-211, 214-218, 223, 224, 235, 245, **246, 260, 262, 309, 310, 317, 346, 350, 357, 360, 365**
Thiophene, as ligand, 235, 240
Thioredoxin, 357, 358
Thiosemicarbazones (*see also* individual names), 42, 44, 45, 199-219, 223, 301, 358
  copper complexes, 45, 198-218, 301
  3-ethoxy-2-oxobutyraldehyde bis-, 44, 45, 199-211, 224
  1-formylisoquinoline, 358
  D-glucosone bis-, 42, 48
  glyoxal bis-, 207, 210
  α-N-heterocyclic carbaldehyde-, 212-218
4-Thiouracil, 93
Thiourea, phenyl-, 6
Threonine (and residues), 99, 102
Thrombocytosis, 291
Thymidine, 37, 94, 95
  ^{3}H, 39
Thymidine 5'-triphosphate, *see* TTP
Thymidylate synthetase, 249, 250
Thymine (and moiety), 26, 92, 98, 109, 111, 171, 223, 268, 309
  platinum blue, 342
Thymus, 97, 272, 315
Thyroid gland, 149
Tin, 3, 68
Tissue
  blood, 349
  bone, 49
  distribution of Ru, 261-263
Toxicity, 15, 16, 18, 19, 27, 49, 52, 153, 200, 219, 242, 262, 266-268, 295, 297, 299, 300
[Toxicity]
  cardiac, 184
  curare-like, 10
  cyto-, 173, 174, 198, 199, 202, 204, 207, 208, 210-214, 218, 220, 224, 295, 317, 326, 362
  gastrointestinal, 142
  hematologic, 142, 184
  kidney, 130, 131, 140, 148, 150, 153, 176-178, 184, 186, 189, 198, 271, 322, 340, 342, 349
  level, 8
  neurologic, 184, 264, 273
  of cisplatin, 2, 5, 141-143, 148, 176, 184, 342, 352
  of metal ions (complexes), 198, 296, 352
  of Ru complexes, 269, 270
  oto-, 142, 340, 342
  selective, 7, 47, 48, 163-172
Toyocamycin, 359
Tracer experiments, *see* Radioactive tracer experiments
Trans effect (*see also* Cis effect), 73-75, 79, 80
Transferases
  amido-, 359, 360
  amidophosphoribosyl-, 359, 360
  sialyl-, 341
  terminal deoxynucleotidyl, 353
Transition metal ions (*see also* individual names), 225, 241, 259, 311, 340
  3d, 28, 29, 316
  4d, 29
  5d, 28, 29
  heavier, 28, 198
Transport
  of calcium, 264, 342
  of cisplatin, 150-152
  of electrons, *see* Electron transfer
  of p-aminohippuric acid, 153
  phosphate, 342
  through membranes, *see* Membrane
Trenimon, 289
Triazenes, 287, 290, 291, 293
Triethylenemelamine, *see* Melamine

Tris, *see* Tris(hydroxymethyl)-methylamine
2,3,S-Tris(aziridino)-1,4-benzoquinone, *see* Trenimon
Tris(1-aziridino)phosphine
  oxide, 286, 289, 296, 297
  sulfide, 289, 291, 297, 298
Tris(hydroxymethyl)methylamine, 106, 107
Tritium, *see* Hydrogen
Trypsin, 344
  α-chymo-, 104, 106, 339, 343, 348
Tryptophan (and residues), 99, 100, 102, 106, 344
dTTP, 351, 358
  arabinosyl-, 356
Tubercidin, 250, 356, 357, 359, 362
Tumors (*see also* Carcinomas, Leukemias, Lymphomas, Melanomas, Sarcomas, and individual names), 134, 319
  ADJ/PC6, 136, 138
  ADJ/PC6A, 4, 15-17, 32, 33, 35, 40, 42, 48, 165, 168
  B16 melanoma, 22, 136
  benign, 175, 182
  bladder, 25, 157, 187
  brain, 159, 189, 291
  cervix, 25, 186, 187
  colon, 22, 137, 189, 341
  Ehrlich ascites, 4-6, 35-40, 43, 44, 49, 50, 136, 150-152, 207, 209, 210, 213, 216, 217, 219, 220, 263, 272, 298
  embryonal testicular, 134
  Ewing's, 134
  head, 25, 177, 185, 186, 263, 340
  imaging nuclides, 49, 232, 233, 261-264, 271-274, 311
  kidney, 189
  Landschutz ascites, 5, 41
  lung, 25, 137, 166, 175, 188, 263, 291
  mammary, 22, 200, 341
  mice, 3-6, 42, 47, 48, 54, 134, 138, 152, 154, 155, 157-159, 164, 165, 200-205, 210, 213, 270, 272, 297, 298, 341, 353
  neck, 25, 177, 185, 186, 263, 340
[Tumors]
  ovarian, 25, 130-132, 137, 177, 183, 184, 291, 340
  P388, 22, 47
  prostate, 25, 188
  rat, 4, 5, 45, 46, 51, 148, 149, 159, 165, 200, 201, 224, 264, 341, 342
  skin, 5, 175, 185
  stomach, 189
  testes, 24, 130-132, 177, 179-183, 291, 307, 340, 341
  Wilm's, 134, 291
Tuna ferrocytochrome c, 105, 107
Tungsten(III), 29
Tyrosinase, 350
Tyrosine (and residues), 99, 102, 106, 358
  benzoyl-L--ethylester, 348

U

UDP, 358
Ultrafiltration, 339
Ultraviolet, *see* UV
UMP, 95
Uracil (and moiety), 92, 109, 111, 113, 114
  ^{14}C, 114
  5-fluoro-, 133, 154-157, 184, 187, 362
  mustard, 288, 291
  4-thio-, 93
Uranium(VI), 13
Uranylacetate, 13
Urea(s)
  blood, 141
  hydroxy-, 156, 157, 358, 359
  nitroso-, 287, 290, 291, 293, 295
  phenylthio-, 6
Uric acid, 364, 365
  6-thio-, 364
Uridine, 35, 94, 95
  6-aza-, 362
  deaza-, 360, 362
  5-fluoro-, 362
Uridine 5'-diphosphate, *see* UDP
Uridine 5'-monophosphate, *see* UMP
Uridine 5'-triphosphate, *see* UTP

Urine, 40, 212, 342
  excretion of cisplatin, 25, 139, 145, 176
  excretion of Ru, 261, 262, 271
Uterus, 272
UTP, 96, 359
UV
  absorption (or spectra) (*see also* Absorption spectra), 117, 221, 222, 247, 256, 312, 346
  spectroscopy techniques, 24, 94, 117
UV radiation, 11

V

Valine (and residues), 101, 103, 107
  leucyl, 101
Valylleucine, 101
Vanadium, 3
Vinblastine, 131, 133, 156, 179-181, 186, 341
Vincristine, 156, 181, 182, 188
Viruses, 11, 147, 353, 356
  avian myeloblastosis, 353
Visible spectra, *see* Absorption bands and spectra
Vitamins (*see also* individual names)
  B_1, *see* Thiamine
  B_{12}, 6
Voltammetry, cyclic, 246

W

Walker-256 carcinosarcoma, 32-34, 41, 45, 46, 48-51, 136, 138, 139, 149, 165, 200-204, 209, 341, 352
Wilm's tumor, 134, 291
Wilson disease, 198

X

Xanthine, 248, 251, 253, 254, 364
  allo-, 365
  1,3-dimethyl-, 242, 248, 252-254, 267
  1,9-dimethyl-, 252
  3,9-dimethyl-, 252
  hypo-, *see* Hypoxanthine
  oxidase, 364, 365
X-ray radiation or therapy (*see also* Radiation and Radiotherapy), 11, 158-160
  ^{60}Co source, 158
X-ray
  extended absorption fine structure (EXAFS), 112, 116, 117
  fluorescence, 339
  photoelectron spectroscopy, 98, 112, 117
  structural studies (or analysis), 73, 88, 93, 94, 96, 103, 104, 109, 110, 116, 117, 205, 220, 221, 247-250, 316, 317, 324, 326, 339, 343

Y

Yeast
  alcohol dehydrogenase, 345
  RNA polymerase, 353
  $tRNA^{Phe}$, 96
Yoshi 864 drug, 156, 157
Yoshida sarcoma, 5, 340

Z

Zeise's complex, 86, 87, 101
Zinc(II), 5, 28, 44, 47, 48, 198, 203, 204, 309, 311, 330, 340, 344, 353-355, 359, 363
  bleomycin, 318
  -deficient diet, 202
  tallysomycin, 322